P9-CKE-363

The Psychology of Safety

The Psychology of SAFETY

How to improve behaviors and attitudes on the job

E. Scott Geller, Ph.D.
Center for Applied Behavior Systems
Department of Psychology
Virginia Polytechnic Institute and State University

Chilton Book Company
Radnor, Pennsylvania

Published in Radnor, Pennsylvania 19089, by Chilton Book Company

Designed by Anthony Jacobson
Manufactured in the United States of America

Library of Congress Cataloging-in-Publication Data
Geller, E. Scott, 1942–
The psychology of safey : how to improve behaviors and attitudes on the job / E. Scott Geller.
p. cm.
Includes bibliographical references and index.
ISBN 0-8019-8733-4
1. Industrial safety—Psychological aspects. I. Title.
T55.3.B43G45 1996
658.3'82'01—dc20 96-7994
CIP

1 2 3 4 5 6 7 8 9 0 5 4 3 2 1 0 9 8 7 6

Past

- to my Mom, Margaret J. Scott, and Dad, Edward I. Geller, who taught me the value of learning and reinforced my need to achieve
- to B. F. Skinner and W. Edwards Deming, who developed and researched the most applicable principles in this text, and inspired me to teach them

Present

- to my wife, Carol Ann, and mother-in-law, Betty Jane, whose continuous support for 25 years made preparation to write this book possible
- to the students and associates in our university Center for Applied Behavior Systems whose data collection and analysis provided practical examples for the principles

Future

- to my daughters Krista and Karly, who I hope will someday experience the sense of accomplishment that I feel at completing this text
- to my five senior partners in Safety Performance Solutions, Inc., who I hope will continuously improve at assisting the achievement of Total Safety Cultures worldwide

Contents

Part Four: Actively Caring for Safety

Part Five: Putting It All Together

Preface

Psychology influences every aspect of our lives, including our safety and health. And psychology can be used to benefit almost every aspect of our lives, including our safety and health.

So what is "psychology" anyway?

My *American Heritage Dictionary,* Second College Edition, defines psychology as "1. The science of mental processes and behavior, 2. The emotional and behavioral characteristics of an individual, group, or activity" (p. 1000). Similarly, the two definitions in the *New Merriam-Webster Dictionary* are "1. The science of mind and behavior, 2. The mental and behavioral characteristics of an individual or group" (p.587).

In both dictionaries, the first definition of "psychology" uses the term "science" and refers to behavioral and mental processes. *Behaviors* are the outside, objective and observable aspects of people; *mental* or *mind* reflects our inside, subjective and unobservable characteristics. *Science* implies the application of the scientific method, or the objective and systematic analysis and interpretation of reliable observations of natural or experimental phenomena.

So what should you expect from a book on the psychology of safety? Obviously, such a book should show how psychology influences the safety and health of people. And to be useful, it should explain ways to apply psychology to improve safety and health. This is, in fact, my purpose for writing this book—to teach you how to use psychology to both explain and reduce personal injury.

As a science of mind and behavior, psychology is actually a vast field of numerous sub-disciplines. Areas covered in a standard college course in introductory psychology, for example, include: research methods, physiological foundations, sensation and perception, language and thinking, consciousness and memory, learning, motivation and emotion, human development, intelligence, personality, psychological disorders, treatment of mental disorders, social thought and behavior, environmental psychology, and industrial/organizational psychology and human factors engineering. This book does not cover all of these areas of psychology, only those directly relevant to understanding and influencing safety-related behaviors and attitudes. In addition, my coverage of information within any one sub-discipline of psychology is not comprehensive, but focuses on those aspects directly relevant to reducing injury in organizational and community settings.

This information will help you to improve safety and health in any setting, from your home to the workplace and every community location in between. You can apply the knowledge gained from reading this book in all aspects of your daily life. Most organized safety-improvement efforts occur in work environments, however, because that's where the exposure to hazardous conditions and at-risk behavior is most obvious. As a result, most (but not all) of my illustrations and examples use an industrial context. My hope is that you will see direct relevance of the principles and procedures to domains beyond the workplace.

A psychology of safety must be based on rigorous research, not common sense or intuition. This is what science is all about. Much of the psychology in self-help paperbacks, audiotapes, and motivational speeches is not founded on scientific investigation, but is presented because it sounds good and it "sells." The psychology in this book was not selected on the basis of armchair hunches but rather from the relevant research literature. In sum, the information in this book is consistent with a literal definition of its title: *the psychology of safety.*

The human element of occupational health and safety has recently been an extremely popular topic at national and regional safety conferences. Safety leaders realize that reducing injuries below current levels requires increased attention to human factors. Engineering interventions and government policy have made their mark. Now it's time to work with the human dynamics of injury prevention, *the psychology of safety.*

Most attempts to deal with the human aspects of safety have been limited in scope. Many trainers and consultants claim to have answers to the human side of safety, but their solutions are too often impractical, shortsighted, or illusory. And to support their particular program, consultants, authors, and conference speakers often give unfair and inaccurate criticism of alternative methods.

Tools from behavior-based safety have been criticized in an attempt to justify a focus on people's attitudes or values. In contrast, promoters of behavior-based safety have ridiculed a focus on attitudes as being too subjective, unscientific, and unrealistic. And both behavior- and attitude-oriented approaches to injury prevention have been faulted in order to vindicate a systems or culture-based approach. The truth of the matter is that both behaviors and attitudes require attention in order to develop large-scale improvement in people's safety and health.

There are a number of books on the market that offer advice regarding the human element of occupational safety. Unfortunately, many of these texts offer a limited perspective. And I've found none comprehensive and practical enough to show how to integrate behavior- and attitude-based perspectives for a system-wide total culture transformation. This book was written to do just that, and in this regard, it is one of a kind.

Simply put, behavioral science principles provide the basic tools and procedures for building an improved safety system. But the people in a work culture need to accept and use these behavior-based techniques appropriately. This is where a broader perspective is needed, including insight regarding more subjective concepts like attitude, value, and thought processes. Recall that psychology includes the scientific study of both mind and behavior. Therefore, a practical book on the psychology of safety needs to teach science-based and feasible approaches to change what people think (attitude) and do (behavior) in order to achieve a Total Safety Culture.

I refer to a Total Safety Culture throughout this text as the ultimate vision of a safety-improvement mission. In a Total Safety Culture everyone feels responsible for safety and pursues it on a daily basis. At work, employees go beyond "the call of duty" to identify environmental hazards and at-risk behaviors. Then they intervene to correct them. And safe work practices are supported with proper recognition procedures. In a Total Safety Culture, safety is not a priority that gets shifted according to situational demands. Rather, safety is a value linked to all situational priorities.

Obviously, building a Total Safety Culture requires a long-term continuous improvement process. It involves cultivating constructive change in both the behaviors and attitudes of everyone in the culture. This book provides you with principles and procedures to make this happen. Applying what you read here might not result in a Total Safety Culture. But it's sure to make a beneficial difference in your own safety and health, and in the safety and health of others you choose to help.

I refer to helping others as *actively caring*. This book shows you how to increase the quality and quantity of your own and others' actively caring behavior. Indeed, actively caring is the key to safety improvement. The more people actively care for the safety and health of others, the less remote is the achievement of our ultimate vision—a Total Safety Culture.

WHO SHOULD READ THIS BOOK?

My editor has warned me that one book can serve only a limited audience. I know he's right, but at the same time a practical book on reducing injuries is relevant for everyone. All of us risk personal injury of some sort during the course of our days, and all of us can do something to reduce that risk to ourselves and others. Therefore, a book that teaches practical ways to do this is pertinent reading for everyone.

The average person, however, won't spend valuable time reading a book on ways to reduce personal risk for injury. Indeed, most people don't believe they are at risk for personal injury, so why should they read a book about improving safety? So while I believe everyone *should* read this book, a text on the psychology of safety is destined for a select audience: an elite group of people who are concerned about the rate of injuries in their organization or community and want to do something about it.

In order to reach more people, two versions of this book have been published. I've been referring to the title of this hardcover version, *The Psychology of Safety: How to Improve Behaviors and Attitudes on the Job,* but there is also an abbreviated softcover version: *Working Safe: How to Help People Actively Care for Health and Safety.*

Only the hardcover edition includes substantial research descriptions and citations to support the various principles and procedures. This version is for the safety professional, practitioner and researcher, who wants to understand the science behind the theory and techniques. The numerous citations to the research literature provide direction for readers to strengthen their profound knowledge in various domains.

I also recommend the hardcover version to students in colleges and universities. A practical text on the psychology of safety could be used as required or recommended reading in a number of undergraduate or graduate courses. This text is ideal for courses on human factors engineering, safety management, or organizational performance management. Many engineering and psychology departments do not offer courses with safety or human factors in their titles. However, this text is quite suitable for such standard courses as applied psychology, organizational psychology, management systems, engineering psychology, applied engineering, and even introductory psychology.

If assigned as supplementary reading for an undergraduate college course, the softcover version of this book might be preferable. The absence of references and research descriptions makes this version easier to get through. The course instructor could read the more thorough hardcover edition, and explain the supportive research during class lecture or group discussion. Similarly, the safety director of an organization will find the hardcover version useful to justify a number of safety improvement interventions. Then the softcover version can be distributed throughout the workforce as an introduction to the basic philosophy and principles behind a mission to involve employees in cultivating a Total Safety Culture.

The writing style and format is different than any professional text I have written or read. Most authors of professional books, including me, have been taught a particular academic or research style of writing that is not particularly enjoyable to read. When did you last pick up a nonfiction technical book for recreational or "fun" reading?

I have coauthored three other professional textbooks about important topics relevant to a broad audience. Their titles suggest important contents: *Preserving the Environment: New Strategies for Behavior Change* (coauthored with Drs. R. A. Winett, & P. B. Everett, published in 1982 by Pergamon Press), *Behavior Analysis Training for Occupational Safety* (coauthored with Drs. G. R. Lehman & M. J. Kalsher and published in 1989 by Make-A-Difference, Inc.), and *Motivating Health Behavior* (coauthored by Drs. J. P. Elder, M. F. Hovell, & J. A. Mayer, published in 1994 by Delmar Publishers, Inc.). Each of these texts followed standard professional guidelines (including the use of third person, past verb tense, and professional jargon), and as a result they are not very enjoyable to read. As with most professional textbooks, they are read by a select group: teachers or investigators preparing lectures, scholarship, or research designs; and students who are tested on content. These texts are not interesting enough to appeal to the countless people, including professional practitioners and organizational leaders, who could make a difference in protecting the environment and improving people's health and safety. Thus, these books have not had nearly the impact warranted by their contents.

To make this book accessible to a wide variety of readers, it is written in an exciting, easy-to-read style without jargon, thanks to invaluable editorial coaching by Dave Johnson, editor of *Industrial Safety and Hygiene News.*

Each chapter includes several original drawings by George Wills to illustrate concepts and add some humor to the learning process. I intersperse these drawings in my professional addresses and workshops, and audiences find them both enjoyable and enlightening. I predict that some of you will page through the book and look for these illustrations. That's a useful beginning to learning concepts and techniques for improving the human dynamics of safety. Then read the explanatory text for a second useful step toward making a difference with this information. If you then discuss the principles and procedures with others, you'll be on your way to putting this information to work in your organization, community, or home.

A TESTIMONY

Throughout this book, I include personal anecdotes to supplement the rationale of a principle or the description of a technique or process. I'd like to end this preface with one such anecdote.

In August, 1994, the Hercules Portland Plant stopped chemical production for two consecutive days so all 64 employees at the facility could receive a two-day workshop on the information presented in this book. Management had received a request for this all-employee workshop from a team of hourly workers who previously attended my two-day professional development conference sponsored by the Mt. St. Helena Section of the American Society of Safety Engineers.

Rick Moreno, a Hercules warehouse operator and hazardous materials unloader for more than 20 years, wrote the following reaction to my workshop. He read it to his coworkers at the start of the Hercules seminar. It set the stage for a most constructive and gratifying two days of education and training. If you approach the information in this book with some of the enthusiasm and optimism reflected in Rick's words, you can't help but make a difference in someone's safety and health.

> *Knowledge is precious. It's like trying to carry water in your cupped hands to a thirsty friend. Ideas that were crystal clear upon hearing them, tend to slip from your memory like water through the creases of your hands, and*

while you may have brought back enough water to wet your friend's lips, he will not enjoy the full drink that you were able to take.

And so it is with this analogy of the Total Safety Culture. Those who were there can only wet your lips with this new concept. Not a class or a program, but a safe well way to live your life that spills into other avenues of our environment.

It has no limit or boundaries as in this year, this plant. It's more like we're on our way and something wonderful is going to happen.

And even though no answers are promised or given, the avenues in which to find our own answers for our own problems will be within our reach. . . That is why it is important that everyone has the opportunity to take a full drink of the Total Safety Culture instead of having our lips wet. Something wonderful is going to happen.

This book is for you, Rick Moreno—and the many others who want to understand the psychology of safety and reduce personal injuries. Hopefully, this material will be used as a source of principles and procedures that you can return to for guidance and benchmarks along your innovative journey toward building a safer culture of more actively caring people.

E. Scott Geller
February 1996

Acknowledgments

I purchased an attractive print of a newborn colt from an artist at Galeria San Juan, Puerto Rico, in December 1992. While the artist, Jan D'Esopo, was signing my print, I asked her how long it took to complete the original.

"Twenty-five minutes or 25 years," she replied, "depending on how you look at it."

"What do you mean?" I asked.

"Well, it took me only 25 minutes to fill the canvas, but it took me 25 years of training and experience to prepare for the artistry."

I feel similarly about completing this book. While I started to write the original draft about a year ago, I've been preparing to write this book since entering the College of Wooster in Wooster, Ohio, in 1960. Almost all exams at this small liberal arts college required written discussion (rather than selecting an answer from a list of alternatives). Therefore, I received early experience and feedback at integrating concepts and research findings from a variety of sources. I was introduced to the scientific method at Wooster, and applied it to my own behavioral science research during both my junior and senior years.

Throughout five years of graduate education at Southern Illinois University in Carbondale, I developed sincere respect and appreciation for the scientific method as the key to profound knowledge. My primary areas of graduate study were learning, personality, social dynamics, and human information processing and decision making. The chairman of both my thesis and dissertation committees, Dr. Gordon F. Pitz, gave me special coaching in research methodology and data analysis, and refined my skills for professional writing.

In 1968, I was introduced to the principles and procedures of applied behavior analysis (the foundation of behavior-based safety) from one graduate course and a few visits to Anna State Hospital in Anna, Illinois, where two eminent scholars (Drs. Ted Ayllon and Nate Azrin) were conducting seminal research in this field. These learning experiences (brief in comparison with all my other education) convinced me that behavior-focused psychology could make large-scale improvements in people's lives. This insight was to have dramatic influence on my future teaching, research, and scholarship.

I started my professional career in 1969 as Assistant Professor of Psychology at Virginia Polytechnic Institute and State University (Virginia Tech). With assistance from undergraduate and graduate students, I developed a productive laboratory and research program in cognitive psychology. My tenure and promotion to Associate Professor was based entirely upon my professional scholarship in this domain. However, in the mid-1970s I became concerned that this laboratory work had limited potential for helping people. And this conflicted with my personal mission statement to make beneficial differences in people's quality of life. Therefore I turned to another line of research.

Given my conviction that behavior-based psychology has the greatest potential for solving organizational and community problems, I focused my research on finding ways to make this happen. Inspired by the first Earth Day in April 1970, my students and I developed, evaluated, and refined a number of community-based techniques for increasing environment-constructive behaviors and decreasing environment-destructive behaviors. This prolific research program culminated with the 1982 Pergamon Press publication of *Preserving the environment: New strategies for behavior change*, which I coauthored with Drs. Richard A. Winett and Peter B. Everett.

Besides targeting environmental protection, my students and I applied behavior-based psychology to a number of problem areas, including prison administration, school discipline, community theft, transportation management, and alcohol-impaired driving. In the mid-1970s we began researching strategies for increasing the use of vehicle safety belts. This led to a focus on the application of behavior-based psychology to prevent unintentional injury in organizational and community settings.

Perhaps this brief history of my professional education and research experience legitimizes my authorship of a text on the psychology of safety. However, my purpose is not so much to establish credibility as to acknowledge the vast number of individuals—teachers, researchers, colleagues, and students—who prepared me for writing this book. Critical to this preparation were our numerous research projects (since 1970), and this could not have been possible without dedicated contributions from hundreds of university students. My graduate students managed most of these field studies, and I'm truly grateful for their valuable talents and loyal efforts.

Financial support from a number of corporations and government agencies made our 25 years of intervention research possible. We received significant research funds from the Alcohol, Drug Abuse, and Mental Health Administration; the Alcoholic Beverage Medical Research Foundation; Anheuser-Busch Companies, Inc.; Centers for Disease Control and Prevention; Domino's Pizza, Inc.; Exxon Chemical Company; General Motors Research Laboratories; Hoechst-Celanese; the Motor Vehicle Manufacturers Association; Motors Insurance Corporation; the National Highway Traffic Safety Administration; the National Institute on Alcohol Abuse and Alcoholism; the National Institute for Occupational Safety and Health; the National Science Foundation; Sara Lee Knit Products; the U.S. Department of Education; the U.S. Department of Energy; the U.S. Department of Health, Education, and Welfare; the U.S. Department of Transportation; and the Virginia Departments of Agriculture and Commerce,

Litter Control, Motor Vehicles, and Welfare and Institutions. Profound knowledge is only possible through programmatic research, and these organizations made it possible for my students and me to develop and systematically evaluate ways to improve attitudes and behaviors throughout organizations and communities.

I am also indebted to the numerous guiding and motivating communications I have received from corporate and community safety professionals worldwide. Daily contacts with these individuals shaped my research and scholarship, and challenged me to improve the connection between research and application. They also provided valuable positive reinforcement to prevent burnout. It would take pages to name all of these friends and acquaintances, and then I would necessarily miss many. You know who you are—Thank you!

The advice, feedback, and friendship of two individuals—Harry Glaser and Dave Johnson—have been invaluable for my preparation to write this text. I first met Harry Glaser in September 1992, after I gave a keynote address at a professional development conference for the American Society of Safety Engineers. As Executive Vice President of Tel-A-Train, Inc., Harry decided that a video-training series on the human dynamics I presented in my talk would be useful. That was the start of an ongoing collaboration in developing videotape scripts, training manuals, and facilitator guides, all of which proved to be valuable preparation for writing this book. In particular, my relationship with Harry Glaser improved my ability to communicate the practical implications of academic research and scholarship.

Also vital to bridging the gap between research and application has been my long-term alliance and synergism with Dave Johnson, Editor of *Industrial Safety and Hygiene News.* Dave and I began learning from each other in the Spring of 1990, when I submitted my first article for his magazine. That year I submitted five articles on the psychology of safety, and Dave did substantial editing on each. Every time one of my articles was published, I learned something about communicating more effectively the bottom line of a psychological principle or procedure.

As author of more than 200 research articles and former editor of the premier research journal in the applied behavioral sciences *(Journal of Applied Behavior Analysis),* I knew quite well how to write for a research audience in psychology. But Dave Johnson showed me that when it comes to writing for safety professionals and the general public, I had a lot to learn. And in this regard, I continue to learn from both Dave Johnson and Harry Glaser.

Dave Johnson has invited me to submit articles to *ISHN* periodically; in 1994 and 1995, I submitted an article every month. To date I've submitted 47 articles to *ISHN* and each profited immensely from Dave's suggestions and feedback. Preparing these articles laid the groundwork for this text. And Dave served as editor of this book, dedicating long hours to improving the clarity and readability of my writing. Thus, the talent and insight of Dave Johnson have been incorporated throughout this book, and I am eternally beholden to him.

Before sending my chapter drafts to Dave, I sent them to colleagues for feedback. Several of these individuals gave me useful suggestions, most notably Charlie Hart (Safety Director, Exxon Chemical Company, Mt. Belvieu, Texas), Carl Metzgar (trainer and consultant in safety and loss control, Winston-Salem, North Carolina), Mike Gilmore and Steve Roberts (Senior Partners for Safety Performance Solutions, Blacksburg, Virginia).

The drawings were created by George Wills (Blacksburg, Virginia). I think they add vitality and fun to the written presentation. I hope you agree. But without the craft and dedication of Jason Fortney, the illustrations could not have been electronically scanned for use by the publisher. In fact, Jason coordinated the final processing of this entire text, combining tables and diagrams (which he refined) with George Wills' illustrations, and the word processing from Gayle Kennedy and Candice Chevaillier.

I also sincerely appreciate the daily support and encouragement I received from my graduate students in 1995: Ted Boyce, Steve Clarke, Jason DePasquale, Kent Glindemann, and Chuck Pettinger; and from my senior partners at Safety Performance Solutions: Sherry Casali, Anne French, Mike Gilmore, Molly McClintock, and Steve Roberts. I'm sincerely grateful for the "actively caring" attention and support I received from the Chilton Book Company staff, particularly Susan Clarey for sharing the vision of this text from the start (after attending one of my two-day seminars), and Kathy Conover for patiently guiding me through the continuous improvement process. All of these people, plus many, many more, have contributed to 35 years of preparation for this *Psychology of Safety*. I thank you all very much. I'm hopeful that the synergy from all your contributions will help readers make rewarding and long-term differences in people's lives.

Part One

Orientation and Approach

Chapter 1

Choosing the Right Approach

The basic purpose of this book is outlined in Chapter 1—to explore the human dynamics of occupational health and safety, and show how they can be managed to significantly improve safety performance. The principles and practical procedures you'll learn are not based on common sense or intuition, but rather on reliable scientific investigation. Many recommendations seem counter to "pop psychology" and traditional approaches to safety. So keep an open mind while you read about the psychology of safety.

Safety professionals, team leaders, and concerned workers today scramble to find the "best" safety approach for their workplace. Typically, whatever offers the cheapest "quick fix" sells. This is not surprising, given the "lean and mean" atmosphere of the times. Programs that offer the most benefit with least effort sound best, but will they really work to improve safety over the long term?

This text will help you ask the right questions to determine whether a particular approach to safety improvement will work. More importantly, this text describes the basic ingredients needed to improve organizational and community safety. In fact, you'll find sufficient information to improve any safety process. Learning the principles and procedures described here will enable you to make a beneficial, long-term difference in the safety and health of your workplace, home, and community.

Selecting the Most Appropriate Program

With so many different approaches to safety improvement available, how can we select the best? My first thought is to ask, "What does the research indicate?" In other words, are there objective data available from program comparisons to shed light on our dilemma? Unfortunately, there are few systematic comparisons of alternative safety models. However, this doesn't stop consultants from showing us impressive results regarding the success of their approaches. Nor does it prevent them from implying (or boldly stating) that we can obtain similar fantastic results by simply following their patented "steps to success."

Keep in mind this marketing information usually comes from selected client case studies. Very few of these "success stories" were collected objectively and reliably enough to meet the rigorous standards of a professional research journal. When consultants try to sell you an approach to safety with this kind of data, ask them if they have published their results in a peer-reviewed journal. If they can show you a published report of their impressive results or a professional presentation of a program very similar to theirs, then give their approach special consideration in your selection process.

Most of the published research on safety improvement systematically evaluates whether a particular program worked in a particular situation, but it does not compare one approach with another. In other words, this research tells us whether a particular strategy is better than nothing, but offers no information regarding the relative impact of two or more different strategies to safety improvement. Such research has limited usefulness when selecting between different approaches.

An exception can be found in a 1993 review article in *Safety Science,* where Stephen Guastello (1993) summarized systematically the evaluation data from 53 different research reports of safety programs. Dr. Guastello provided rare and useful information for deciding how to improve safety. You can assume the evaluations were both reliable and valid, since each report appeared in a scientific peer-reviewed journal. All of the studies selected for his summary were conducted in a workplace setting since 1977, and each study evaluated program impact with outcome data (including number and severity of injuries).

As listed in Figure 1.1, ten different approaches to safety improvement were represented in the 53 research articles summarized by Dr. Guastello. They are ranked according to the mean percentage decrease in injury rates as detailed by Dr. Guastello in his careful analysis of the published reports. Since half of the percentages were based on three or fewer research reports, the program ranking should be considered

APPROACH	NUMBER OF STUDIES	NUMBER OF SUBJECTS	AVERAGE REDUCTION (%)
1. Behavior-Based	7	2,444	59.6
2. Ergonomics	3	na	51.6
3. Engineering Change	4	na	29.0
4. Group Problem Solving	1	76	20.0
5. Government Action (Finland)	2	na	18.3
6. Management Audits	4	na	17.0
7. Stress Management	2	1,300	15.0
8. Poster Campaign	2	6,100	14.0
9. Personnel Selection	26	19,177	3.7
10. Near-Miss Reporting	2	na	0.0

na = relevant information not available

Figure 1.1. Research comparisons reveal informative ranking of approaches to reduce work injuries. (Adapted from S. J. Guastello, 1993)

preliminary. More research on program impact is clearly needed, as are systematic comparisons.

From my reading of Dr. Guastello's article, I believe it's safe to say the behavior-based and comprehensive ergonomics approaches lead the field. Personnel selection, the most popular method (26 studies targeted a total of 19,177 employees), is among the least effective. With the exception of "near miss" reporting, the other program techniques are clearly in the middle of the ranking, with insufficient evidence to favor one over another. To appreciate this ranking of program effectiveness, it is helpful to define the program labels given in Figure 1.1. Here are brief descriptions of these approaches to reduce workplace injuries:

1. **BEHAVIOR-BASED.** Programs in this category consisted of employee training regarding particular safe and at-risk behaviors, systematic observation and recording of the targeted behaviors, and feedback to workers regarding the frequency or percentage of safe versus at-risk behavior. Some of these programs included goal setting and/or incentives to encourage the observation and feedback process. See Petersen (1989) for a comprehensive review of behavior-based studies in the research literature and for more evidence that this approach to industrial safety deserves top billing.

2. **COMPREHENSIVE ERGONOMICS.** The ergonomics (or human factors) approach to safety refers essentially to any adjustment of working conditions or equipment in order to reduce the frequency or probability of an environmental hazard or at-risk behavior (Kroemer, 1991). An essential ingredient in these programs was a diagnostic survey or environmental audit by employees which led to specific recommendations for eliminating hazards that put employees at risk or promoted at-risk behaviors. Dr. Guastello noted a direct relationship between injury reductions and the amount of time devoted to dealing with the ergonomic recommendations of a diagnostic survey. See Guastello (1989) for further discussion of the development and application of an ergonomics diagnostic survey.

3. **ENGINEERING CHANGES.** This category included the introduction of robots or the comprehensive redesign of facilities to eliminate certain at-risk behaviors. It was noted, however, that the robotic interventions introduced the potential for new types of workplace injuries, like a robot catching an operator in its work envelope and impaling him or her against a structure. Thus, robotic innovations usually require additional engineering intervention such as equipment guards, emergency kill switches, radar-type sensors, and workplace re-design to prevent injury from robots.

4. **GROUP PROBLEM SOLVING.** For this approach, operations personnel met voluntarily to discuss safety issues and problems, and to develop action plans for safety improvement (Saarela, 1990). This approach is analogous to quality circles where employees who perform similar types of work meet regularly to solve problems of product quality, productivity, and cost.

5. **GOVERNMENT ACTION (IN FINLAND).** Two Finnish government agencies responsible for labor production targeted the most problematic occupational groups and implemented certain action strategies. These included: a) disseminating information to work supervisors regarding the cause of workplace injuries and methods to reduce them; b) setting standards for safe machine repair and use; and c) conducting periodic work site inspections. See Bjurstrom (1989) for more specifics regarding this Finnish national intervention.

6. MANAGEMENT AUDITS. For the programs in this category, designated managers were trained to administer a standard International Safety Rating System (ISRS). This system evaluates workplaces based on 20 components of industrial safety. These include: leadership and administration, management training, planned inspections, task and procedures analysis, accident investigations, task observations, emergency preparedness, organizational rules, accident analysis, employee training, personal protective equipment, health control, program evaluation, engineering controls, and off-the-job safety. Managers conduct the comprehensive audit annually to develop improvement strategies for the next year. Specially certified ISRS personnel visited target sites and recognized a plant with up to five "stars" for exemplary safety performance. See Eisner and Leger (1988) or Pringle and Brown (1990) for more specifics on application and impact of the ISRS.

7. STRESS MANAGEMENT. These programs taught employees how to cope with stressors, or sources of work stress (Ivancevich, Matteson, Freedman, & Phillips, 1990; Murphy, 1984). Exercise was often a key action strategy promoted as a way to prevent stress-related injuries in physically demanding jobs (cf. Cady, Thomas, & Karwasky, 1985). I discuss the topic of stress as it relates to injury prevention in Chapter 6.

8. POSTER CAMPAIGNS. The two published studies in this category evaluated the accident reduction impact of posting signs that urged workers at a shipyard to avoid certain at-risk behaviors and to follow certain safe behaviors. Most signs were posted at relevant locations and gave specific behavioral instructions like "Take material for only one workday," "Gather hoses immediately after use," "Wear your safety helmet," and "Check railing and platform couplings (on scaffolds)."

For one study, safety personnel at the shipyard gave work teams weekly feedback regarding compliance with sign instructions (Saarela, Saari, & Aaltonen, 1989). In the other study, environmental audits, group discussions, and structured interviews were used to develop the poster messages (Saarela, 1989). Thus, it's possible factors other than the posters themselves contributed to the moderate short-term impact of this intervention approach.

9. PERSONNEL SELECTION. This popular but ineffective approach to injury prevention is based on the intuitive notion of "accident proneness." The strategy is to identify aspects of accident proneness among job applicants and then screen out people with critical levels of certain characteristics. Accident proneness characteristics targeted for measurement and screening have included: anxiety, distractibility, tension, insecurity, beliefs about injury control, general expectancies about personal control of life events, social adjustment, reliability, impulsivity, sensation seeking, boredom susceptibility, and self-reported alcohol use.

Although measuring and screening for accident proneness sounds like a "quick fix" approach to injury prevention, this method has several problems you will readily realize as you read more in this book about the psychology of safety. Briefly, this technique has not worked reliably to prevent workplace injuries because: a) the instruments or procedures available to measure the proneness characteristics are unreliable or invalid; b) the characteristics do not carry across settings, so a person might show them at home but not at work or vice versa; and c) a person with a higher desire to take risks (such as a sensation seeker) might be more inclined to take appropriate precautions (like using personal protective equipment) to avoid potential injury.

Also, although individuals have demonstrated different risk levels, many have exhibited these inconsistently—the risk-taking is likely to be influenced by environmental conditions. Additionally, finding correlations between certain person characteristics and injury rates does not mean the proneness factors caused changes in the injury rate (cf. Rundmo, 1992). Other factors, including cultural factors or environmental events, could cause both the person characteristics and the accident proneness. See Geller (1994a,b) for additional details regarding problems with this approach to injury prevention.

10. "NEAR-MISS" REPORTING. This approach involved increased reporting and investigation of incidents that did not result in an injury but certainly could have under slightly different circumstances. One program in this category increased the number of corrective suggestions generated, but did not reduce injury rate. The other scientific publication in this category reported a 56-percent reduction in injury severity as a result of increased reporting of near hits[1], but the overall number of injuries did not change.

THE CRITICAL HUMAN ELEMENT

Every safety approach listed in Figure 1.1 requires you to consider the human element, or the psychology of safety. Indeed, the most successful approaches, behavior-based safety and comprehensive ergonomics, directly address the human aspects of safety. The behavior-based approach targets human behavior and relies on interpersonal observation and feedback for intervention. The success of comprehensive ergonomics depends on employees observing interaction between behaviors and work situations, and then recommending feasible changes in behavior, equipment, or environmental conditions for safety achievement. See the text edited by Oborne *et al* (1993) for a comprehensive discussion of the psychological aspects of ergonomics.

Achieving success in safety today requires concerted efforts in the realm of psychology. Safety professionals are hungry for insights. In recent years, many seminars at national and regional safety conferences purporting to teach aspects of the psychology of safety have attracted standing-room-only crowds. Just look at these titles from recent conferences of the National Safety Council or the American Society of Safety Engineers:

- "Managing Safe Behavior for Lasting Change"
- "Humanizing the Total Safety Program"
- "The Human Element in Achieving a Total Safety Culture"
- "The Psychology of Injury Prevention"
- "Behavior-Based Safety Management: Parallels with the Quality Process"
- "Behavioral Management Techniques for Continuous Improvement"
- "Improving Safety Through Innovative Behavioral and Cultural Approaches"
- "Total Safety Leadership—Empowering All Employees"

1. "Near miss" is used routinely in the workplace to refer to an incident that did not result in an injury. Since a literal translation of this term would mean the injury actually occurred, "near hit" is used throughout this text instead of "near miss."

- "Managing Safety Using a Behavior-Based Approach"
- "Potholes in the Road to Behavioral Safety"
- "Achieving Total Safety Through Employee Involvement"

I attended each of these seminars and found numerous inconsistencies between presentations dealing with the same topic. Sometimes, I noted erroneous and frivolous statements, inaccurate or incomplete reference to psychological theory or research, and invalid or irresponsible comparisons between various approaches to dealing with the psychology of safety. It seemed a primary aim of several presenters was to "sell" their own particular program or consulting services by overstating the benefits of their approach and giving an incomplete or naive discussion of alternative methods or procedures.

THE FOLLY OF CHOOSING WHAT SOUNDS GOOD

The theory, research, and tools in psychology are so vast and often so complex that it can be overwhelming to decide which particular approach or strategy to use. As a result, we are easily biased by common-sense words that sound good. Valid theory, principles, and procedures founded on solid research evidence are often ignored. Today, there is an apparent endless market of self-help books, audiotapes, and videotapes addressing concepts seemingly relevant to the psychology of injury prevention. In recent years, I have listened to the following audiotapes, representing only a fraction of "pop psychology" tapes with topics relevant to the psychology of safety:

- "Coping with Difficult People" by R. M. Branson
- "Personal Excellence" by K. Blanchard
- "How to Build High Self-Esteem" by J. Canfield
- "The 7 Habits of Highly Effective People" by S. R. Covey
- "The Science of Personal Achievement" by N. Hill
- "Increasing Human Effectiveness" by R. Moawad
- "Lead the Field" by E. Nightingale
- "Unlimited Power" by A. Robbins
- "The Psychology of Achievement" by B. Tracey
- "The Psychology of Winning" by D. Waitley
- "Self-Esteem" by J. White
- "Goal Setting" by Z. Zigler

Which, if any, of these "pop psychology" audiotapes gives safety professionals the "truth"—the most effective and practical tools for dealing with the human aspects of safety? Many of the strategies to promote personal growth and achievement, including attitude and behavior change, were selected and listened to with trust and optimism because they sound good—not because there is solid scientific evidence that the strategies work. Many of the same anecdotes and quotes from famous people were repeated across audiotapes. Also, strategies suggested for developing self-esteem and building personal success were quite similar, with goal-setting and self-affirmations (such as repeating "I am the greatest" to oneself) leading the list. Does this repetition of good-sounding self-help strategies make them effective?

Some of the most cost-effective strategies for managing behaviors and attitudes at the personal and organizational level are not even mentioned in many of the "pop psychology" books, audiotapes, and videotapes. This might be the case not only because authors and presenters are unaware of the latest research, but also because many of the best techniques for individual and group improvement do not sound good—at least at first. The primary purpose of this text is to teach the most effective approaches for dealing with the human aspects of occupational safety and health. These principles and procedures were not selected because they sound good, but because their validity has been supported with sound research.

THE FALLACY OF RELYING ON COMMON SENSE

There is no doubt that good common sense can go a long way in selecting effective techniques for benefiting human achievement. I have met many people, including supervisors, line workers, safety professionals, and motivational speakers who seem to have special intuition or common sense for selecting approaches to help people improve. Indeed, Tom Peters, Anthony Robbins, Brian Tracey, Denis Waitley, and many others who successfully market techniques for increasing human potential and achievement are particularly skillful at selecting those principles and procedures backed by research. But it is important to realize that common sense is subjective, based on a person's everyday selective experiences and interpretation of those experiences.

As mentioned, I prefer principles and procedures based on scientific knowledge, which comes from the experience of the researcher. At a four-day "Quality Enhancement Seminar" in 1991, I heard W. Edwards Deming assert, "Experience teaches us nothing; that's why American business is in such a mess." Dr. Deming called for theory to guide objective and reliable observations, and to integrate the results from these systematic data-based experiences. Thus, while common sense is gained through biased subjective experience, scientific knowledge is gained through theory-driven objective experimentation.

Dr. Aubrey C. Daniels, a world-renowned educator and consultant in the field of organizational performance management, asserts in his 1994 book, *Bringing Out the Best in People,* that he is "on a crusade to stamp out the use of common sense in business. Contrary to popular belief there is not too little common sense in business, there is too much." (Daniels, 1994, p. 10). Dr. Daniels lists the following distinctions between common sense and scientific knowledge, reflecting the need to be cautious when relying on only common sense to deal with human aspects of occupational health and safety:

Common-sense knowledge is acquired in ordinary business and living, while scientific knowledge must be pursued deliberately and systematically.

Common-sense knowledge is individual; scientific knowledge is universal.

Common-sense knowledge accepts the obvious; scientific knowledge questions the obvious.

Common-sense knowledge is vague; scientific knowledge is precise.

Common sense cannot be counted on to produce consistent results; applications of scientific knowledge yield the same results every time.

Common sense is gained through uncontrolled experience; scientific knowledge is gained through controlled experimentation.

I've heard or read a number of psychology-related statements from motivational speakers and consultants that sound like good common sense but in fact contradict scientific knowledge. Some of these statements appear so many times in the "pop psychology" literature that they are accepted as basic truths, when in fact they cannot be substantiated with objective evidence.

How many times, for example, have you heard this statement: "Do something 21 times (or for one month) and it becomes a habit." That statement sounds reasonable, unless you know some basics about learning and the maintenance of behavior. This book will enable you to understand why such a statement about habit formation is clearly false. At this point, I hope you are open to questioning the validity of good-sounding statements that are not supported by sound research.

Research in psychology, for example, does not generally support the following common statements related to the psychology of occupational health and safety:

- Practice makes perfect.
- Spare the rod and spoil the child.
- Attitudes need to be changed before behavior will change.
- Incentives and rewards are detrimental to intrinsic motivation.
- Proper education is sufficient to motivate behavior change.
- Human nature motivates safe and healthy behavior.
- People will naturally help in a crisis.
- Rewards for not having injuries reduce injuries.
- All injuries are preventable.
- Zero injuries should be a safety goal.
- Manage only that which can be measured.

These and other common safety beliefs will be refuted in this book by scientific knowledge obtained from systematic research. Sometimes case studies will illustrate the practicality and benefits of a particular principle or procedure, but the validity of the information was not founded on case studies. The approaches presented in this text were originally discovered and verified with systematic and repeated scientific research in laboratory and field settings.

Behavior: The Scientific Approach

Many "pop psychology" self-help books, audiotapes, and motivational speeches give minimal if any attention to behavior-based approaches to personal achievement. "Behavioral control" and "behavior modification" do not sound good. The term "behavior" has negative connotations, as in "let's talk about your behavior at the party last night." Dr. B. F. Skinner, the founder of behavioral science and its many practical applications, was one of the most misunderstood and underappreciated scientists and scholars of this century, primarily because the behavior management principles he taught did not sound good. Two particularly insightful but underappreciated and misunderstood books by Skinner are *Walden Two* (1948) and *Beyond Freedom and Dignity* (1971). Professor Skinner and his followers have shown over and over again that behavior is motivated by its consequences, and thus behavior can be changed by controlling the events that follow behavior. But this principle of "control by consequences" doesn't sound as good as "control by positive thinking and free will." Therefore, the scientific principles and procedures from behavioral science have been underappreciated and underused.

This book teaches you how to apply behavioral science for safety achievement. The research recommends that we start with behavior. But the demonstrated validity of a behavior-based approach does not mean the better-sounding, person-based approaches should not be used. It is important to consider the feelings and attitudes of employees, since it takes people to implement the tools of behavior management. This book will teach you how certain feeling states critical for safety achievement—self-esteem, empowerment, and belongingness—can be increased by applying behavioral science. It is possible to establish interpersonal interactions and behavioral consequences in the workplace to increase important feelings and attitudes. I will show you how increasing these feeling states benefits behavior and helps to achieve safety excellence.

As illustrated in Figure 1.2, an attitude of frustration or an internal state of distress can certainly influence driving behaviors, and vice versa. Indeed, internal (unobserved) states of mind continually influence observable behaviors, while changes in observable behaviors continually affect changes in person states or attitudes. Thus, it is possible to "think a person into safe behaviors" (through education, counselling, and consensus-building exercises), and it is possible to "act a person into safe thinking" (through behavior management techniques).

In an industrial setting, it is most cost effective to target behaviors first through behavior management interventions (to be reviewed in the following chapters) implemented by employees themselves. Small changes in behavior can result in attitude change, fol-

Figure 1.2. Behavior influences attitude and attitude influences behavior.

lowed by more behavior change and more desired attitude change. This spiraling of behavior feeding attitude, attitude feeding behavior, behaviors feeding attitudes and so on can lead to employees becoming totally committed to safety achievement, as reflected in their daily behavior. And all of this could start with a relatively insignificant behavior change in one employee—a "small win."

In Conclusion

In this chapter I have outlined the basic orientation and purpose of this text, which is to teach principles for understanding the human aspects of occupational health and safety, and to illustrate practical procedures for applying these principles to achieve significant improvements in organizational and community-wide safety.

The principles and procedures are not based on common sense or intuition, but rather on reliable scientific investigation. Some will contradict common folklore in "pop psychology" and require shifts in traditional approaches to the management of organizational safety. So approach this material with an open mind. Be ready to relinquish fads, fancies, and folklore for innovations based on unpopular but research-supported theory.

I promise this "psychology of safety" is based on the latest and most reliable scientific knowledge; and I promise greater safety achievement in your organization if you follow the principles and procedures presented here. Reference to the research literature is given throughout this text to verify the concepts, principles, and procedures discussed. Read some of these yourself to experience indirectly the rewards of scientific inquiry and distance yourself from the frivolity of common sense.

REFERENCES

Bjurstrom, L. M. (1989). Priority to key areas and management by results in the national accident prevention policy. In A. Mital (Ed.), *Advances in industrial ergonomics and safety*, Taylor and Francis, London, pp. 801-808.

Cady, L. D., Thomas, P. C., & Karwasky, R. J. (1985). Program for increasing health and physical fitness of fire fighters. *Journal of Occupational Medicine*, 27, 110-114.

Daniels, A. C. (1994). *Bringing out the best in people*. New York: McGraw-Hill, Inc.

Deming, W. E. (1991, May). *Quality, productivity, and competitive position*. Four-day workshop presented in Cincinnati Ohio by Quality Enhancement Seminars, Inc.

Eisner, H. S., and Leger, J. P. (1988). The international safety rating system in South African mining. *Journal of Occupational Accidents*, 10, 141-160.

Geller, E. S. (1994a). What's in a perception survey? *Industrial Safety and Hygiene News*, November, 11-12.

———. (1994b). Survey reliability vs. validity. *Industrial Safety and Hygiene News*. December, 12-13.

Guastello, S. J. (1989). Catastrophe modeling of the accident process: Evaluation of an accident reduction program using the Occupational Hazards Survey. *Accident Analysis and Prevention*, 21, 61-77.

———. (1993). Do we really know how well our occupational accident prevention programs work? *Safety Science*, 16, 445-463.

Ivancevich, J. M., Matteson, M. T., Freedman, S. M., & Phillips, J. S. (1990). Worksite stress management interventions. *American Psychologist*, 45, 252-261.

Kroemer, K. H. (1991). Ergonomics. *Encyclopedia of Human Biology,* 3, 473-480.

Murphy, L. R. (1984). Occupational stress management: A review and appraisal. *Journal of Occupational Psychology,* 57, 1-16.

Oborne, D. J., Branton, R., Leal, F., Shipley, P., & Stewart, T. (1993) (Eds.). *Person-centred ergonomics: A Brontonian view of human factors.* Washington, D.C.: Taylor & Francis.

Petersen, D. (1989). *Safe behavior reinforcement.* New York: Aloray, Inc.

Pringle, D. R. S., & Brown, A. E. (1990). International safety rating system: New Zealand's experience with a successful strategy. *Journal of Occupational Accidents,* 12, 41.

Rundmo, T. (1992). Risk perception and safety on offshore petroleum platforms—Part II: Perceived risk, job stress and accidents. *Safety Science,* 15, 53-68.

Saarela, K. L. (1989). A poster campaign for improving safety on shipyard scaffolds. *Journal of Safety Research,* 20, 177-185.

———. (1990). An intervention program utilizing small groups: A comparative study. *Journal of Safety Research,* 21, 149-156.

Saarela, K. L., Saari, J., & Aaltonen, M. (1989). The effects of an informal safety campaign in the ship building industry. *Journal of Occupational Accidents,* 10, 255-266.

Skinner, B. F. (1971). *Beyond freedom and dignity.* New York: Alfred A Knopf.

Skinner, B. F. (1948). *Walden Two.* New York: MacMillan Publishing Co., Inc.

Chapter 2
Start with Theory

In this chapter we'll consider the value of theory in guiding our approaches to safety and health improvement. You'll see how a vision for a Total Safety Culture is a necessary guide to achieve safety excellence. A basic principle here is that safety performance results from the dynamic interaction of environmental, behavioral, and personal factors. Achieving a Total Safety Culture requires attention to each of these. I make a case for integrating person-based and behavior-based psychology in order to address most effectively the human dynamics of injury prevention.

As you know, some safety efforts suffer from a "flavor of the month" syndrome. New procedures or intervention programs are tried seemingly at random, without an apparent vision, plan, or supporting set of principles. When the mission and principles are not clear, employees' acceptance and involvement suffer.

Without a guiding theory or set of principles it is difficult to design and refine procedures to stay on course. At a four-day workshop on "Quality, Productivity, and Competitive Position," I heard the late W. Edwards Deming (1991) make this statement more than once to emphasize the value of theory:

> *Experience teaches us nothing;*
> *that's why American business is in such a mess.*

Steven R. Covey (1984, 1990) emphasized the same point in his popular books, *The Seven Habits of Highly Effective People, Principle-Centered Leadership,* and *First Things First,* co-authored by A. Roger Merrill and Rebecca R. Merrill (1994).

A theory or set of guiding principles makes it possible to evaluate the consistency and validity of program goals and intervention strategies. By summarizing the appropriate theory or principles into a mission statement, you have a standard for judging the value of your company's procedures, policies, and performance expectations.

It's important to develop a set of comprehensive principles on which to base safety procedures and policies. Then teach these principles to your employees so they are understood, accepted, and appreciated. This buy-in is certainly strengthened when employees or associates help select the safety principles to follow and summarize them in a company mission statement. This is theory-based safety. A critical challenge, of course, is to choose the most relevant theories or principles for your company culture and purpose, and develop an appropriate and feasible mission statement that reflects the right theory.

A Mission Statement

Several years ago I worked with employees of a major chemical company to develop the general mission statement for safety given in Figure 2.1. This vision for a Total Safety Culture serves as a guideline or standard for the material presented throughout this book, in the same way a corporate mission statement serves as a yardstick for gauging the development and implementation of policies and procedures.

This mission statement might not be suited for all organizations, but it is based on appropriate and comprehensive theory, supported by scientific data from research in psychology. Before developing this statement, employees learned basic psychological theories most relevant to improving occupational safety. These principles are illustrated throughout this text, along with operational (real-world) definitions.

THE VALUE OF THEORY AS "MAP"

I'd like to relate an experience to show how a theory can be seen as a map to guide us to a destination. The mission statement in Figure 2.1 reflects a destination for safety within the realm of psychology. This story also reflects the difficulty in finding the best theory among numerous possibilities.

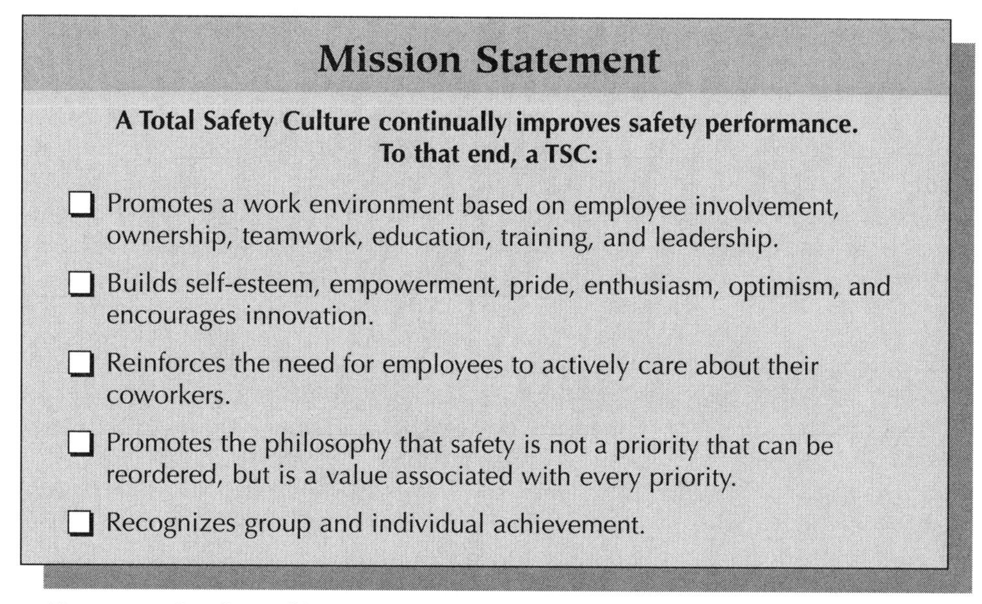

Figure 2.1. A safety achievement mission statement for the principles and procedures covered in this book.

I had the opportunity to conduct a training program at a company in Palatka, Florida. My client sent me step-by-step instructions to take me from Interstate 95 to Palatka. That was my map, limited in scope for sure, but sufficient I presumed to get the job done—to get me to the Holiday Inn in Palatka. But while at the National Car Rental desk, an attendant said my client's directions were incorrect and showed me the "correct way" with National's map of Jacksonville.

I was quick to give up my earlier theory (from my client's handwritten instructions) for this more professional display. After all, I now had a professionally printed map and directions from someone in the business of helping customers with travel plans—a consultant, so to speak. But National's map showed details for a limited area, and Palatka was not on the map. I couldn't verify the attendant's directions with the map, nor could I compare these directions with my client's very different instructions. Without a complete perspective, I chose the theory that looked best. And I got lost.

After traveling 15 miles, I began to question the "National Theory" and wondered whether my client's scribbling had been correct after all. But I stuck by my decision, and drove another ten miles before exiting the highway in search of further instruction.

I certainly needed to reach my destination that night, but motivation without appropriate direction can do more harm than good. In other words, a motivated worker cannot reach safety goals with the wrong theory or principles.

It was late Sunday night and the gas station off the exit ramp was closed, but another vehicle had also just stopped in the parking lot. I drove closer and announced to four tough-looking, grubby characters in a pick-up truck loaded with motorcycles that I was lost and wondered if they knew how to get to Palatka. None of these men had heard of Palatka, but one pulled out a detailed map of Florida and eventually found the town of Palatka. I couldn't see the details in the dark, but I accepted this new "theory" anyway, with no personal verification.

The packaging of this theory was not impressive, but my back was against the wall. I was desperate for a solution to my problem and had no other place to turn. As I left the parking lot with a new theory, I wondered whether I was now on the right track. Perhaps the theory obtained from the National Car rental attendants was correct and I had missed an exit. Whom should I believe? Fortunately, I looked beyond the slick packaging and went with the guy who had the more comprehensive perspective (the larger map). This theory got me to the Holiday Inn Palatka.

RELEVANCE TO OCCUPATIONAL SAFETY

That evening I thought about my experience and its relevance to safety. It reminded me of the dilemma facing many safety professionals when they choose approaches, programs, and consultants to help solve people problems related to safety. Theories, research, and tools in psychology are so vast and often so complex that it can be an overwhelming task to select a theory or set of principles to follow. As discussed in Chapter 1, there is a huge market of self-help books, audiotapes, and videotapes addressing concepts seemingly relevant to understanding and managing the human dynamics of safety. But many of the anecdotes, principles, and procedures given in "pop psychology" books and audiotapes are founded on limited or no scientific data. In fact, more of the material was probably used because it sounded good rather than because systematic research found it valid.

The theory that got me to Palatka was not the most professional or believable, nor was it "packaged" impressively. This doesn't mean you should avoid the slick, well-marketed approaches to occupational safety. I only wish these factors were given much less weight than scientific data.

It is relevant, though, that the more comprehensive map enabled me to find my destination. I have found that many of the human approaches to improving safety are limited in scope or theoretical foundation. Many are sold or taught as packaged programs or step-by-step procedures for any workplace culture.

In the long run, it's more useful to teach comprehensive theory and principles. On this foundation, culture-relevant procedures and interventions can be crafted by employees who will "own" and thus follow them. As the old saying goes, "Give a man a fish and you feed him for a day, teach him how to fish and you feed him for a lifetime."

At breakfast, I told the human resource manager and the safety director, the one who gave me the handwritten instructions, about my problems finding Palatka. Interestingly, each had a different theory on the best way to travel between the Jacksonville airport and Palatka. The safety director stuck with his initial instructions. The human resource manager recommended the route I eventually took. Their discussion was not enlightening; in fact it got me more confused because I didn't have a visual picture or schema (a comprehensive map) in which to fit the various approaches (or routes) they were discussing. In other words, I did not have a framework or paradigm to organize their verbal descriptions. Without a relevant theory, my experience taught me nothing, except the need for an appropriate theory—in this case a map.

A theory should serve as the map that provides direction to meet a specific safety challenge. Obviously it's important to teach the basic theory to everyone who must meet the challenge. Then it's a good idea to have an employee task force summarize the theory in a safety mission statement. When the workforce understands the theory and accepts the summary mission statement, intervention processes based on the theory will not be viewed as "flavor of the month," but as an action plan to bring the theory to life.

When employees appreciate and affirm the theory, they will get involved in designing and implementing the action steps. They will also suggest ways to refine or expand action plans and theory on the basis of systematic observations or scientific evidence. This is the best kind of continuous improvement.

A BASIC MISSION AND THEORY

The mission statement in Figure 2.1 reflects the ultimate in safety—a Total Safety Culture, in which:

- Everyone feels responsible for safety and does something about it on a daily basis.
- People go beyond the call of duty to identify unsafe conditions and at-risk behaviors, and they intervene to correct them.
- Safe work practices are supported intermittently with rewarding feedback from both peers and managers.
- People "actively care" continuously for the safety of themselves and others.
- Safety is not considered a priority that can be conveniently shifted depending on the demands of the situation; rather safety is considered a value linked with every priority of a given situation.

This Total Safety Culture mission is much easier said than done, but it is achievable through a variety of safety processes rooted in the disciplines of engineering and psychology. Generally, a Total Safety Culture requires continual attention to three domains:

1. Environment factors (including equipment, tools, physical layout, procedures, standards, and temperature);

2. Person factors (including people's attitudes, beliefs, and personalities); and

3. Behavior factors (including safe and at-risk work practices, as well as going beyond the call of duty to intervene on behalf of another person's safety).

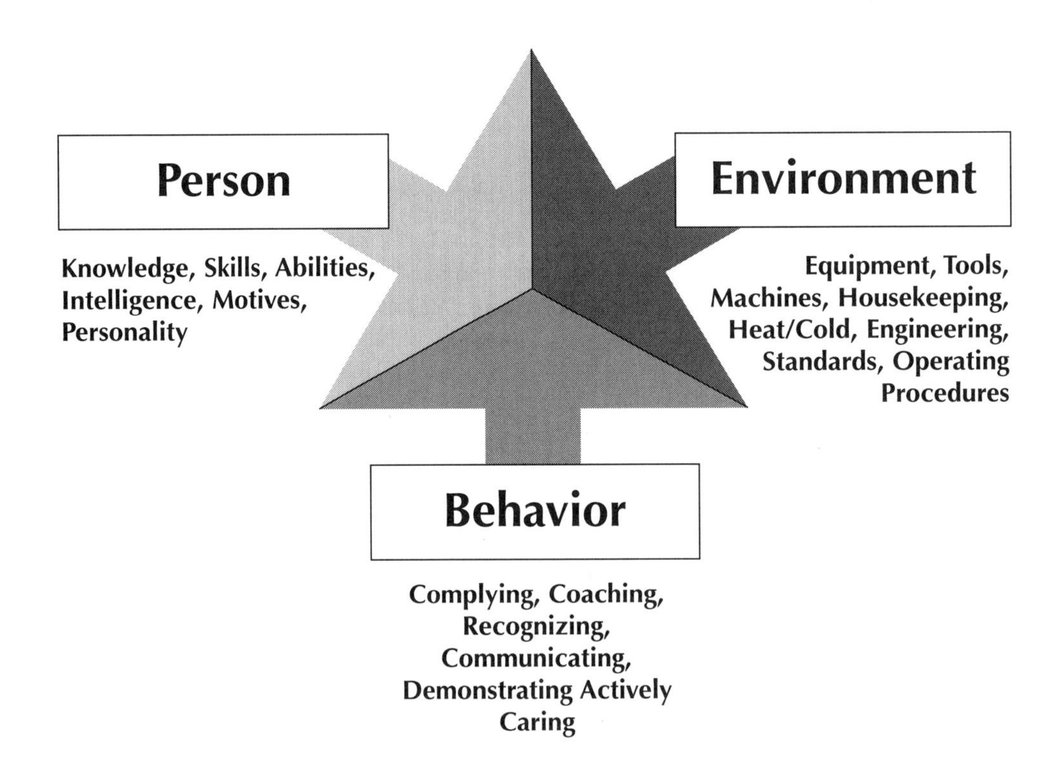

Figure 2.2. A Total Safety Culture requires continual attention to three types of factors.

This triangle of safety-related factors has been termed "The Safety Triad" (Geller, 1989; Geller, Lehman, & Kalsher, 1989) and is illustrated in Figure 2.2.

These three factors are dynamic and interactive. Changes in one factor eventually impact the other two. For example, behaviors that reduce the probability of injury often involve environmental change, and lead to attitudes consistent with the safe behaviors. This is especially true if the behaviors are viewed as voluntary. In other words, when people choose to act safely, they act themselves into safe thinking. These behaviors often result in some environmental change.

The behavior and person factors represent the human dynamics of occupational safety and are addressed in this book. The basic principle here is that behavior-based and person-based factors need to be addressed in order to achieve a Total Safety Culture. These two divergent approaches to understanding and managing the human element represent the psychology of injury prevention.

Behavior-Based versus Person-Based Approaches

Paying attention to only behavior-based factors (the observable activities of people) or to only person-based factors (unobservable feeling states or attitudes of people) is like using a limited map to find a destination, as with my attempt to find Palatka, Florida. The mission to achieve a Total Safety Culture requires a comprehensive framework—a complete map of the relevant psychological territory. Figure 2.3 illustrates the complex interaction of environment, person, and behavior factors.

There are myriad opinions and recommendations on how the psychology of safety can be used to produce beneficial changes in people and organizations. Most can be classified into one of two basic approaches: person-based and behavior-based. In fact, most of the numerous psychotherapies available to treat developmental disabilities and psychological disorders, from neurosis to psychosis, can be classified as essentially person-based or behavior-based.

That is, most psychotherapies focus on changing people either from the inside ("thinking people into acting differently") or from the outside ("acting people into thinking differently"). Person-based approaches

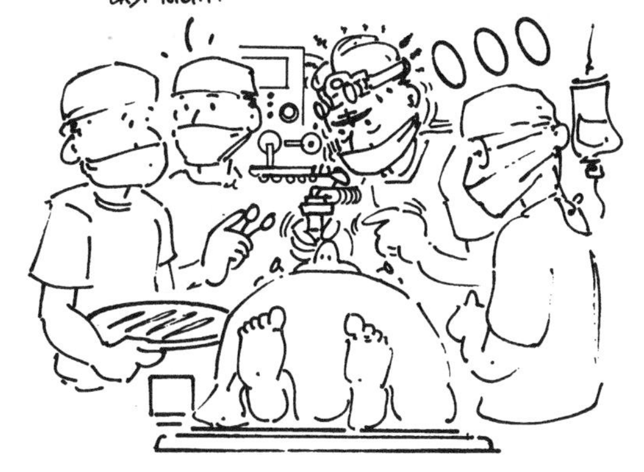

Figure 2.3. Performance results from the dynamic interaction of environment, behavior, and person factors.

attack individual attitudes or thinking processes directly. They teach clients new thinking strategies or give them insight into the origin of their abnormal or unhealthy thoughts, attitudes, or feelings. In contrast, behavior-based approaches attack the clients' behaviors directly. They change relationships between behaviors and their consequences.

Many clinical psychologists use both person-based and behavior-based techniques with their clients, depending upon the nature of the problem. Sometimes the same client is treated with person-based and behavior-based intervention strategies. I am convinced both are relevant in certain ways for improving health and safety. This text will show you how to integrate relevant principles from these two psychological approaches in order to achieve a Total Safety Culture.

THE PERSON-BASED APPROACH

Imagine you see two employees pushing each other in a parking lot, as a crowd gathers around to watch. Is this aggressive behavior, horseplay, or mutual instruction for self defense? Are the employees physically attacking each other to inflict harm, or does this physical contact indicate a special friendship and mutual understanding of the line between aggression and play? Perhaps if you watch longer, and pay attention to verbal behavior, you'll decide whether this is aggression, horseplay, or a teaching/learning demonstration.

However, a truly accurate account might require you to assess each individual's personal feelings, attitudes, or intentions. It's possible, in fact, that one person was being hostile while the other was just having fun, or the contact started as horseplay and progressed to aggression.

This scenario illustrates a basic premise of the person-based approach. Focusing only on observable behavior does not explain enough. People are much more than their behaviors. Concepts like intention, creativity, intrinsic motivation, subjective interpretation, self-esteem, and mental attitude are essential to understanding and appreciating the human dynamics of a problem. Thus, a person-based approach in the workplace applies surveys, personal interviews and focus-group discussions to find out how individuals feel about certain situations, conditions, behaviors, or personal interactions.

A wide range of therapies fall within the general framework of person-based, from the psychoanalytic techniques of Sigmund Freud, Alfred Adler, and Carl Jung to the client-centered humanism developed and practiced by Carl Rogers, Abraham Maslow, and Viktor Frankl (cf. Wandersman, Popper, & Ricks, 1976). Humanism is the most popular person-based approach today, as evidenced by the current market of "pop psychology" videotapes, audiotapes, and self-help books. Some current popular industrial psychology tools—such as the Myers-Briggs Type Indicator and other trait measures of personality, motivation, or risk-taking propensity—stem from psychoanalytic theory and practice. The key principles of humanism found in most "pop psychology" approaches to increasing personal achievement are:

1. Everyone is unique in numerous ways. The special characteristics of individuals cannot be understood or appreciated by applying general principles or con-

cepts, such as the behavior-based principles of performance management, or the permanent personality trait perspective of psychoanalysis.

2. Individuals have far more potential to achieve than they typically realize and should not feel hampered by past experiences or present liabilities.

3. The present state of an individual in terms of feeling, thinking, and believing is a critical determinant of personal success.

4. One's self-concept influences mental and physical health, as well as personal effectiveness and achievement.

5. Ineffectiveness and abnormal behavior result from large discrepancies between one's real self ("who I am") and ideal self ("who I would like to be").

6. Individual motives vary widely and come from within a person.

Readers familiar with the writings of W. Edwards Deming (1986, 1993) and Stephen R. Covey (1989, 1990) will recognize these eminent industrial consultants as humanists, or advocates of a person-based approach.

THE BEHAVIOR-BASED APPROACH

The behavior-based approach to applied psychology is founded on behavioral science as conceptualized and researched by B. F. Skinner (1938, 1974). In his experimental analysis of behavior, Skinner rejected for scientific study unobservable factors such as self-esteem, intentions, and attitudes. He researched only observable behavior and its social, environmental, and physiological determinants. The behavior-based approach starts by identifying observable behaviors targeted for change and the environmental conditions or contingencies that can be manipulated to influence the target behavior(s) in desired directions. (Contingencies are relationships between designated target behaviors and their supporting consequences.)

Behaviorism has effectively solved environmental, safety, and health problems in organizations and throughout entire communities. You first define the problem in terms of relevant observable behavior, and then design and implement an intervention process to decrease behaviors causing the problem and/or increase behaviors that can alleviate the problem (Elder *et al,* 1994; Geller *et al* 1982; Goldstein & Krasner, 1987; Greene *et al,* 1987). The behavior-based approach is reflected in the research and scholarship of several safety consultants and is becoming increasingly popular for industrial applications (Geller *et al,* 1989; Krause *et al,* 1990; McSween, 1995; Petersen, 1989).

CONSIDERING COST-EFFECTIVENESS

When people act in certain ways they usually adjust their mental attitude and self-talk to parallel their actions (Festinger, 1957); and when people change their attitudes, values, or thinking strategies, certain behaviors change as a result (Fishbein & Ajzen, 1975). Thus, person-based and behavior-based approaches to changing people can influence both attitudes and behaviors, either directly or indirectly. Most parents, teachers, first-line supervisors, and safety managers use both approaches in their attempts to change a person's knowledge, skills, attitudes, or behaviors:

When we lecture, counsel, or educate others in a one-on-one or group situation, we are essentially using a person-based approach

When we recognize, correct or discipline others for what they have done, we are operating from a behavior-based perspective

Unfortunately, we're not always effective with our person-based or behavior-based change techniques, and often we don't know whether our intervention worked as intended. In order to apply person-based techniques to psychotherapy, clinical psychologists receive specialized therapy or counseling training for four years or more, followed by an internship of at least a year. This intensive training is needed because tapping into an individual's perceptions, attitudes, and thinking styles is a demanding and complex process. Also, internal dimensions of people are extremely difficult to measure reliably, making it cumbersome to assess therapeutic progress and obtain straightforward feedback regarding therapy skills. Consequently, the person-based therapy process can be very time-consuming, requiring numerous one-on-one sessions between professional therapist and client.

In contrast, behavior-based psychotherapy was designed to be administered by individuals with minimal professional training. From the start, the idea was to reach people where problems occur—in the home, school, rehabilitation institute, and workplace, for example—and teach parents, teachers, supervisors, friends, or co-workers the behavior-change techniques most likely to work under the circumstances (Ullman & Krasner, 1965).

More than three decades of research have shown convincingly that this on-site approach is cost effective, primarily because behavior-change techniques are straightforward and relatively easy to administer, and because intervention progress can be readily monitored by the ongoing observation of target behaviors. By obtaining objective feedback on the impact of intervention techniques, a behavior-based process can be continually refined or altered.

Integrating Approaches

A common perspective, even among psychologists, is that humanists and behaviorists are complete opposites (Newman, 1992; Wandersmann *et al,* 1976). Behaviorists are considered cold, objective, and mechanistic, operating with minimal concern for people's feelings. In contrast, humanists are thought of as warm, subjective and caring, with limited concern for directly changing another person's behavior or attitude. The basic humanistic approach is termed "nondirective" or "client-centered." Therapists, counselors, or coaches do not directly change their clients, but rather provide empathy and a caring and supportive environment for enabling clients to change themselves, from the inside out.

Given the foundations of humanism and behaviorism, it is easy to build barriers between person-based and behavior-based perspectives and assume you must follow one or the other when designing an intervention process. In fact, many consultants in the safety management field market themselves as using one or the other approach, but not both. It is my firm belief that these approaches need to be integrated in order to truly understand the psychology of safety and build a Total Safety Culture.

In Conclusion

Theory or basic principles are needed to organize research findings and guide our approaches to improve the safety and health of an organization. Similarly, a vision for a Total Safety Culture incorporated into a mission statement is needed to guide us in developing action plans to achieve safety excellence.

When employees understand and accept the mission statement and guiding principles, they become more involved in the mission. The action plan will not be viewed as "one more flavor of the month," but as relevant to the right principles and useful for achieving shared goals. Indeed, the workforce will help design and implement the action plans. This is crucial for developing a Total Safety Culture.

A basic principle introduced in this chapter is that the safety performance of an organization results from the dynamic interaction of environment, behavior, and person factors. The behavior and person dimensions represent the human aspect of industrial safety, and reflect two divergent approaches to understanding the psychology of injury prevention. The behavior-based approach is more cost-effective than the person-based approach in effecting large-scale change. But it cannot be effective unless the work culture believes in the behavior-based principles and willingly applies them to achieve the mutual safety mission. This involves person-based psychology, which means a Total Safety Culture requires us to integrate person-based and behavior-based psychology.

REFERENCES

Covey, S. R. (1989). *The seven habits of highly effective people: Restoring the character ethic.* New York: Simon and Schuster.

Covey, S. R. (1990). *Principle-centered leadership.* New York: Simon and Schuster.

Covey, S. R., Merril, A. R., & Merril, R. R. (1994). *First things first.* New York: Simon and Schuster.

Deming, W. E. (1986). *Out of the crisis.* Cambridge, MA: Massachusetts Institute of Technology, Center for Advanced Engineering Study.

———. (1991, May). *Quality, productivity, and competitive position.* Four-day workshop presented in Cincinnati, Ohio by Quality Enhancement Seminars, Inc.

———. (1993). *The new economics for industry, government, education.* Cambridge, MA: Massachusetts Institute of Technology, Center for Advanced Engineering Study.

Elder, J. P., Geller, E. S., Hovell, M. F., & Mayer, J. A. (1994). *Motivating health behavior.* New York: Delmar Publishers, Inc.

Festinger, L. (1957). *A theory of cognitive dissonance.* Stanford, CA: Stanford University Press.

Fishbein, M., & Ajzen, I. (1975). *Belief, attitude, intention, and behavior: An introduction to theory and research.* Reading, MA: Addison-Wesley.

Geller, E. S. (1989). Managing occupational safety in the auto industry. *Journal of Organizational Behavior Management,* 10, (1), 181-185.

Geller, E. S., Lehman, G. R., & Kalsher, M. R. (1989). *Behavior analysis training for occupational safety.* Newport, VA: Make-A-Difference, Inc.

Geller, E. S., Winett, R. A., & Everett, P. B. (1982). *Preserving the environment: New strategies for behavior change.* Elmsford, NY: Pergamon Press.

Goldstein, A. P., & Krasner, L. (1987). *Modern applied psychology.* New York: Pergamon Press.

Greene, B. F., Winett, R. A., Van Houten, R., Geller, E. S., & Iwata, B. A. (1987) (Eds.). *Behavior analysis in the community: Readings from the Journal of Applied Behavior Analysis.* Lawrence, KS: University of Kansas Press.

Krause, T. R., Hidley, J. H., & Hodson, S. J. (1990). *The behavior-based safety process.* New York: Van Nostrand Reinhold.

McSween, T. E. (1995). *The value-based safety process.* New York: Van Nostrand Reinhold.

Newman, B. (1992). *The reluctant alliance: Behaviorism and humanism.* Buffalo, NY: Prometheus Books.

Petersen, D. (1989). *Safe behavior reinforcement.* New York: Alorey, Inc.

Skinner, B. F. (1938). *The behavior of organisms.* Acton, MA: Copley Publishing Group.

———. (1974). *About behaviorism.* New York: Alfred A. Knopf.

Ullman, L. P., & Krasner, L. (1965). (Eds.) *Case studies in behavior modification.* New York: Holt, Rinehart, & Winston.

Wandersman, A., Popper, P., & Ricks, D. (1976). *Humanism and behaviorism: Dialogue and growth.* New York: Pergamon Press.

Chapter 3

Paradigm Shifts for Total Safety

This chapter outlines ten new perspectives we need to adopt in order to exceed current levels of safety excellence and reach our ultimate goal—a Total Safety Culture. The traditional three Es of safety management—engineering, education, and enforcement—have only gotten us so far. A Total Safety Culture requires understanding and applying three additional Es—empowerment, ergonomics, and evaluation.

Safety in industry has improved dramatically in this century. Let's examine the evolution of accident prevention to see how this was accomplished. The first systematic research began in the early 1900s, and focused on finding the psychological causes of accidents. It assumed people were responsible for most accidents and injuries, usually through mental errors caused by anxiety, attitude, fear, stress, personality, or emotional state (Guarnieri, 1992). Reducing accidents was typically attempted by "readjusting" attitude or personality, usually through supervisor counseling or discipline (Heinrich, 1931).

This so-called "psychological approach" held that certain individuals were "accident prone." By removing these workers from risky jobs or by disciplining them to correct their attitude or personality problems, it was thought that accidents could be reduced. But as I discussed in Chapter 1, this focus on accident-proneness has not been effective, partly because reliable and valid measurement procedures are not available. Also, the person factors contributing to accident proneness are probably not consistent characteristics or traits in people, but vary from time to time and situation to situation.

The "Old" Three Es

Enthusiasm for the early "psychological approach" waned because of the difficulty of measuring its impact (Barry, 1975). In addition, the seminal research and scholarship of William Haddon (1963, 1968) suggested that engineering changes held the most promise for large-scale, long-term reductions in injury severity.

As the first administrator of the National Highway Safety Bureau (now the National Highway Traffic Safety Administration), Dr. Haddon was able to turn his theory and research into the first federal automobile safety standards. Haddon believed injury is caused by delivering excess energy to the body, and that injury prevention depends on controlling that energy. The prevention focus now shifted to engineering and epidemiology, and resulted in developing personal protective equipment (PPE) for work and recreational environments, as well as standards and policy regarding the use of PPE. In vehicles, Haddon's basic theory eventually led to collapsible steering wheels, padded dashboards, head restraints, and airbags in automobiles.

This brief history of the safety movement in the U.S. explains why engineering is the dominant paradigm in industrial health and safety (Peterson, 1991; Winn & Probert, 1995), with secondary emphasis on two additional *Es*—education and enforcement. Over the past several decades, the basic protocol for reducing injury has been to:

1. Design the safest equipment, environmental settings, or protective devices
2. Educate people regarding the use of the engineering interventions
3. Use discipline to enforce compliance with recommended safe work practices

Thanks to this paradigm, most safety professionals are safety *engineers,* who commonly advocate that "Safety is a condition of employment."

The three *Es* have dramatically reduced injury severity in the workplace, at home, and on the road. Let's take a look at motor vehicle safety for a minute. The Government Accounting Office has estimated conservatively that the early automobile safety standards ushered through Congress by Dr. Haddon had saved at least 28,000 American lives by 1974 (Guarnieri, 1992). In addition, the state laws passed in the 1980s requiring use of vehicle safety belts and child safety seats have saved countless more lives. Many more lives would be saved and injuries avoided if more people buckled up and used child safety seats for their children.

The current rate of safety belt use in the U.S. is about 67 percent (Novak, 1995), a dramatic improvement from the 15 percent prior to statewide interventions, including belt-use laws, campaigns to educate people about the value of safety-belt use, and large-scale enforcement blitzes by local and state police officers.

There is still much room for improvement, especially considering that most of the riskiest drivers still don't buckle up (Evans *et al,* 1982; Wagenaar, 1984; Waller, 1987). Each year since 1990, the U.S. Department of Transportation has set nationwide belt-use goals of 70 percent, but to date this goal has not been met—at least over the long term. It seems the effectiveness of current methods to increase the use of this particular type of PPE has plateaued or asymptoted below 70 percent.

Now let's turn our attention to industry. I've worked with many corporate safety professionals over the years who say their plants' safety performance has reached a plateau. Yes, their overall safety record is vastly better than it once was. But continuous improvement is elusive. A frantic search for ways to take safety to the next level has not paid off. The old "three *Es*" paradigm will not get us there.

A certain percentage of people keep falling through the cracks. Keep on doing what you're doing and you'll keep on getting what you're getting. As I heard Dr. Deming (1991) say many times, "Goals without method, what could be worse?"

The New Three Es

I'm going to introduce you to three new *E*s in this book. *E*rgonomics. *E*mpowerment. *E*valuation. I certainly don't suggest abandoning tradition. We need to maintain a focus on engineering, education, and enforcement strategies. But to get beyond current plateaus and reach new heights in safety excellence, we must attend more competently to the psychology of injury prevention. These three new *E*s suggest specific directions or principles.

ERGONOMICS

As discussed in Chapter 1, ergonomics requires careful study of relationships between environment and behaviors, as well as developing action plans (such as equipment work orders, safer operating procedures, training exercises) to avoid possible acute or chronic injury from the environment/behavior interaction. This requires consistent and voluntary participation by those who perform the behaviors in the various work environments. These are usually line operators or hourly workers in an organization, and their participation will happen when these individuals are empowered.

Throughout this book I discuss ways to develop an empowered work culture, and I explain procedures for involving employees in ergonomic interventions.

EMPOWERMENT

Some operational definitions of the traditional three *E*s for safety (especially enforcement) have been detrimental to employee empowerment. Many supervisors have translated "enforcement" into a strict punishment approach, and the result has turned off many employees to safety programs. These workers may do what is required, but no more. Some individuals who feel especially controlled by safety regulations might try and beat the system, and success will likely bring a sense of gratification or freedom. This is predictable from theory and research in the area of psychological reactance (Brehm, 1966, 1972), and is illustrated in Figure 3.1.

Figure 3.1. Some top-down rules have undesirable side effects.

EVALUATION

The third new E essential to achieving a Total Safety Culture is evaluation. Without appropriate feedback or evaluation, practice does not make perfect. Thus, we need the right kind of evaluation processes. Later in this book, especially in Chapter 15, I detail procedures for conducting the right kind of comprehensive evaluation. Right now, what's important to understand is that some traditional methods of evaluation actually decrease or stifle empowerment. This calls for changing some safety measurement paradigms.

Remember the need for a guiding theory or set of principles? Basic theory from person-based and behavior-based psychology suggests a shift to new safety paradigms. And, these paradigm shifts provide a new set of guiding principles for achieving new heights in safety excellence.

Shifting Paradigms

I have heard many definitions of "paradigm," some humorous, some academic, and some practical. From my perspective, this is one of those superfluous academic terms that is completely unnecessary. We have simple and straightforward words in the English language to cover every definition of paradigm. Perhaps that's why I often get humorous or sarcastic reactions from audiences when I ask "What is a paradigm?"

"Isn't that twenty cents?" shouts one participant ("pair-a-dimes"). Another replies, "A paradigm is what I use on the farm to dig post holes" ("pair-a-dig-ems"). When I was a graduate student of psychology in the mid-1960s, paradigm was used to refer to a particular experimental procedure or methodology in psychological research. For this discussion here, I consulted three different dictionaries (*Webster's New Universal Unabridged, The American Heritage Dictionary,* and *The Scribner-Bantam English Dictionary),* and came up with a consensus definition for paradigm. It's a pattern, example, or model. However, words can change their meaning through usage, as discussed brilliantly by S. I. Hayakawa (1978) in his instructive and provocative text *Language in Thought and Action.* In business, paradigm has been equated with psychological terms such as perception, attitude, cognition, belief, and value.

The popular 1989 video "Discovering the Future: The Business of Paradigms" by Joel Barker (see also Barker, 1992) was certainly responsible for some of the new applications of the term "paradigm." A number of articles and speeches in the safety field have supported and precipitated this change. Indeed, Dan Petersen's keynote speech at the 1993 Professional Development Conference of the American Society of Safety Engineers was entitled, "Dealing with Safety's Paradigm Shift," and followed up his earlier 1991 article in *Professional Safety* entitled "Safety's Paradigm Shift." Here Dr. Petersen claimed that safety has shifted its focus to large-scale culture change through employee involvement. But some safety professionals assert that a revolutionary change in ideas, beliefs, and approaches—the new definition of paradigm—has not yet occurred in safety (Winn, 1992; Winn & Probert, 1995).

The aim of this chapter is not to dissect the meaning of paradigm, nor to debate whether one or more paradigm shifts have occurred in industrial safety. Instead, I want to define ten basic changes in belief, attitude, or perception that are needed to develop the ultimate Total Safety Culture. These shifts require new principles, approaches, or procedures, and will result in different behaviors and attitudes among top managers and hourly workers. Empowerment will increase throughout the work culture.

The shift in how paradigm is commonly defined does contain an important lesson. When we adopt and use new definitions, our "mind-set" or perception changes. In other words, as I indicated in the previous chapter, we act ourselves into a new way of thinking or perceiving. This is a primary theme of this book. When employees get involved in more effective procedures to control safety, they develop a more constructive and optimistic attitude toward safety and the achievement of a Total Safety Culture. Let's consider the shifts in principles, procedures, beliefs, attitudes, or perceptions needed for the three new Es—ergonomics, empowerment, and evaluation—and for achieving a Total Safety Culture.

1. FROM OSHA REGULATIONS TO CORPORATE RESPONSIBILITY. Many safety activities and programs in U.S. industry are driven by OSHA (the U.S. Occupational Safety and Health Administration) or MSHA (the Mine Safety and Health Administration) rather than by the employers and employees who can benefit from a safety process. In other words, many in industry do "safety stuff" because the government requires it—not because it was their idea and initiative.

But people are more motivated and willing to go beyond the call of duty when they are achieving their own self-initiated goals. Ownership, commitment, and proactive behaviors are less likely when you're working to avoid missing goals or deadlines set by someone else. This statement is intuitive. Just compare your own motivation when working for personal gain versus someone else's gain, or when working to earn a reward versus to avoid a penalty.

The language used to define safety programs and activities influences personal participation. Remember, we can act ourselves into an attitude. So it makes sense to talk about safety as a company mission that is owned and achieved by the very people it benefits. A safety process is not intended to benefit federal regulators. Let's work to achieve a Total Safety Culture for the right reasons.

2. FROM FAILURE ORIENTED TO ACHIEVEMENT ORIENTED. If you strive to meet someone else's goals rather than your own, you'll probably develop an attitude of "working to avoid failure" rather than "working to achieve success." But we are more motivated by achieving success than avoiding failure. If you have a choice between earning positive reinforcers (rewards) or avoiding punishers (penalties), you'll probably choose the positive reinforcement situation. Plus, if you feel controlled by punishment or negative reinforcement, you often procrastinate and take a reactive rather than a proactive stance (Skinner, 1971).

This principle helps explain why more continuous and proactive attention goes to productivity and quality than to safety. Productivity and quality goals are typically stated in achievement terms, and gains are tracked and recorded as individual or team accomplishments, sometimes followed by rewards or recognition awards.

Figure 3.2. Safety reward programs should pass the "dead man's" test.

In contrast, safety goals are most often stated in negative reinforcement terms. How many times have you heard: "We will reach our safety goal after another month without a lost-time injury." And "keeping score" in safety means tracking and recording losses or injuries.

Measuring safety with only records of injuries not only limits evaluation to a reactive stance, it also sets up a negative motivational system that is apt to take a back seat to the positive system used for productivity and quality. Giving safety an achievement perspective (like production and quality) requires a different scoring system, as indicated by the next paradigm shift.

3. FROM OUTCOME FOCUSED TO BEHAVIOR FOCUSED. Companies are frequently ranked according to their OSHA recordables and lost-time injuries. Within companies, work groups or individual workers earn safety awards according to outcomes—those with the lowest numbers win. Offering incentives for fewer injuries, for instance, can often reduce the reported numbers while not improving safety. Pressure to reduce outcomes without changing the process (or ongoing behaviors) often causes employees to cover up their injuries. How many times have you heard of an injured employee being driven to work each day to sign in and then promptly returned to the hospital or home to recuperate? This keeps the outcome numbers low but does more harm than good to the corporate culture. Likewise, failure to report even a minor first-aid case prohibits key personnel from correcting the factors that led to the incident. A scoring system based on what people do for safety (as in a behavior-based process) not only attacks a root cause of most work injuries, it can also be achievement oriented. This puts safety in the same motivational framework as productivity and quality.

A misguided emphasis on outcomes rather than process is illustrated in Figure 3.2. Although the idea of a dead person receiving a safety award is clearly ridiculous, this type of incentive/reward process is quite common in American industry. A 1993 survey of more than 400 companies in Wisconsin revealed 58 percent used rewards to motivate safety; of these, more than 85 percent based rewards on outcomes such as OSHA recordables rather than process (Koepnick, 1993). These programs often bring down numbers by influencing the reporting of injuries, but rarely do they benefit the safety processes which control results.

In Chapter 10, I explain principles for establishing an incentive/reward process to motivate the kinds of safety processes that influence outcomes. For now, just recognize and appreciate the advantage of focusing on achieving process improvements over working to avoid failure. This is especially true if a failure-oriented goal is remote, such as a plant-wide reduction in injuries, and thus might be perceived as uncontrollable.

Safety can be on equal footing with productivity and quality if it is recorded and tracked with an achievement score perceived by employees as directly controllable and obtainable. This occurs with a focus on the safety processes that can decrease an organization's injury rate, as well as an ongoing measurement system that continuously tracks safety accomplishments and displays them to the workforce.

As discussed in Chapter 2, safety accomplishments occur in three general areas (environment, behavior, and person), with environmental successes easiest to record and track. Environmental achievements for safety range widely, from purchasing safer equipment, to correcting environmental hazards and demonstrating improved environmental audits.

Person factors are influenced by numerous situations, such as safety education, safety celebrations, increased safety personnel. It's possible to estimate achievements in this domain by counting the occurrences of these events. A more direct assessment can occur through periodic perception surveys, interviews, or focus-group discussions (as detailed later in Chapter 15). These measurements can be rather time-consuming, though, and the reliability and validity of results from intermittent subjective surveys are equivocal. Plus, finding an improvement in perception does not necessarily imply an increase in safe work practices—the human dynamic most directly linked to reducing work injuries.

Work practices can be observed, recorded, and tracked objectively (Geller *et al*, 1989; Krauss *et al*, 1990, McSween, 1995). When daily displays of behavioral records show increases in safe behaviors and decreases in at-risk behaviors, the workforce can celebrate the success of an improved safety process. In Part 3 of this book, I detail principles and procedures for accomplishing this.

4. FROM TOP-DOWN CONTROL TO BOTTOM-UP INVOLVEMENT. As I discussed when introducing three new *E*s, a Total Safety Culture requires continual involvement from operations personnel such as hourly workers. After all, these are the people who know where safety hazards are located and when the at-risk behaviors occur. Also, they can have the most influence in supporting safe behaviors and correcting at-risk behaviors and conditions. In fact, the ongoing processes involved in developing a Total Safety Culture need to be supported from the top but driven from the bottom. This is more than employee participation; it is employee ownership, commitment, and empowerment.

As discussed in Chapter 1, research has shown that safe work practices can be increased and work injuries decreased with behavior-based interventions (Geller, 1990; Sulzer-Azaroff, 1982, 1987). But this research invariably involved outside agents such as consultants to help implement and evaluate the tactics, and the projects were usually short-term and small-scale. Large-scale and long-term behavior change requires employees themselves to apply the techniques throughout their workplace. For this to happen, employees must understand the relevant behavioral science principles and feel good about using them to prevent work injuries.

Understanding and feeling good about something brings us to considering again those person factors such as knowledge, intentions, attitudes, expectancies, and mood states. Certain dispositions or mood states, for example, influence an individual's propensity to help another person, and it's possible to increase these person fac-

Figure 3.3. Some work environments create barriers to synergy.

tors through changing environment and behavior factors (see reviews by Carlson, Charlin, & Miller, 1988; and Geller, 1994).

This supports the general principle I introduced in Chapter 2: A Total Safety Culture requires integrating both behavior-based and person-based approaches to understand and influence the human dynamics of a corporation. To show you how to do this is my primary aim with this book.

5. FROM RUGGED INDIVIDUALISM TO TEAMWORK. An employee-driven safety process requires teamwork founded on interpersonal trust, synergy, and win/win contingencies. However, from childhood most of us have been taught an individualistic, win/lose perspective, supported by such popular slogans as "You have to blow your own horn," "Nice guys finish last," "No one can fill your shoes like you," and "It's the squeaky wheel that gets the grease." Grades in school, the legal system, and many sports also orient us to think individualism and win/lose. This is why a true team approach to safety does not come easily.

Figure 3.3 illustrates a competitive situation quite common in the workplace. Although some office environments were originally designed to promote more open communication and group interaction, physical and psychological barriers have often been erected to maintain privacy and an individualistic atmosphere. This partially results from work systems that offer more rewards for individual than group achievement. Processes and systems can be implemented to promote group behaviors and interdependence over individual behaviors and independence. These processes and contingencies are emphasized throughout this book, since a Total Safety Culture requires more teamwork than rugged individualism.

6. FROM A PIECEMEAL TO A SYSTEMS APPROACH. The long-term improvements of a Total Safety Culture can only be achieved with a systems approach, including balanced attention to all aspects of the corporate culture. Dr. Deming (1986, 1993) emphasized that total quality can only be achieved through a systems approach, and of course the same is true for safety. As I discussed earlier in Chapter 2, three basic domains need attention when designing and evaluating safety processes and when investigating the root causes of near hits and injuries:

1. Environment factors such as equipment, tools, machines, housekeeping, heat/cold, engineering;

2. Person factors such as employees' knowledge, skills, abilities, intelligence, motives, personality; and

3. Behavior factors such as complying, coaching, recognizing, communicating, and "actively caring."

Two of these system variables involve human factors. Each generally receives less attention than the environment, mostly because it's more difficult to visibly measure the outcomes of efforts to change the human factors. Some human factors programs focus on behaviors (as in behavior-based safety); others focus on attitudes (as in a person-based approach). A Total Safety Culture integrates these two approaches.

7. FROM FAULT FINDING TO FACT FINDING. Blaming an individual or group of individuals for an injury-producing incident is not consistent with a systems approach to safety. Instead, an injury or near hit provides an opportunity to investigate facts from all aspects of the system that could have contributed to the incident. Immediate environment, person, and behavior factors should be explored for root causes, and numerous historical factors should also be considered. For example:

How common was the at-risk behavior?

How many individuals observed the at-risk behavior without intervening?

And what aspects of operations and the management system supported the at-risk behavior?

Several years ago I helped a company investigate an injury which occurred when an employee slipped on a metal plate covering a large hole on scaffolding three stories above a concrete floor. The worker fell through the hole, and would have fallen three stories were he not able to throw out his arms and catch himself on the sides of the hole. He suffered painful cuts and scrapes, but obviously the injury could have been much worse.

Whose fault was it? What was the root cause of this incident? It was tempting to finger the welder who failed to secure the metal plate over the hole as the culprit—the root cause. This would be piecemeal and fault finding. Instead a fact finding, systems approach was followed. What were the management demands, for example, that led to careless work by the welder? How many individuals had stepped on the plate, noticed that it was loose, and did not report their near hit? (The investigation revealed that numerous employees had been aware of the loose plate.) What environment or person factors prevented people from reporting their near hit with the loose plate? What processes should be put in place to facilitate observing, reporting, and correcting environmental hazards like the one contributing to this injury?

This is obviously only a partial list of questions related to a systems-level, fact-finding investigation. But I hope it's clear that answering these questions will be far more constructive than finding an individual to blame and perhaps punishing him or her for the mishap.

In this case, I doubt there was one root cause. Indeed, efforts to search for the root cause of an injury can be fruitless and lead to more fault finding. Many environment, behavior, and person factors contribute to the system that causes workplace injuries. Piecemeal corrective action is narrow-minded and short-sighted; systems-level corrective action can have a large-scale, long-term impact.

8. FROM REACTIVE TO PROACTIVE. Investigating events preceding an incident, be it a near hit or an injury, demonstrates the need to think and act proactively. Unfortunately, a proactive stance is extremely difficult to maintain, especially in a corporate culture that is increasingly complex and demanding. There is a higher and higher price tag on "free time." With barely enough time to react sufficiently to crises each day, how can we find time to be proactive?

9. FROM QUICK FIX TO CONTINUOUS IMPROVEMENT. There is no quick-fix answer to this challenging question. "Proactive" can be substituted for "reactive" only with a systems perspective and an optimistic attitude of continuous improvement through increased employee involvement. Understanding the psychology of safety can be a great aid here. The principles and procedures described in this book will enable you to influence incremental changes in work practices and attitudes that can prevent an injury. This represents a proactive, continuous improvement paradigm, which will surely improve your safety performance.

10. FROM PRIORITY TO VALUE. "Safety is a priority." This is probably the most common safety slogan found in workplaces and voiced by safety leaders. I've seen signs, pens, buttons, hats, T-shirts, and note pads with this message. No wonder safety and health professionals are surprised when I say that safety should not be a priority. To justify my case, I offer the following explanation:

Think about a typical workday morning. We all follow a prioritized agenda, often a standard routine, before traveling to work. Some people eat a hearty breakfast, read the morning newspaper, take a shower, and wash dishes. Others wake up early enough to go for a morning jog before work. Some grab a roll and a cup of coffee, and leave their home in disarray until they get back in the evening.

In each of these scenarios the agenda—the priorities—are different. Yet there is one common activity. It's not a priority but a basic value. Do you know what it is? One morning you wake up late. Perhaps your alarm-clock failed. You have only 15 minutes to prepare for work. Your morning routine changes drastically. Priorities must be rearranged. You might skip breakfast, a shower, or a shave. Yet every morning schedule still has one item in common. It's not a priority, capable of being dropped from a routine due to time constraints or a new agenda. No, this particular morning activity represents a value which we've been taught as infants, and it's never compromised. Have you guessed it by now? Yes, this common link in everyone's morning routine, regardless of time constraints, is "getting dressed."

This simple scenario shows how circumstances can alter behavior and priorities. Actually, labeling a behavior a "priority" implies that its order in a hierarchy of daily activities can be rearranged. How often does this happen at work? Does safety sometimes take a "back seat" when the emphasis is on other priorities such as production quantity or quality?

Enduring Values

It's human nature to shift priorities, or behavioral hierarchies, according to situational demands or contingencies. But values remain constant. The early morning anecdote illustrates that the activity of "getting dressed" is a value that is never dropped from the routine. Shouldn't "working safely" hold the same status as "getting dressed"? Safe work practices should occur regardless of the demands of a particular day. Safety should be a value linked with every activity or priority in a work routine.

Safe work should be the enduring norm, whether the current focus is on quantity, quality, or cost-effectiveness as the "number one priority."

The ultimate aim of a Total Safety Culture is to make safety an integral aspect of all performance, regardless of the task. Safety should be more than the behaviors of "using personal protective equipment," more than "locking out power" and "checking equipment for potential hazards," and more than "practicing good housekeeping." Safety should be an unwritten rule, a social norm, that workers follow regardless of the situation. It should become a value that is never questioned—never compromised.

This of course is much easier said than done. How do you even begin to work for such lofty aims? Figure 3.4 summarizes the relationships between intentions, behaviors, attitudes, and values. It outlines a starting point and general process for developing safety as a corporate value. A key point is that attitudes and values follow from behavior. This brings us to behavior management techniques. They are the starting point for acting a person into safe thinking.

This is how it works: When you follow safe procedures consistently for every job and attribute your behavior to a voluntary decision, you begin thinking safe. Eventually, working safe becomes part of your value system.

Figure 3.4 illustrates how attitudes and values influence intentions and behaviors directly. But as discussed in Chapter 2, it's not cost-effective to manage attitudes and values directly to "think people into safe acting." Notice in the figure the different thicknesses of rectangles enclosing the terms. The thicker the border, the more mea-

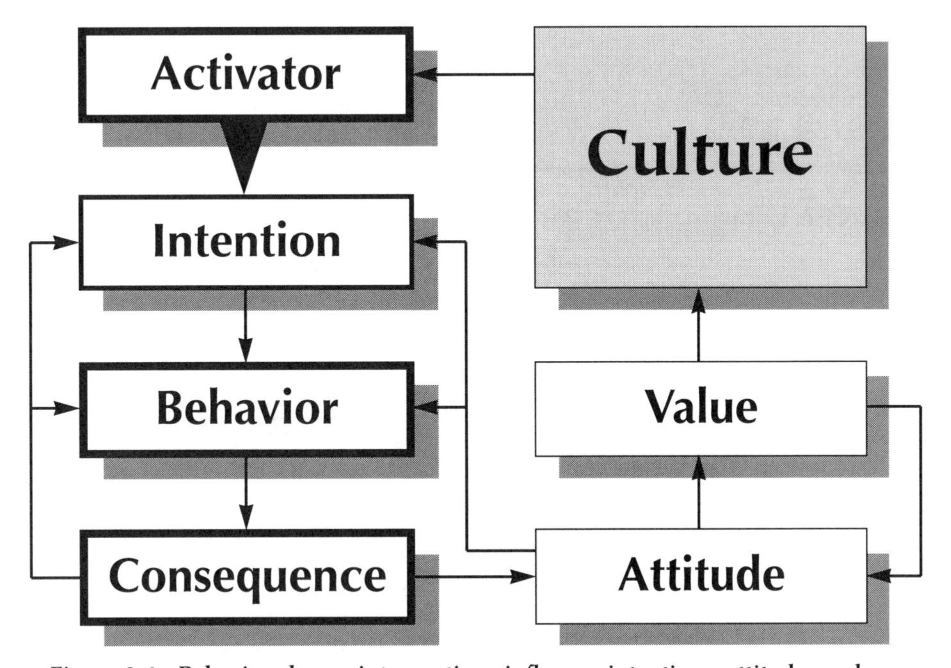

Figure 3.4. Behavior change interventions influence intentions, attitudes, values, and culture.

surable and manageable the concept. Activators (antecedent conditions which direct behavior), behaviors, and consequences (events which follow and motivate behaviors and influence attitudes) are easiest to define, measure, and manage.

In contrast, values and culture are the most difficult to measure reliably and influence directly. This book gives you specific techniques for managing behaviors to promote supportive safety attitudes and values. Put them all together and eventually you will construct an integrated Total Safety Culture.

In Conclusion

This chapter describes ten shifts in perspective needed to go beyond current levels of safety excellence. The first nine could be considered goals for achieving a Total Safety Culture. The tenth—making safety a value—is not something that can be readily measured and tracked. It's the ideal vision for our safety mission.

Here's how the new paradigms fit together: Your safety achievement process should be considered a company responsibility, not a regulatory obligation. It should be achievement oriented with a focus on behavior, supported by all managers and supervisors but driven by the line workers or operators through teamwork. A systems approach is needed, which leads to a fact-finding perspective, a proactive stance, and a commitment to continuous improvement.

These new perspectives reflect new principles to follow, new procedures to develop and implement. This "new safety work" will lead to different perceptions, attitudes, and even values. Ultimately, the tenth paradigm shift can be reached. When safety goes from priority to value, it won't be compromised at work, at home, or on the road. Naturally, numerous injuries will be prevented and lives saved everyday. This vision should motivate each of us to be active in the safety achievement process. This book helps you define your role.

REFERENCES

Barry, P. Z. (1975). Individual versus community orientation in the prevention of injuries. *Preventive Medicine, 4,* 47-56.

Brehm, J. W. (1972). *Responses to loss of freedom: A theory of psychological reactance.* New York: General Learning Press.

———. (1966). *A theory of psychological reactance.* New York: Academic Press.

Carlson, M., Charlin, V., & Miller, N. (1988). Positive mood and helping behavior: A test of six hypotheses. *Journal of Personality and Social Psychology, 55,* 211-229.

Deming, W. E. (1986). *Out of the crisis.* Cambridge, MA: Massachusetts Institute of Technology, Center for Advanced Engineering Study.

———. (1991, May). *Quality, productivity, and competitive position.* Four-day workshop presented in Cincinnati Ohio by Quality Enhancement Seminars, Inc.

———. (1993). *The new economics for industry, government, education.* Cambridge, MA: Massachusetts Institute of Technology, Center for Advanced Engineering Study.

Evans, L., Wasielewski, P., & von Buseck, C. R. (1982). Compulsory seat belt usage and driver risk-taking behavior. *Human Factors, 24,* 41-48.

Geller, E. S. (1994). The human element in integrated environmental management. In Cairns, J., Crawford, T. V., & Salwasser, H. (Eds.). *Implementing integrated environmental management.* Blacksburg, VA: University Press.

———. (1990). Performance management and occupational safety: Start with a safety belt program. *Journal of Organizational Behavior Management,* 11(1), 149-174.

Geller, E. S., Lehman, G. R., & Kalsher, M. J. (1989). *Behavior analysis training for occupational safety.* Newport, VA: Make-A-Difference, Inc.

Guarnieri, M. (1992). Landmarks in history of safety. *Journal of Safety Research,* 23, 151-158.

Haddon, W., Jr. (1968). The changing approach to the epidemiology, prevention, and amelioration of trauma: The transition to approaches etiologically rather than descriptively based. *American Journal of Public Health,* 58, 1431-1438.

Haddon, W., Jr. (1963). A note concerning accident theory and research with special reference to motor vehicle accidents. *Annals of the New York Academy of Sciences,* 107, 635-646.

Hayakawa, S. I. (1978). *Language in thought and action,* Fourth Edition. New York: Harcourt Brace Jovanovich, Publishers.

Heinrich, W. W. (1931). *Industrial accident prevention.* New York, NY: McGraw-Hill.

Koepnick, W. (1993, October). *Do safety incentive programs really work?* Presentation at the National Safety Council Congress & Exposition, Chicago, Illinois.

Krause, T. R., Hidley, J. H., & Hodson, S. J. (1990). *The behavior-based safety process.* New York: Van Nostrand Reinhold.

McSween, T. E. (1995). *The values-based safety process: Improving your safety culture with a behavioral approach.* New York: Van Nostrand Reinhold.

Novak, L. (1995, May). National Highway Traffic Safety Administration; personal communication with E.S. Geller.

Petersen, D. (1991). Safety's paradigm shift. *Professional Safety,* 36(8), 47-49.

Skinner, B. F. (1971). *Beyond freedom and dignity.* New York: Alfred A Knopf.

Sulzer-Azaroff, B. (1987). The modification of occupational safety behavior. *The Journal of Occupational Accidents,* 9, 177-197.

———. (1982). Behavioral approaches to occupational health and safety. In L. W. Frederiksen (Ed.). *Handbook of organizational behavior management.* (pp.505-538). New York: John Wiley & Sons.

Wagenaar, A. C. (1984). *Restraint usage among crash-involved motor vehicle occupants.* Report UMTRI-84-2 Ann Arbor, MI: University of Michigan Transportation Research Institute.

Waller, J. A. (1987). Injury: Conceptual shifts and prevention implication. *Annual Review of Public Health,* 8, 21-49.

Winn, G. L. (1992). In the crucible: Testing for a real paradigm shift. *Professional Safety,* 37(12), 30-33.

Winn, G. L., & Probert, L. L. (1995). Philosopher's stone: It may take another Monongah. *Professional Safety,* 40(5), 18-22.

Part Two

Human Barriers to Safety

Chapter 4

The Complexity of Human Nature

Safety is usually a continuous fight with human nature. This chapter explains why. Understanding this basic point will lead to less victim-blaming and fault-finding when investigating an injury. Instead, we'll be able to find factors in the system that can be changed in advance to prevent injuries at work, at home, and throughout the community.

"All injuries are preventable."
"It's human nature to work safely."
"Safety is just common sense."
"Safety is a condition of employment."

Read these familiar statements and you get the idea that working safely is easy or natural. Nothing could be further from the truth.

In fact, it's often more convenient, more comfortable, more expedient, and more common to take risks than to work safely. And past experience usually supports our decisions to choose the at-risk behavior, whether we're working, traveling, or playing. So we're often engaged in a continuous fight with human nature to motivate ourselves and others to avoid those risky behaviors and maintain safe ones.

Let's consider what holds us back from choosing the safe way, whether it's following safe operating procedures, driving our automobile, or using personal protective equipment.

When I ask safety professionals, corporate executives, or hourly workers what causes work-related injuries, I get long and varied lists of factors. Actually, each list is quite similar. After all, everyone experiences events, attitudes, demands, distractions, responsibilities, and circumstances that get in the way of performing a task safely.

Most of us have been in situations where we weren't sure how to perform safely.

Perhaps we lacked training. Maybe the surrounding environment was not as safe as it could be.

Demands from a supervisor, co-worker, or friend put pressure on us to take a short cut or risk. Maybe it was inconvenient or uncomfortable to follow all of the safety procedures.

It's possible a physical condition (fatigue, boredom, drug impairment) influenced at-risk performance. There are other factors. Have you ever been unsafely distracted by external stimuli, like another person's presence? Or by an internal state, like personal thoughts or emotions?

Can you remember a time when you just didn't feel like taking the extra time to be safe? And I'm sure you've experienced the "macho" attitude, from yourself or others, that "It will never happen to me." Fortunately, it's rare that an injury follows unsafe behavior. The "It can't happen to me" attitude is usually supported or rewarded by our actual experiences. Risk-taking is rarely punished with an injury or even a near hit, instead it's consistently rewarded with convenience, comfort, or time saved. This creates something of a vicious cycle. The rewards of risky behavior mean you're likely to take more chances. As you gain experience at work you often master dangerous shortcuts. Since these at-risk behaviors are not followed by a near hit or injury, they remain unpunished, and they persist.

This basic principle of human nature reinforced throughout our lives runs counter to the safety efforts of individuals, groups, organizations, and communities. It explains why promoting safety and health is the most difficult on-going challenge at work. The reality is that injuries really do happen to the "other guy."

Learning to be At Risk

Remember when you first learned to drive a car? I bet this was an important but stressful occasion. Even with the right amount of driver training from your parents or a professional instructor, you felt a bit nervous getting behind the wheel for the first time.

At first, you were probably very careful to follow all the safe procedures you learned. Both hands on the wheel—the ten o'clock and two o'clock positions. Both eyes on the road at all times. You always used your turn signal. Always stopped when traffic lights turned yellow. If a safety belt was in the car, you used it. Conversations with passengers were avoided, as well as distractions from a radio or cassette tape. This was all right and proper, of course, because driving is a relatively complex and risky task, requiring the driver's undivided attention. This time human nature was on the safe side.

But how quickly you took driving for granted! Your complete concentration was no longer needed as tasks became automatic and "second nature." Many precautionary behaviors fell by the wayside. You began driving with one hand on the wheel. Your other hand held a drink, a cigarette, or a passenger's hand. Distractions were soon permitted—loud music, emotional conversations (sometimes on a telephone), and even "love making." I've even seen some people read a book, a letter, or a map while driving. All this while blowing past the speed limit, running a "yellow" traffic light, or following too close behind another vehicle.

We continue these risky driving behaviors every day because we perceive them as being "cool"—they are fun, convenient, and save us time. We never think of crashing, and thank goodness it usually doesn't happen to us.

In a short time behind the wheel we've gone from controlled processing to automatic processing (Schneider & Shiffrin, 1977; Shiffrin & Dumais, 1981). Various risky practices are adopted for fun, comfort, or convenience. These consequences reward the risky behavior and sustain it. This is human nature on the side of at-risk behavior, and can be explained by basic principles of behavioral science.

Dimensions of Human Nature

I always categorize factors contributing to a work injury into three areas: (1) environment factors, (2) person factors, and (3) behavior factors. This is the "Safety Triad" (Geller *et al,* 1989) introduced in Chapter 2. The most common reaction to an injury is to correct something about the environment— modify or fix equipment, tools, housekeeping, or an environmental hazard.

Often the accident report includes some mention of personal factors, like the employee's knowledge, skills, ability, intelligence, motives, or personality. These factors are typically translated into general recommendations, such as "The employee will be disciplined," or "The employee will be retrained."

This kind of vague attention to critical human aspects of a work injury shows how frustrating and difficult it is to deal with the psychology of safety—the person and behavior sides of the Safety Triad. The human factors contributing to injury are indeed complex, often unpredictable and uncontrollable. This justifies my conclusion that not *all* injuries can be prevented.

The acronym **BASIC ID** reflects the complexity and uncontrollability of human nature. As depicted in Figure 4.1, each letter represents one of seven human dimensions.

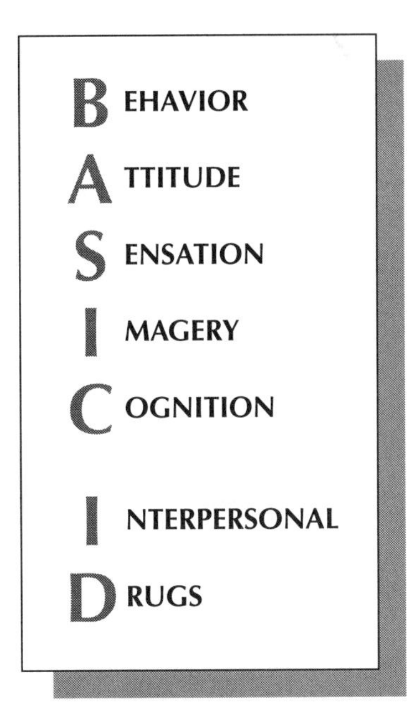

Figure 4.1. The acronym BASIC ID reflects the complexity of people and potential contributions to injury.

Many clinical psychologists use a similar acronym as a reminder that helping people improve their psychological state requires attention to each area (Lazarus, 1971, 1976). A simple scenario underscores the need in safety to understand these dimensions of human nature:

> *Dave, an experienced and skilled craftsman, works rapidly to make an equipment adjustment while the machinery continues to operate. As he works, production-line employees watch and wait to resume their work. Dave realizes all too well that the sooner he finishes his task, the sooner his co-work-*

ers can resume quality production. So he doesn't shut down and lock out the equipment power. After all, he's adjusted this equipment numerous times before without locking out and he's never gotten injured.

A morning argument with his teenage daughter pervades Dave's thoughts as he works, and suddenly he experiences a near hit. His late timing nearly results in his hand being crushed in a pinch point.

Removing his hand just in time, Dave feels weak in his knees and begins to perspire. This stress reaction is accompanied by a vivid image of a crushed right hand. After gathering his composure, Dave walks to the switch panel, shuts down and locks out the power, then lights up a cigarette. He thinks about this scary event for the rest of the day, and talks about the near hit to fellow workers during his breaks.

This brief episode illustrates each of the psychological dimensions represented by BASIC ID (see Figure 4.1), and demonstrates the complexity of human activity.

Behavior is illustrated by observable actions such as adjusting equipment, pulling a hand away from the moving machinery, lighting up a cigarette, and talking to co-workers.

Dave's **attitude** about work was fairly neutral at the start of the day, but immediately following his near hit he felt a rush of emotion. His attitude toward "energy control and power lock out" changed dramatically, and his commitment to locking out increased after relating his "near hit" to friends.

Sensation is evidenced by Dave's dependence on visual acuity, hand-eye coordination, and a keen sense of timing when adjusting the machinery. His ability to react quickly to the dangerous situation prevented severe pain and potential loss of valuable touch sensation.

Imagery occurred after the "near hit" when Dave visualized a crushed hand in his mind's eye, and this contributed to the significance and distress of the incident. Dave will probably experience this mental image periodically for some time to come. And this will motivate him to perform appropriate lock-out procedures, at least for the immediate future.

Cognition or "mental speech" about the morning argument with his daughter may have contributed to the timing error that resulted in the near hit. Dave will probably remind himself of this episode in the future, and these cognitions may help trigger proper lock-out behavior.

Interpersonal refers to the other people in Dave's life who contributed to his near hit, and will be influential in determining whether he initiates and maintains appropriate lock-out practices.

For example, it was the interpersonal discussion with his daughter that occupied his thoughts or cognitions before the near hit. The presence of production-line workers influenced Dave through subtle peer pressure to quickly adjust equipment without lock-out practices. These onlookers may have distracted Dave from the task, or they could have motivated him to show off his adjustment skills. After Dave's near hit, his interpersonal discussions were therapeutic, helping him relieve his distress and increase his personal commitment to occupational safety.

Drugs in the form of caffeine from morning coffee may have contributed to Dave's timing error. The extra cigarettes Dave smokes as a "natural" reaction to distress also had physiological consequences, which could have been reflected in Dave's subsequent behavior, attitude, sensation, or cognition. Dave's lesson shows how human nature interacts with environmental factors to cause at-risk work practices, near hits, and sometimes personal injuries. It's relatively easy to control the environmental factors. And, as I'll explain in Part Three on behavior-based safety, it's feasible to measure and control the behavior factors. However, the complex person factors, described by the BASIC ID acronym, are quite elusive. The field of psychology provides insights here, and this information can benefit occupational safety and health programs.

Let's further discuss person-based dimensions that make safety achievement so challenging.

COGNITIVE FAILURES

"All injuries are preventable."

I've heard this said so many times that it seems to be a slogan or personal affirmation that safety pros use to keep themselves motivated. I suspect some readers will resist any challenge to this ideal. I certainly appreciate their optimism, and there's no harm if such perfectionism is kept to oneself. But sharing this belief with others can actually inhibit achieving a Total Safety Culture.

You see, if a common workplace slogan declares all injuries preventable, workers may be reluctant to admit they were injured, or had a near hit. After all, if all injuries are preventable and I have an injury, I must be a real "jerk" for getting hurt.

Combine this slogan with a *goal* of zero injuries and a *reward* for not having an injury and human nature will dictate covering up an injury if possible. Also, as I'll discuss in the next chapter, claiming that all injuries are preventable can reduce the perceived risk of the situation. This can create the notion that "it won't happen to me," a perception that can increase the probability of at-risk behavior and an eventual injury.

Eliminate the "all injuries are preventable" slogan from your safety discussions. The most important reason to drop it is that most people don't believe it, anyway. They have been in situations where all the factors contributing to the near hit or injury could not have been anticipated, controlled, or prevented. The most uncontrollable factors are the person-based or internal subjective dimensions of people. Consider, for example, the role of cognitive failures.

Just what is a "cognitive failure"? I liken it to a "brain cramp." And research by Broadbent *et al*, 1982, demonstrated that people who report greater frequency of "cognitive failures" are more likely to experience an injury. The items listed in Figure 4.2 were used by Professor Broadbent and his associates to measure cognitive failures. Respondents were merely asked to indicate on a five-point scale the extent to which they agreed with each statement (from "strongly disagree" to "strongly agree"). Professor Broadbent's measurement instrument offers an operational definition for this person dimension that apparently influences injury frequency.

- I sometimes forget why I went from one part of the house to another.
- I often fail to notice signposts on the road.
- I sometimes bump into things or people.
- I often forget whether I've turned off a light or the coffeepot, or locked the door.
- I sometimes forget which way to turn on a road I know well but rarely use.
- I sometimes fail to see what I want in a supermarket (although it's right there).
- I often forget where I put something like a newspaper or a book.
- I often daydream when I ought to be listening to something.
- At home, I often start doing one thing and get distracted by something else (unintentionally).
- I sometimes forget what I came to the store to buy.
- I often drop things.
- I often find myself putting the wrong things in the wrong place when I'm done with them—like putting milk in the cereal cupboard.
- I frequently confuse right and left when giving directions.

Figure 4.2. Items from the survey developed by Broadbent *et al* (1982) to measure cognitive failures.

Scientific protocol won't allow us to conclude from the research by Broadbent *et al* that cognitive failures cause injury, but based on personal experience it sure seems reasonable to interpret a cause-and-effect relationship. It's likely every reader has experienced one or more of the "brain cramps" listed in Figure 4.2. Surely you have walked into a room to get something and forgotten why you were there. And how often have you left home for work more than once in a single morning because you forgot something? The same sort of breakdown in cognitive functioning can cause an injury. Does Figure 4.3 reflect potential reality?

Figure 4.3. A cognitive failure or "brain cramp" can cause a workplace injury.

Being actively aware of possible cognitive failures (listed in Figure 4.2) can help us develop strategies to decrease how often we experience them. For example, reducing fatigue and stress (discussed in Chapter 6) can make a marked difference. But I doubt that any personal strategy can eliminate all cognitive failures. Furthermore, it's unlikely you can have much impact on the cognitive failures of others.

So, if you believe as I do that "brain cramps" can cause personal injury, and that it's impossible to control completely these sudden, unexpected mental lapses, then it's illogical to presume that all injuries are preventable. There's a benefit to realizing that personal factors like cognitive failures are extremely hard to control—it should make us less likely to take risk for granted, even in familiar situations. We should be motivated to take greater precautions or to work more defensively. But this isn't easy, and interpersonal factors are one reason why. Now let's look more closely at the second ***I*** of BASIC ID.

INTERPERSONAL FACTORS

Our interpersonal relationships dramatically influence our thoughts, attitudes, and actions. How much of your time each day goes into efforts to gain the approval of others? Of course, we sometimes attempt to avoid the disapproval of others, be they a parent, spouse, work supervisor, or department head. As discussed in Chapter 3, we don't feel as good—or as *free*—when working to avoid failure or disapproval as when working to achieve success or approval. In both cases, other people are the cause of our motivation and behavior.

Social psychologists study systematically how we all influence each other's attitudes and behaviors. I want to discuss two reliable phenomena from social psychology research that strongly affect our daily activities. These phenomena are no doubt responsible for many at-risk behaviors and injuries, yet they are often overlooked.

PEER PRESSURE

Research conducted by Solomon Asch and associates in the mid-1950s found more than one out of three intelligent and well-intentioned college students were willing to publicly deny reality in order to follow the obviously inaccurate judgments of their peers. Dr. Asch's classic studies of conformity (1955, 1956, 1958) involved six to nine individuals sitting around a table judging which of three comparison lines was the same length as the standard. Figure 4.4 depicts one of these judgment situations.

All but the last individual to voice an opinion were research associates posing as subjects. Sometimes the research associates uniformly gave obviously incorrect judgments. The last person to decide was the real subject of the experiment. About one-third of the time the subject denied the obvious truth in order to go along with the group consensus.

This and similar procedures were used in numerous social psychology experiments to study factors influencing the extent of conformity. For example, a subject's willingness to deny reality in order to conform with the group was bolstered by increasing group size (Asch, 1955) and the apparent competence or status of group members (Crutchfield, 1955; Endler & Hartley, 1973). On the other hand, the presence of one dissenter or nonconformist in the group was enough to significantly

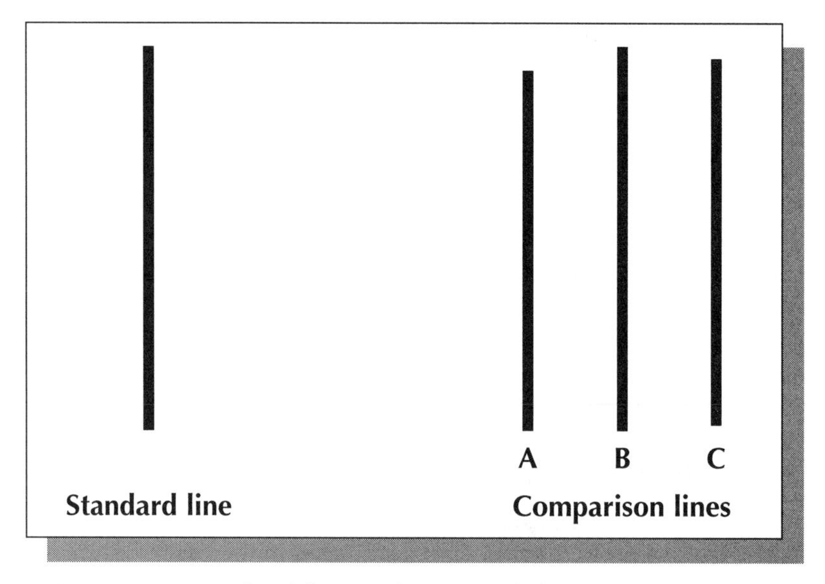

Figure 4.4. A sample of the stimulus materials for the Asch conformity experiment on group influence.

decrease conformity—it increased a subject's willingness to choose the correct line even when only 1 of 15 prior decisions reflected the correct choice (Nemeth, 1986).

The phenomenon of social conformity is certainly not new to any reader. We see examples of it every day, from the clothes people wear to how they communicate both verbally and in writing. We can't overlook the power of conformity in influencing at-risk behavior. Plus, we've learned that peer pressure increases when more people are involved, and when the group members are seen as relatively competent or experienced. Can you relate, for example, to the workplace conformity situation illustrated in Figure 4.5?

Figure 4.5. Social conformity can be risky.

It's important to remember, though, that one dissenter—a leader willing to ignore peer pressure and do the right thing—is often enough to prevent another person from succumbing to potentially dangerous conformity at work.

POWER OF AUTHORITY

Imagine you're among nearly 1000 participants in one of professor Stanley Milgram's 20 obedience studies at Yale University in the 1960s. You and another individual are led to a laboratory to participate in a human learning experiment.

First, you draw slips of paper out of a hat to determine randomly who will be the "teacher" and who the "learner."

Figure 4.6. Subjects readily gave electric shocks to peers.

You get to be the teacher; the learner is taken to an adjacent room and strapped to a chair wired through the wall to an electric shock machine containing 30 switches with labels ranging from "15 volts—light shock" to "450 volts—severe shock." You sit behind this shock generator and are instructed to punish the learner for errors in the learning task by delivering brief electric shocks, starting with the 15-volt switch and moving up to the next higher voltage with each of the learner's errors. The scenario is depicted in Figure 4.6. Complying with the experimenter's instructions, you hear the learner moan as you flick the third, fourth, and fifth switches. When flicking the eighth switch (labeled "120 volts"), the learner screams "These shocks are painful"; and when the tenth switch is activated the learner shouts "Get me out of here!"

At this point, you might think about stopping, but the experimenter prompts you, "Please continue— the experiment requires that you continue."

Increasing the shock intensity with each of the learner's errors, you reach the 330-volt level. Now you hear shrieks of pain—the learner pounds on the wall, then becomes silent. Still, the experimenter urges you to flick the 450-volt switch when the learner fails to respond to the next question.

At what point will you refuse to obey the instructions? Milgram asked this question to a group of people, including 40 psychiatrists, before conducting the experiment. They thought the sadistic game would stop soon after the learner indicated the shock was painful. So Milgram and his associates were surprised that 65 percent of his actual subjects, ranging in age from 20 to 50, went along with the experimenter's request right up to the last 450-volt switch (Milgram, 1963, 1974).

Why did they keep following along? Did they figure out that the learner was a confederate of the experimenter and did not really receive the shocks? Did they realize they were being deceived in order to test their obedience?

No, the subjects sweated, trembled, and bit their lips when giving the shocks. Some laughed nervously. Others openly questioned the instructions. But most did as they were told.

Milgram and associates learned more about the power of authority in further studies. Full obedience exceeded 65 percent, with as many as 93 percent flicking the highest shock switch, when:

The authority figure—the one giving the orders—was in the room with the subject;

The authority was supported by a prestigious institution, such as Yale University;

The shocks were given by a group of "teachers" in disguise and remaining anonymous (Zimbardo, 1970);

There was no evidence of resistance—no other subject was observed disobeying the experimenter;

The victim was depersonalized or distanced from the subject in another room.

Stanley Milgram drew this lesson from the research: "Ordinary people, simply doing their jobs, and without any particular hostility on their part, can become agents in a terrible destructive process" (Milgram, 1974, p.6).

Let's apply this research to the workplace. As a result of social obedience or social conformity, people might perform risky acts or overlook obvious safety hazards, and put themselves and others in danger. To say, "I was just following orders" reflects the obedience phenomenon, and "Everyone else does it" implies social conformity or peer pressure.

To achieve a Total Safety Culture, we need to realize the power of these two interpersonal factors. Interventions capable of overcoming peer pressure and blind obedience are detailed in Part Three. What I want to stress at this point is the vital role of leadership. One person can make a difference—decreasing both destructive conformity and obedience—by deviating from the norm and setting a safe example. And when a critical mass of individuals boards the "safety bandwagon," you get constructive conformity and obedience that supports a Total Safety Culture.

In Conclusion

We need to understand a problem as completely as possible and from many perspectives before we can solve it. In this chapter, we explored dimensions of the safety problem by considering the complexity of people. I attempted to convince you that human nature does not usually support safety. The natural relationships between behavior and its motivating consequences usually result in some form of convenient, time-saving—and risky—behavior. Consequently, to achieve a Total Safety Culture, you should prepare for an ongoing fight with human nature.

Human barriers to safety are represented by a popular acronym from clinical psychology (BASIC ID). The ***C*** (cognitions) and second ***I*** (interpersonal) dimensions of this acronym in particular explain the special challenges of achieving a Total Safety Culture. The phenomenon of cognitive failures shows the shallowness—in fact the potential danger—of the popular safety slogan, "All injuries are preventable." Conformity and obedience, two powerful phenomena from social psychological research, further help us understand the individual, group, and system factors responsible for at-risk behavior and injury.

The human barriers to safety discussed here should lead us to be more defensive and alert in hazardous environments. They also show how difficult it is to find the root cause, if indeed there is such a thing, of a "near hit" or injury. Another psychological challenge to safety is explored in the next chapter when we discuss the ***S*** (sensation) of BASIC ID.

REFERENCES

Asch, S. E. (1958). Effects of group pressure upon modification and distortion of judgments. In E. E. Maccoby, T. M. Newcomb, & E.L. Hartley (Eds.). *Readings in social psychology* (3rd ed.). New York: Holt, Rinehart & Winston.

———. (1956). Studies of independence and conformity: A minority of one against a unanimous majority. *Psychological Monographs, 70*, 1-70.

———. (1955). Opinions and social pressure. *Scientific American*, 193, 31-35.

Broadbent, D., Cooper, P. F., Fitzgerald, P., & Parker, K. (1982). The cognitive failures questionnaire (CFQ) and its correlates. *British Journal of Clinical Psychology*, 21, 1-16.

Crutchfield, R. S. (1955). Conformity and character. *American Psychologist*, 10, 191-198.

Endler, N. S., & Hartley, S. (1973). Relative competence, reinforcement and conformity. *European Journal of Social Psychology*, 3, 63-72.

Geller, E. S., Lehman, G. R., & Kalsher, M. J. (1989). *Behavior analysis training for occupational safety*. Newport, VA: Make-A-Difference, Inc.

Lazarus, A. A. (1976). *Multimodal behavior therapy*. New York: Springer Publishing Company, Inc.

———. (1971). *Behavior therapy and beyond*. New York: McGraw-Hill.

Milgram, S. (1963). Behavioral studies of obedience. *Journal of Abnormal and Social Psychology*, 67, 371-378.

———. (1974). *Obedience to authority*. New York: Harper Collins.

Nemeth, C. (1986). Differential contribution of majority and minority influence. *Psychological Review*, 93, 23-32.

Schneider, W., & Shiffrin, R. M. (1977). Controlled and automatic information processing. I: Detection, search, and attention. *Psychological Review*, 84, 1-66.

Shiffrin, R. M., & Dumais, S. T. (1981). The development of automatism. In J. R. Anderson (Ed.), *Cognitive skills and their acquisition*. Hillsdale, NJ: Eilbaum.

Zimbardo, P. G. (1970). The human choice: Individuation, reason, and order versus deindividuation, impulse, and chaos. In W. J. Arnold & D. Levine (Eds.), *Nebraska Symposium on Motivation*. Lincoln, NB: University of Nebraska Press.

Chapter 5

Sensation, Perception, and Perceived Risk

It's critically important to understand that perceptions of risk vary among individuals. We can't dramatically improve safety until people increase their perception of risk in various situations, and reduce their overall tolerance for risk.

In this chapter we'll explore the notion of selective sensation or perception, and then relate the concept to perceived risk and injury control. Several factors will be discussed that impact whether employees react to workplace hazards with alarm, apathy, or something in between. Taken together, these factors shape personal perceptions of risk and illustrate why the job of improving safety is so challenging.

The ***S*** of the BASIC ID acronym introduced in Chapter 4 refers to **sensation**—a human dimension that influences our thinking, attitudes, emotions, and behavior. In grade school we learned there are five basic senses we use daily to experience our world (we see, hear, smell, taste, and touch). Later we learned that our senses do not take in all of the information available in our immediate surroundings. Instead, we intentionally and unintentionally tune in and tune out certain features of our environment; thus, some potential experiences are never realized.

This is a complex process. To experience life on a selective basis we begin by using our five senses. But from there we:

Define (or encode) the information received;
Interpret its meaning or relevance to us;
Decide whether the information is worth remembering or responding to;
Plan and execute a response (if called for).

At any time in this chain of information processing and decision making we can—and do—impose our own individual bias, which is shaped by our past experiences, personality, intentions, aspirations, and expectations. You can see how our everyday sensations are dramatically influenced consciously and unconsciously by a number of person factors unique to the situation and the individual sensing the situation. Psychologists refer to such biased sensation as **perception**.

There is also a term called "selective perception" that is commonly used to refer to our biased sensations. But since all perceptions result from our intentional or unintentional distortion of sensations, adding the adjective "selective" to perception is unnecessary and actually redundant. We experience our surroundings through the natural selection of our sensations. This process is simply referred to as perception.

Selective Sensation or Perception

At the 1994 Professional Development Conference of the American Society of Safety Engineers (ASSE), the following instructions were printed in half of the 40-page handouts distributed to the audience of more than 350 individuals at the start of my two-hour presentation.

> *You are going to look briefly at a picture and then answer some questions about it. The picture is a rough sketch of a poster of a couple at a costume ball. Do not dwell on the picture. Look at it only long enough to "take it all in" at once. After this, you will answer "yes" or "no" to a series of questions.*

After the participants read the instructions, I presented the illustration depicted in Figure 5.1 for about five seconds. If you would like to experience the biased visual sensation (or perception) demonstrated to the ASSE audience, please read the instructions given above and then look at Figure 5.1 for approximately five seconds.

Figure 5.1. Ambiguous drawing illustrates selective sensation. Please read instructions, then look at the drawing for five seconds. Afterward, answer the five questions.

Now, answer the questions that I asked my ASSE audience:

1. Did you see a man in the picture?

2. Did you see a woman in the picture?

3. Did you see a animal in the picture? If so, what kind of animal did you see?

4. What other details did you detect in the brief exposure to the drawing?__ A woman's purse?__ A man's cane?__ A trainer's whip?__ A fish?__ A ball?__ A curtain?__ A tent?

Practically everyone in the audience raised a hand to answer "yes" to the first question, and I suspect you also see a man in Figure 5.1. But only about half of the audience acknowledged seeing a woman in the drawing, and many said they had seen an animal.

When I asked what type of animal, the common response heard across the room was "seal." This drew many laughs, and the laughter got louder when I asked what else was quickly perceived in the illustration.

Several people saw a woman's purse and a man's cane; others said they had seen a trainer's whip, a fish, and a beach ball. Some remembered seeing a curtain. Others saw part of a circus tent. What did you see in Figure 5.1?

"What's going on here?" I asked the ASSE audience. Why are we getting these diverse reactions to one simple picture? Some people speculated about environmental factors in the seminar room, including lighting, spatial orientation and visual distance from the presentation screen. Others thought that individual differences, including gender, age, occupation, and personal experiences even "last night," could be responsible. Finally, someone asked whether the instructions printed in their handouts could have influenced the different perceptions. This was in fact the case.

Every handout included the same exact instructions except for a few words. Half included all the words given above; the rest had the words "trained seal act" substituted for "couple at a costume ball." This was enough to make a marked difference in perceptions. Perhaps this makes perfect sense to you. Critical words in the instructions created expectations for a particular visual experience. I had set up my audience. Was your perception of Figure 5.1 influenced by this "set up"?

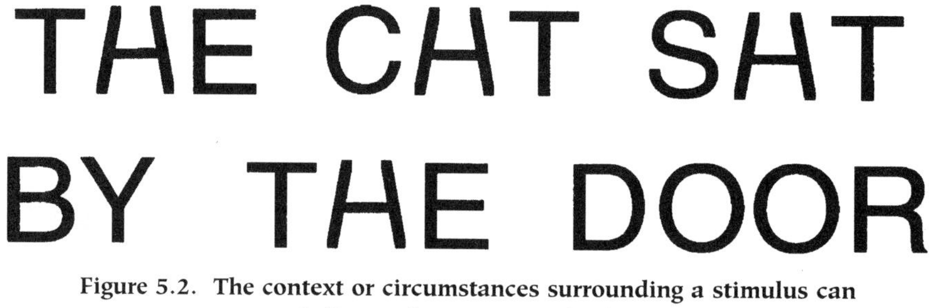

THE CAT SAT
BY THE DOOR

Figure 5.2. The context or circumstances surrounding a stimulus can influence how we perceive it.

Figure 5.3. The environmental context influences personal perception of the man in uniform.

BIASED BY CONTEXT

Now take a look at Figure 5.2. I suspect you have no difficulty reading the sentence as "The cat sat by the door," even though the symbol for "H" is exactly the same as the symbol for "A." It's a matter of context. The symbol was positioned in a way that influenced your labeling (or encoding) of the symbol. Likewise, the context or environmental surroundings in our visual field influence how we see particular stimuli.

The same is true for our other senses—hearing, smelling, tasting, and touching. How we experience food, which involves the sensations of smell, touch, and taste, can be dramatically influenced by the atmosphere in which it is served. This is a basic rule of the restaurant business. Of course other factors also bias food sensations, including hunger and past experiences with the same and similar food.

Let's take a look at Figure 5.3. What label do you give the man in the drawing on the left? The setting or context certainly influences your decision. The sign, keys, and uniform are cues that the man is probably a doorman. The environmental context in the drawing on the right leads to a different perception and label for the same person. Here he is a policeman enforcing a safety policy.

Now let's take our discussion of perception and apply it to the workplace. Here, perceptions of people can be shaped by equipment, housekeeping, job titles, and work attire. In fact, our own job title or work assignment can influence perceptions of ourselves, as well as affect our perceptions of others. This can dramatically influence how we interact with others, if indeed we choose to interact at all.

It's important to recognize this contextual bias. Pick out someone you communicate with at work, and think how your relationship would be different in another setting. Would you still feel superior or inferior? Also, as depicted in Figure 5.4, the work setting has a way of turning individuals into numbers, depersonalizing them. This impression certainly can be misleading, and might cause you to overlook someone's

Figue 5.4. The environmental context influences personal perception and behavior.

potential. In another setting, the same individual might feel empowered and be perceived as a leader.

BIASED BY OUR PAST

Perhaps every reader realizes that our past experiences influence our present perceptions. In Chapter 3, we considered shifts in methods and perceptions needed to achieve a Total Safety Culture. When I give workshops on paradigm shifts, someone invariably expresses concern about resistance. "He (or she) keeps playing old tapes and is not open to new ideas," is a common refrain. Past experiences are biasing present perceptions. Actually, there is a long trail of intertwined factors here. Past experiences filter through a personal evaluation process that is influenced by person factors, including many past perceived experiences. The cumulative collection of these previous perceived experiences biases every new experience, and makes it indeed difficult to "teach an old dog new tricks."

Some participants arrive at my seminars and workshops with a "closed mind" and a "have to be here" attitude. Others start with an "open mind" and an "opportunity to learn" outlook. This is another example of the power of personal perception—how much one learns at these seminars depends on perceptions going in.

Perhaps you'll find it worthwhile to copy Figure 5.5 and use it for a group demonstration. You can show how current impressions are affected by prior perceptions by asking participants to call out what they see as you reveal each drawing. The drawings must be uncovered in a particular order. Show the top row of pictures first, revealing each successive one from left to right. The last picture will probably be identified as the face of an elderly man.

With the top row covered, then show the successive animal pictures of the second row. Now the last picture will likely be identified as a rat or mouse. Even after knowing the purpose of the demonstration, you can view serially the row pictures in Figure 5.5 and see how your perception of the last drawing changes depending on whether you previously looked at human faces or animals.

Figure 5.5. Prior perceptual experience influences current perception. (Adapted from Bugelski & Alimpay, 1961)

Now I'd like you to read the sentence given in Figure 5.6 with the intent of understanding what it means. The sentence might seem to make little sense, but treat it as a sentence in a memo you have received from a colleague or supervisor. Some of those memos seem meaningless, too. Your past experience at reading memos, as well as your mood at the time, can influence how you perceive and react to a memo.

After reading the sentence in Figure 5.6, go back and quickly count the number of times the letter *F* occurs in the sentence. Record your answer. When I show this sentence to workshop participants and ask the same question, most will answer "three." A few will shout out "six," usually because they have seen the demonstration before. "Six" is indeed the correct answer, but even after knowing this, a number of people cannot find more than three *F*s in the sentence. Why?

When I was first introduced to this exercise many years ago, I showed the sentence to my two young daughters, and they both found six *F*s immediately. Karly was in kindergarten and Krista was a second grader. Neither could understand the words.

My wife had the same difficulty as I, and could only find three. I remember looking at the sentence again and again, trying to find six *F*s but to no avail. My past experience at speed reading had conditioned me to simply overlook small, unessential words like "of." My history had biased my perception. I bet you had a similar expe-

FINISHED FILES ARE THE RESULT OF YEARS OF SCIENTIFIC STUDY COMBINED WITH THE EXPERIENCE OF MANY YEARS.

Figure 5.6. Past experience can teach us to overlook details.

rience, unless you had seen this demonstration before. If so, then *that* experience biased your current perception of Figure 5.6.

If you need more evidence that past experience influences current perceptions, take a look at the woman in Figure 5.7.

Figure 5.7. Viewing the face from a different orientation (by turning the book upside-down) will influence a different perception.

Notice anything strange, other than that the picture is upside-down? Is this face relatively attractive, or at least normal? Now turn the book upside-down and view the woman's face right-side-up. Has your perception changed? Why didn't you notice her awkward (actually ugly) mouth when the picture was upside-down? Perhaps both context and prior experience (or learning) biased your initial perception. I bet this perceptual bias will persist even after you realize the cause of the distortion, and after viewing the face several times in both positions. A biased perception can be difficult to correct. It's not easy to fight human nature.

RELEVANCE TO ACHIEVING A TOTAL SAFETY CULTURE

Is the relevance of this discussion to occupational safety and health obvious? Perhaps by understanding factors that lead to diverse perceptions we can become more tolerant of individuals who don't appear to share our opinion or viewpoint. Perhaps the person factors discussed here increase your appreciation and respect for diversity and support the basic need to actively listen. "Seek first to understand, before being understood" is Stephen Covey's fifth habit for highly effective people (Covey, 1989, p.235).

It's also possible that this discussion and the exercises on personal perception have reduced your tendency to blame individuals for an injury or to look for a single root cause of an undesirable incident. Before we react to an incident or injury with our own viewpoint, recommendation, or corrective action, we need to ask others about their perceptions.

Hopefully, I have not reduced your optimism toward achieving a Total Safety Culture. But maybe I've alerted you to challenges not previously considered. If I haven't convinced you yet to stop claiming "All injuries are preventable," the next section should do the trick.

Perceived Risk

People are generally underwhelmed or unimpressed by risks or safety hazards at work. Why? Our experiences on the job lead us to perceive a relatively low level of risk. This is strange. After all, it's quite probable someone will eventually be hurt on the job when you factor in the number of hours workers are exposed to various hazards.

In Chapter 4, I discussed one major reason for low perceptions of risk in the workplace. It's elemental, really—we usually get away with risky behavior. As each day goes by without receiving an injury, or even a near hit, we become more accepting of the common belief, "It's not going to happen to me." Now let's further explore why we're generally not impressed by safety hazards at work.

REAL VERSUS PERCEIVED RISK

The real risk associated with a particular hazard or behavior is determined by the magnitude of loss if an mishap occurs, and the probability that the loss or accident will indeed occur. For example, the risk that comes from driving during any one trip can be estimated by calculating the probability of a vehicle crash on one trip and multiplying this value by the magnitude of injury from a crash. Of course, the injury potential or mortality rate from a vehicle crash is influenced by many other factors, including size of vehicle(s) involved, speed of vehicle(s), road conditions, and whether the vehicle occupants were using safety belts.

On any single trip the chance of a vehicle crash is minuscule; but in a lifetime of driving the probability is quite high, varying from .30 to .50 depending upon factors such as geographic location, trip frequency and duration, and characteristics of the driver such as age, gender, reaction time, or mental state (Evans, 1991). Obviously, the risk of driving an automobile is difficult to assess, although it has been estimated that 55 percent of all fatalities and 65 percent of all injuries would have been prevented if a combination shoulder and lap belt had been used (*Federal Register,* 1984).

Estimating the risk of injury from working with certain equipment is even more difficult to determine, because work situations vary so dramatically. Plus the risk can be eliminated completely by the use of appropriate protective clothing or equipment. Still, many people don't appreciate the value of using personal protective equipment or following safe operating procedures. Their perception of risk is generally much lower than actual risk. This thinking pervades society. Automobile accidents are the nation's leading cause of lost productivity, greater than AIDS, cancer, and heart disease (National Academy Press, 1985; Waller, 1986). But how many of us take driving for granted?

The risk of a fatality from driving a vehicle or working in a factory is much higher than from the environmental contamination of radiation, asbestos, or industrial chemicals. Yet look at the protests over asbestos in schools and neighborhood chemical plants.

LOWER RISK	HIGHER RISK
• exposure is voluntary	• exposure is mandatory
• hazard is familiar	• hazard is unusual
• hazard is forgettable	• hazard is memorable
• hazard is cumulative	• hazard is catastrophic
• collective statistics	• individual statistics
• hazard is understood	• hazard is unknown
• hazard is controllable	• hazard is uncontrollable
• hazard affects anyone	• hazard affects vulnerable people
• preventable	• only reducible
• consequential	• inconsequential

Figure 5.8. Factors which influence perceived risk. Those in the left-hand column *reduce* risk perception and are generally associated with the workplace. (Adapted from Sandman, 1991)

Researchers of risk communication have found that various characteristics of a hazard, irrelevant to actual risk, influence people's perceptions (Covello *et al,* 1991; Slovic, 1991). It's important to consider these characteristics, because behavior is determined by perceived rather than actual risk.

Figure 5.8 shows factors that influence our risk perceptions. It is derived from research by Drs. Peter Sandman (1991), Paul Slovic (1991), and their colleagues. The factors listed on the left reduce perceptions of risk and are typically associated with the workplace. The opposing factors in the right-hand column have been found to increase risk perception, and these are not usually experienced in the work setting. As a consequence, our perception of risk on the job is not as high as it should be, and therefore we don't work as defensively as we should. Discussing some of these factors will reveal strategies for increasing our own and others' perception of risk in certain situations.

THE POWER OF CHOICE

Hazards we choose to experience (like driving, skiing, and working) are seen as less risky than ones we feel forced to endure (like food preservatives, environmental pollution, and earthquakes). Of course, the perception of choice is also subjective, varying dramatically among individuals. For example, people who feel they have the freedom to pull up stakes and move whenever they want would likely perceive less risk from a nearby nuclear plant or seismic fault. Likewise, employees who feel they have their pick of places to work generally perceive less risk in a work environment. They are typically more motivated and less distressed. In the next chapter, I'll discuss relationships between perceived choice, stress, and distress.

FAMILIARITY BREEDS COMPLACENCY

Familiarity is probably a more powerful determinant of perceived risk than choice. The more we know about a risk, the less it threatens us. Remember how attentive you were when first learning to drive, or when you were first introduced to the equipment in your workplace? It wasn't long before you lowered your perceptions of risk, and changed your behavior accordingly. When driving, for example, most of us quickly shifted from two hands on the wheel and no distractions to steering with one hand while turning up the radio and carrying on a conversation.

It's so easy to tune out the familiar hazards of the workplace. Safety professionals respond by constantly reminding employees of risks with a steady stream of memos, newsletters, safety meetings, and signs. Still, these efforts can't compete with the impact of unusual, catastrophic, and memorable events broadcast by the media and dramatized on television and in the movies. Publicity of memorable injuries, like those suffered by John Wayne Bobbitt and Nancy Kerrigan in 1994, influences misperception of actual risk.

SYMPATHY FOR VICTIMS

Many people feel sympathy for victims of a publicized incident, even vividly visualizing the injury as if it happened to them. Personalizing these experiences increases perceived risk. At work, employees show much more attention and concern for hazards when injuries or "near hits" are discussed by the coworkers who experienced them, compared to a presentation of statistics. The average person can't relate to group numbers. But there's power in personal stories. I have met many people over the years who accepted individual accounts in lieu of convincing statistics: "The police officer told Uncle Jake he would have been killed if he had been buckled up"; "Aunt Martha is 91 years old and still smokes two packs of cigarettes a day."

This suggests that we should shift the focus of safety meetings away from statistics, emphasizing instead the human element of safety. Safety talks and intervention strategies should center on individual experiences rather than numbers. This might be easier said than done. Encouraging victims to come forward with their stories is often stifled by management systems in many companies that seem to value fault finding over fact finding, piecemeal rather than system approaches to injury investigation, and enforcement and discipline more than recognition and rewards to influence on-the-job behaviors.

UNDERSTOOD AND CONTROLLABLE HAZARDS

Hazards we can explain and control cause much less alarm than hazards that are not understood, and thus are perceived as uncontrollable. This points up a problem with many employee safety education and training programs. Workplace hazards are explained in a way that creates the impression they can be controlled. Indeed, safety professionals often state a vision or goal of "zero injuries," implying complete control over the factors that cause injuries. This actually lowers perceived risk by convincing people the causes of occupational injuries are understood and controllable.

Perhaps it would be better for safety leaders to admit and publicize that only two of the three types of factors contributing to workplace injuries can be managed effectively—environmental/equipment factors and work behaviors. As I've already discussed in preceding chapters, the mysterious inside, unobservable, and subjective world of people dramatically influences the risk of personal injury. These attitudes, expectancies, perceptions, and personality characteristics cannot be measured, managed, or controlled reliably. Internal human factors make it impossible to prevent all injuries. By discussing the complexity of people and their integral contribution to most workplace hazards and injuries, you can increase both the perceived value of ongoing safety interventions and the belief that a Total Safety Culture requires total commitment and involvement of all concerned.

ACCEPTABLE CONSEQUENCES

We're less likely to feel threatened by risk-taking or a risk exposure that has its own rewards. But if few benefits are perceived by an at-risk behavior or environmental condition, outrage—or heightened perceived risk—is likely to be the reaction, along with a concerted effort to prevent or curtail the risk.

Some people, for example, perceive guns, cigarettes, and alcohol as having limited benefit and thus lobby to restrict or eliminate these societal hazards. The availability of and exposure to these hazards will continue, though, as long as a significant number of individuals perceive that the risk benefits outweigh the risk costs. You can see how cost-benefit analyses are easily politicized by subjective perceptions of risk.

The benefits of risky work behaviors are obvious. You might feel cooler and more comfortable not wearing a respirator. It's convenient—you can talk easier. Feeling relieved, you might actually be more productive. The costs of not wearing the mask might be abstract and delayed (if the exposure is not immediately life-threatening). Statistics might point out a chance of getting a lung disease, which won't surface for decades, if ever. Decisions about risk-taking are made every day by workers. By playing the odds and shooting for short-term gains, risky work practices are often accepted and not perceived to be as dangerous as they actually are.

SENSE OF FAIRNESS

Most people believe in a just and fair world (Lerner, 1975, 1977). "What goes around comes around." We believe people generally get what they deserve. When people receive benefits like increased productivity from their risky behavior, the outrage, public attention, or perceived risk is relatively low. On the other hand, when hazards or injuries seem unfair, as when a child is molested or inflicted with a deadly disease, special attention is given. This increased attention results in more perceived risk.

This makes it relatively easy to obtain contributions or voluntary assistance for programs that target vulnerable populations, like learning-disabled children. Workplace injuries, though, are perceived as fair. They are indiscriminately distributed among employees who take risks, and these people deserve what they get. This perception or attitude naturally lowers the outrage we feel when someone gets injured on the job. Lower outrage translates into lower perceived risk.

Figure 5.9. Personal protective equipment can reduce the perception of risk.

Risk Compensation

A discussion of risk perception would not be complete without examining one of the most controversial concepts in the field of safety. In recent years, it has been given different labels, including *risk homeostasis, risk or danger compensation, risk-offsetting behavior,* and *perverse compensation*. Whatever the name, the basic idea is quite simple and straightforward. People are presumed to adjust their behavior to compensate for changes in perceived risk. If a job is made safer with machine guards or the use of personal protective equipment, workers might reduce their perception of risk and thus perform more recklessly. For example, if the individual depicted in Figure 5.9 is taking a risk due to the perceived security of fall protection, we would have support for risk compensation theory (Peltzman, 1975).

The notion of taking more risks to compensate for lower risk perception certainly seems intuitive. I bet every reader has experienced this phenomenon. I clearly remember taking more risks after donning a standard high school football uniform. With helmet and shoulder pads, I'd willingly throw my body in the path of another player or leap to catch a pass. I didn't perform these behaviors until perceiving security from the personal protective equipment (PPE).

Today I experience risk compensation of a different sort on the tennis court. If I get ahead of my opponent by a few games, I take more chances. I'll hit out for a winner or go to the net for a volley. When I get behind my opponent by two or more games, I play more conservatively from the base line. I adjust the risk level of my game depending on the circumstances—my opponent's skills and the score of the match.

Risk compensation has seemingly universal applications. How can the phenomenon be denied?

Figure 5.10 depicts a workplace situation quite analogous to my teenage experiences on the football field. There appears to be limited scientific data to support the use of commercially available back belts (Metzgar, 1995). Could this be partly due to risk compensation? If the use of a back belt leads to employees lifting heavier loads, then the potential protection from this device could be offset by greater risk taking.

Figure 5.10. Back belts can give a false sense of security.

The protective device could give a false sense of security and reduce one's perception of being vulnerable to back injury. The result could be more frequent and heavier lifting, and greater probability of injury. This is why back belt suppliers emphasize the need for training and education in the use of belts.

Obviously, the notion that an individual's behavior could offset the safety benefits of PPE is extremely repugnant to a safety professional. Could this mean that efforts to make environments safer with engineering innovations are useless in the long run? Are safety belts and air bags responsible for increases in vehicle speeds? Does this mean laws and policy to enforce safe behavior actually provoke offsetting at-risk behavior?

Some researchers and scholars are convinced risk compensation is real and is detrimental to injury prevention (Adams, 1985; Peltzman, 1975; Wilde, 1994); others contend the phenomenon doesn't exist. B.J. Lehman and Howard Gage (1995) proclaim, for example, that "this alleged theory [risk compensation] has neither experimental nor analytical scientific basis" (p.37); and Dr. Leonard Evans of General Motors Research Laboratories is quoted as saying "...there are no epicycles and there is no phlogiston...similarly, there is no risk homeostasis" (in Wilde, 1994, ii).

SUPPORT FROM RESEARCH

In fact, there is scientific evidence that risk compensation, or risk homeostasis, is real, as our intuition or common sense tells us. But the offsetting or compensating behavior does not negate the benefits of intervention. Although football players increase at-risk behaviors when suited up, for example, they sustain far fewer injuries than they would without the PPE. This is true even if a lack of protection reduced their risk-taking substantially. More importantly, if people lower the level of risk they are willing to accept (as promoted in a Total Safety Culture), then risk compensation or risk homeostasis is irrelevant. I'll explain this "good news" further, but first let's look more closely at research evidence supporting the phenomenon.

Comparisons between people. The notion of risk compensation made its debut among safety professionals following the theorizing and archival research of University of Chicago economist Dr. Sam Peltzman (1975). Peltzman systematically compared vehicle crash statistics before (1947-1965) versus after (1966-1972) the regulated installation of safety engineering innovations in vehicles, including seat belts, energy-absorbing steering columns, padded instrument panels, penetration-resistant windshields, and dual braking systems. As predicted by risk compensation theory, Peltzman found that these vehicle-manufacturing safety standards had not reduced the frequency of crash fatalities per miles driven. Perhaps the most convincing evidence of risk compensation was that the cars equipped with safety devices were involved in a disproportionately high number of crashes.

Peltzman's article has been criticized on a number of counts, primarily statistical; but it did stimulate follow-up investigations. Dr. John Adams of University College, London, for example, compared traffic fatality rates among countries with and without safety-belt use laws. His annual comparisons (from 1970 to 1978) showed dramatic reductions in fatal vehicle crash rates after countries introduced seat-belt use laws. Taken alone this data would lend strong support to seat-belt legislation. But the drop in fatality rates was even greater in countries without safety-belt use laws (Adams, 1985b). Apparently, the large-scale impact of increased use of vehicle safety belts has not been nearly as beneficial as expected from laboratory crash tests. Risk compensation has been proposed to explain this discrepancy.

There are obviously other possible explanations for the fluctuations in large data bases compiled and analyzed by Peltzman (1975) and Adams (1985a,b)—changes in the economy, improvements in vehicle performance, and media promotion of particular lifestyles, to name a few. Regarding safety-belt mandates, for example, it's generally believed that the safest drivers are the first to buckle up and comply, meaning the most prominent decrease in injuries from vehicle crashes won't occur until the remaining 30 percent buckle up—those currently resisting belt-use laws (Campbell *et al,* 1987). In other words, "those segments of the driving population who are least likely to comply with safe driving laws are precisely those groups that are at highest risk of serious injury" (Waller, 1987, p.43).

Research supports this presumed direct relationship between at-risk behavior and non-compliance with safety policy. Young males (Preusser *et al,* 1985), persons with elevated blood alcohol levels (Wagenaar, 1984), and "tailgaters" who drive dangerously close to the vehicles they follow (Evans *et al,* 1982) are less likely to comply with a belt-use law. These findings could certainly have implications for occupational safety. If the riskiest workers are least likely to comply with rules and policy, traditional top-down enforcement and discipline are not sufficient to achieve a Total Safety Culture. Of course, this is a primary theme of this book. But let's get back to the issue of risk compensation.

Studies that compared risk behaviors across large data sets and found varying characteristics among people who complied with a safety policy versus those who did not certainly weakens the case for risk compensation. Behavioral scientists call this *between-groups research,* and it can only indirectly test the occurrence of risk compensation. Since risk compensation theory predicts that individuals increase their risky behavior after perceiving an increase in safety or security, the theory can only be tested by comparing the same group of individuals under different conditions. Behavioral-science researchers call this a *within-subjects design.*

WITHIN-SUBJECT COMPARISONS. Most within-subject tests of risk compensation theory have been restricted to simulated laboratory investigations (Wilde *et al,* 1985). These observations of different risk conditions are time-consuming and quite difficult to pull off in a real-world situation. Dr. Fredrick Streff and I conducted one such study in 1987. We built an oval clay go-cart track about 100 meters in circumference and equipped a 5-horsepower go-cart with an inertia reel-type combination shoulder-lap safety harness.

Subjects were told to drive the go-cart around the track "quickly, but at a speed that is comfortable for you" (Streff & Geller, 1988, p.280). The 56 subjects were either buckled or unbuckled in the first of two phases of driving trials. After the first phase, the safety condition was switched for half the subjects. That is, the safety belt was no longer used if the drivers had previously been buckled up, or the belt was used by drivers who previously did not use it. The speed and accuracy of each subject's driving trial were systematically measured. Following the first and second phases (consisting of 15 trials each), the subjects completed a brief questionnaire to assess their perceived risk while driving the go-cart.

The between-subject comparisons showed no risk compensation. Subjects who used the safety belt for all trials did not drive faster than subjects who never used the safety belt. Perceptions of risk were not different across these groups of subjects, either.

On the other hand, the within-subject differences did show the predicted changes in risk perception and significant risk compensation. Subjects reported feeling safer when they buckled up, and subsequently drove the go-cart significantly faster than subjects who used the safety belt during both phases. Those who took off their safety belts reported a significant decrease in perceived safety, but this change in risk perception was not reflected in slower driving speeds, compared to drivers who never buckled up in the go-cart.

Our go-cart study was later followed up in the Netherlands using a real car on real roads. Convincing evidence of risk compensation was found (Jansson, 1994). Specifically, habitual, "hard-core" non-users of safety belts buckled up at the request of the experimenter. Compared to measures taken when not using a safety belt, these buckled-up drivers drove faster, followed more closely behind vehicles in front of them, changed lanes at higher speeds, and braked later when approaching an obstacle.

IMPLICATIONS OF RISK COMPENSATION. I'm convinced from personal experience and reading the research literature that risk compensation is a real phenomenon. What does this mean for injury prevention? Professor Wilde (1994) says it means safety excellence cannot be achieved through top-down rules and enforcement. Some people only follow the rules when they are supervised, and might take greater risks when they can get away with it. This behavior is not only predicted by risk compensation theory, but also by the theory of psychological reactance (Brehm, 1966) discussed in Chapter 3 (see Figure 3.1).

According to reactance theory, some people feel a sense of freedom or accomplishment when they don't comply with top-down regulations. They like to beat the system. Skinner (1971) referred to reactive behavior as *countercontrol*—a means by which some people attempt to assert their freedom and dignity when feeling controlled.

Whether dangerous behavior results from psychological reactance or risk compensation, our risk reduction attempts are the same. As the title of his book *Target Risk* indicates, Professor Wilde advocates that safety interventions need to lower the level of risk people are willing to tolerate. This requires a change in values. Wilde (1994) asserts that improvements in safety cannot be "achieved by interventions in the form of training, engineering or enforcement" (p. 213). "The extent of risk taking with respect to safety and health in a given society, therefore, ultimately depends on values that prevail in that society, and not on the available technology" (Wilde, 1994, p. 223).

I hope it's obvious that Wilde's position is consistent with the theme of this text. When people understand and accept the paradigm shifts needed for a Total Safety Culture (see Chapter 3), they are on track to reducing their tolerance for risk. Next they need to believe in the vision of a Total Safety Culture and buy into the mission of achieving it. Then they need to understand and accept the procedures that can achieve this vision. These methods are explained in Part Three of this text. Through a continuous process of applying the right procedures, the work force will feel empowered to actively care for a Total Safety Culture. Finally, they will come to treat safety as a value rather than a priority. I discuss these concepts more fully in Part Four.

In Conclusion

This chapter explored the concept of selective sensation or perception, and related it to perceived risk and injury control. Visual exercises illustrated the impact of past experience and contextual cues on present perception. This allows us to appreciate diversity and realize the value of actively listening during personal interaction. We need to work diligently to understand the perceptions of others before we impulsively jump to conclusions or attempt to exert our influence.

We must realize that perceptions of risk vary dramatically among individuals. And we can't improve safety unless people increase their perception of, and reduce their tolerance for, risk. Changes in risk perception and acceptance will occur when individuals get involved in achieving a Total Safety Culture with the principles and procedures discussed in this book.

Several factors were discussed in this chapter that affect whether employees react to workplace hazards with alarm, apathy, or something in between. Taken together, these factors shape personal perceptions of risk and illustrate why the job of improving safety is so daunting. This justifies more resources for safety and health programs, and intervention plans to motivate continual employee involvement. We will get to a discussion of these interventions in Part Three. But before discussing strategies to fix the problem, we need to understand how stress, distress, and personal attributions contribute to the problem. That's our topic for the next chapter.

REFERENCES

Adams, J.G.U. (1985a). *Risk and freedom*. London: Transport Publishing Projects.

———. (1985b). Smead's law, seat belts and the emperor's new clothes. In L. Evans and R. C. Schwing (Eds.), *Human Behavior and Traffic Safety*, (pp. 193-245). New York: Plenum.

Brehm, J. W. (1966). *A theory of psychological reactance*. New York: Academic Press.

Bugelski, B. R., & Alimpay, D. A. (1961). The role of frequency in developing perceptual sets. *Canadian Journal of Psychology*, 15, 205-211.

Campbell, B. J., Stewart, J. R., & Campbell, F. A. (1987). *1985-1986 experiences with belt laws in the United States*. Chapel Hill, NC: UNC Highway Safety Research Center.

Covell, V. T., Sandman, P. M., & Slovic, P. (1991). Guidelines for communicating information about chemical risks effectively and responsibly. In D. G. Mayo & R. D. Hollander (Eds.), *Acceptable evidence: Science and values in risk management.* (pp. 66-90). New York: Oxford University Press.

Covey, S. R. (1989). *The seven habits of highly effective people: Restoring the character ethic.* New York: Simon and Schuster.

Evans, L. (1991). *Traffic safety and the driver.* New York: Van Nostrand Reinhold.

Evans, L., Wasielewski, P., & von Buseck, C. R. (1982). Compulsory seat-belt usage and driver risk-taking behavior. *Human Factors,* 24, 41-48.

Federal Register 9, (1984). *Federal motor vehicle safety standards: Occupant crash protection, Final Rule,* 48 (no. 138), July. Washington, DC: US Department of Transportation.

Janssen, W. (1994). Seat belt wearing and driving behavior: An instrumented-vehicle study. *Accident Analysis and Prevention,* 26, 249-261.

Lehman, B. J., & Gaze, H. (1995). How much is safety really worth? Countering a false hypothesis. *Professional Safety,* 40(6), 37-40.

Lerner, M. S. (1975). The justice motive in social behavior. *Journal of Social Issues,* 31(3), 1-19.

Lerner, J. J. (1977). The justice motive: Some hypotheses as to its origins and forms. *Journal of Personality,* 45, 1-52.

Metzgar, C. R. (1995). Placebos, back belts and the Hawthorne effect. *Professional Safety,* 40(4), 26-29.

National Academy Press, (1985). *Injury in America: A continuing public health problem.* Washington, D.C.: National Academy Press.

Peltzman, S. (1975). The effects of automobile safety regulation. *Journal of Political Economics,* 83, 677-725.

Preusser, D. F., Williams, A. F., & Lund, A. K. (1985). *The effect of New York's seat belt law on teenage drivers.* Washington, D.C.: Insurance Institute for Highway Safety.

Sandman, P. M. (1991). *Risk = Hazard + Outrage: A formula for effective risk communication.* Videotaped presentation for the American Industrial Hygiene Association. Environmental Communication Research Program, P.O. Box 231, Cook College, Rutgers University, New Brunswick, NJ.

Skinner, B. F. (1971). *Beyond freedom and dignity.* New York: Alfred A. Knopf.

Slovic, P. (1991). Beyond numbers: A broader perspective on risk perception and risk communication. In D. G. Mayo & R. D. Hollander (Eds.), *Deceptable evidence: Science and values in risk management* (pp. 48-65). New York: Oxford University Press.

Streff, F. M., & Geller, E. S. (1988). An experimental test of risk compensation: Between-subject versus within-subject analysis. *Accident Analysis and Prevention,* 20, 277-287.

Wagenaar, A. C. (1984). *Restraint usage among crash-involved motor vehicle occupants.* Report VMTRI-84-2. Ann Arbor, MI: University of Michigan Transportation Research Institute.

Waller, J. A. (1987). Injury: Conceptual shifts and prevention implications. *Annual Review of Public Health,* 8, 21-49.

———. (1986). State liquor laws as enablers for impaired driving and other impaired behaviors. *American Journal of Public Health,* 76, 787-792.

Wilde, G.J.S. (1994). Target risk. Toronto, Ontario, Canada: PDE Publications.

Wilde, G.J.S., Claxton-Oldfield, S. P., & Platenius, P. H. (1985). Risk homeostasis in an experimental context. In Evans, L. and Schwing, R. C. (Eds.), *Human Behavior and Traffic Safety* (pp. 114-149). New York: Plenum Press.

Chapter 6

Stress versus Distress

Stressors can contribute to a near hit or injury; they are barriers to a Total Safety Culture. However, stressors can provoke positive stress rather than negative distress—which can lead to constructive problem solving rather than destructive, at-risk behavior. This chapter explains the important distinction between stress and distress, and defines factors which determine the occurrence of one or the other.

The concept of "attribution" is reviewed as a process we use to reduce personal distress. This is one way accident victims shift blame and distort follow-up investigations. I show, however, how this self-serving bias can actually be used to benefit safety efforts.

Judy was tired and worried. She had just left her six-year-old son at her sister's house with instructions for her to take him to Dr. Slayton's office for a 10:30 a.m. appointment. She had been up much of the night with Robbie, attempting to comfort him. With tears in his eyes, he had complained of a "hurt" in his stomach. This was the third night his cough had periodically awakened her, but last night Robbie's cough was deeper, seemingly coming from his lungs.

Judy arrived at her work station a little later than normal and found it more messy than usual. Grumbling under her breath that the night shift had been "careless, sloppy, and thoughtless," she downed her usual cup of coffee and waited for the production line to crank up, rather than cleaning up the work area. After all, it wasn't her mess. The graveyard shift is not nearly as busy as the day shift. How could they be so sloppy and inconsiderate?

Judy was ready to start her inspection and sorting when she noticed the "load cart" was misaligned. She inserted a wooden handle in the bracket and pulled hard to jerk the cart into place. Suddenly the handle broke, and Judy fell backward against the control panel. Fortunately, she wasn't hurt, and the only damage was the broken handle. Judy discarded it, inserted another one, and put the cart in place.

During lunch Judy called the doctor's office and learned that her son had the flu, and would be fine in a day or two. She completed the day in a much better mood, and without incident.

At the end of her shift, Judy filled out a near-hit report on her morning mishap. She wrote that someone on the graveyard shift had left her work area in disarray, including a misaligned load tray. She also indicated that the design of the cart handle made damage likely; in the past other cart handles had been broken. She recommended a re-design of the handle brackets and immediate discipline for the graveyard shift in her work area.

The fact that Judy filled out a near-hit report is certainly good news. But was this a complete report? Were there some person factors within Judy that could have influenced the incident? Was Judy under stress or distress, and if so, could this have been a contributing factor? It's been estimated that from 75 to 85 percent of all industrial injuries can be partially attributed to inappropriate reactions to stress (Jones, 1984). Furthermore, stress-related headaches are the leading cause of lost work time in the United States (Jones, 1984).

Judy's near-hit report was also clearly biased by common attribution errors researched by social psychologists, and used by all of us at some time to deflect potential criticism and reduce distress. Attribution errors, along with stress and distress, represent potential barriers to achieving a Total Safety Culture.

What is Stress?

In simple terms, stress is a psychological and physiological reaction to events or situations in our environment. Whatever triggers the reaction is called a stressor. So stress is the reaction of our mind and body to *stressors* such as demands, threats, conflicts, frustrations, overloads, or changes.

Figure 6.1. Technology cannot always substitute for personnel.

Figure 6.1 depicts an office worker reacting to work overload—an almost everyday stressor. "Lean and mean" organizations can create very stressful circumstances. The quick-fix solution to cutting costs by reducing headcount can lead to long-term stress and safety problems, unless employees can adjust to increased or shifted responsibities.

Let's return to the basics of stress. When you interpret a situation as being stressful, your body prepares to deal with it. This is the fight-or-flight syndrome, a process controlled through our sympathetic nervous system. Adrenaline rushes into the bloodstream, the heart pumps faster, and breathing increases. Blood flows quickly from our abdomen to our muscles—causing those "butterflies in our stomach." We can also feel tense muscles or nervous strain in our back, neck, legs, and arms (Selye, 1974, 1976).

CONSTRUCTIVE OR DESTRUCTIVE?

The terms stress and distress commonly carry negative connotations. Something unwanted and bad is happening. But the first definition of stress in my *American Heritage Dictionary* (1992) is "importance, significance, or emphasis placed on something" (p. 1205). Similarly, the *New Merriam-Webster Dictionary* (1989) defines stress as "a factor that induces bodily or mental tension . . . a state induced by such a stress . . . urgency, emphasis" (p. 701). In contrast, distress is defined as "anxiety or suffering . . .severe strain resulting from exhaustion or an accident" (*The American Heritage Dictionary,* 1991, p. 410), or "suffering of body or mind; pain, anguish; trouble, misfortune...a condition of desperate need (*The New Merriam-Webster Dictionary,* 1989, p. 224).

Psychological research supports these distinctions between stress and distress. Stress can be positive, giving us heightened awareness, sharpened mental alertness, and an increased readiness to perform. Certain psychological theories presume that some stress is necessary for people to perform. The person who asserts, "I work best under pressure," understands the motivational power of stress. But can too much pressure, too many deadlines, be destructive?

I'm sure most of you have been in situations—or predicaments—where the pressure to perform seemed overwhelming. This is the point where too much pressure can hurt performance, where stress becomes *dis*tress. The relationship between external stimulation or pressure to perform and actual performance is depicted in Figure 6.2. This inverted U-shaped function is known as the Yerkes-Dodson Law (Yerkes & Dodson, 1908).

The Yerkes-Dodson law states that, up to a point, performance will increase as arousal, or pressure to perform well, increases. But the best performance comes when arousal is optimum rather than maximum. Push a person too far and performance starts to deteriorate. In fact, at exceptionally high levels of pressure or tension a person might perform as poorly as when they are hardly stimulated at all. Ask someone who is hysterical, and someone who is about to fall asleep, to do the same job, and you won't be pleased by either result. Hans Selye, the Austrian-born founder of stress research said, "Complete freedom from stress is death" (Selye, 1974, p.32). It is extreme, disorganizing stress we need to avoid. Watch out for distress.

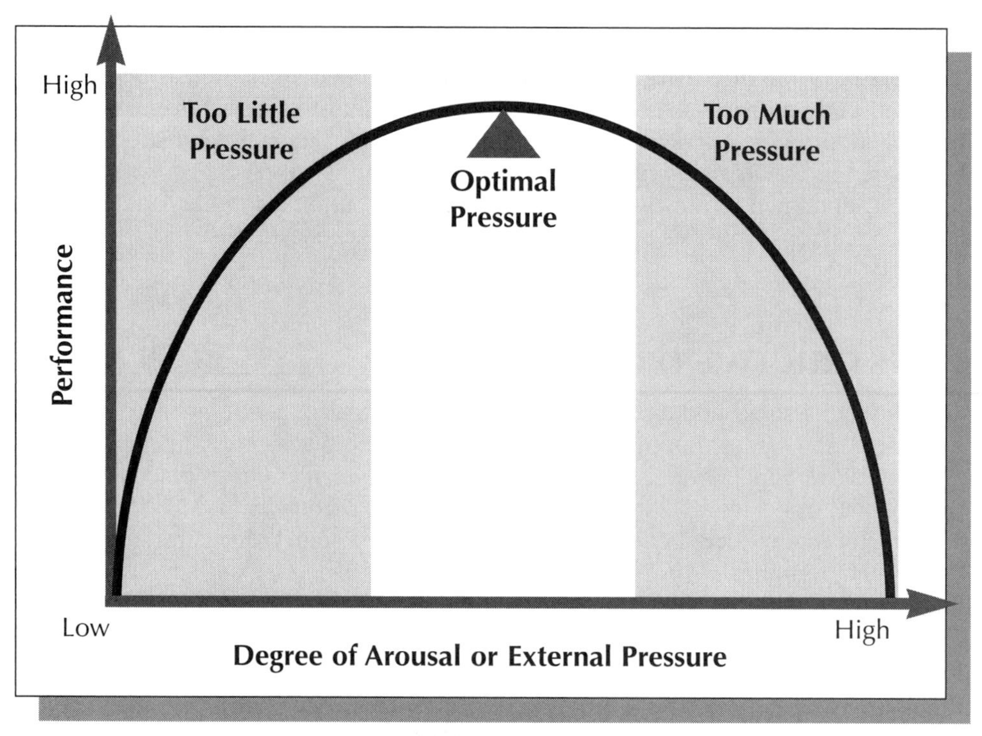

Figure 6.2. Arousal from external pressure (or a stressor) improves performance to an optimal point.

THE EYES OF THE BEHOLDER

Perceptions play an important role in stress and distress. The boss gives a group of employees a deadline; some tighten up inside, others take it in stride. Some circle their calendars and can't take their minds off the due date. Others seem to pay it no mind.

When a stressor is noticed and causes a reaction, the result can be constructive or destructive. If we believe we are *in control*—that we can deal with the overload, frustrations, conflicts, or whatever is triggered by the stressor—we become aroused and motivated to go beyond the call of duty. We actually achieve more. On the other hand, when we believe we can't handle the demands of the stressor, the resulting psychological and physiological reactions are likely to be detrimental to performance, and to our health and safety.

Why does the boss's deadline motivate one person and distress another? It depends on a number of internal "person factors." These include the amount of arousal already present in the individual and the person's degree of preparedness or self-confidence.

Those "butterflies" we feel in our stomach can help or hinder performance, depending on personal perception. When the "butterflies" are aligned for goal-directed behavior, we feel in control of the situation: prepared for it, challenged by it. The stress is positive—it arouses or motivates performance improvement. When the "butterflies" are misaligned and scattered in different directions, we feel unprepared, burdened rather than challenged. Stress is experienced as distress. This arousal can divert our attention, interfere with thinking processes, disrupt performance and reduce our ability and overall motiviation to perform well. The result can be at-risk behavior and a serious injury.

Identifying the Stressors

Stress or distress can be provoked by a wide range of demands and circumstances. Some stressors are acute, sudden life events, such as death or injury to a loved one, marriage, marital separation or divorce, birth of a baby, failure in school or at work, and a job promotion or relocation. Other stressors include the all-too-frequent minor hassles of everyday life, from long lines and excessive traffic to downsized work conditions and worries about personal finances (Holmes & Masuda, 1974).

Prolonged, uncontrollable stressors can lead to "burnout." Common symptoms of burnout include: 1) physical exhaustion resulting in lack of energy, headaches, backaches and general fatigue; 2) emotional exhaustion manifested by loss of appetite, feelings of helplessness, and depression; and 3) mental or attitudinal exhaustion revealed through irritability, cynicism, and a negative outlook on life (Baron, 1995). Obviously, burnout puts people at risk for causing injury to themselves or others.

Our jobs or careers are filled with stressors. Consider for a moment how much time you spend working, or thinking about your work. Some workplace stressors are obvious; others might not be as evident but are just as powerful. Work overload can obviously become a stressor and provoke either stress or distress, but what about "work underload"? Being asked to do too little can produce profound feelings of boredom, which can also lead to distress. Performance appraisals are stressors that can be motivational if perceived as objective and fair, or they can contribute to distress and inferior performance if viewed as subjective and unfair.

Other work-related factors that can be perceived as stressors and lead to distress and eventual burnout include: role conflict or ambiguity; uncertainty about one's job responsibilities; responsibility for others; a crowded, noisy, smelly, or dirty work environment; lack of involvement or participation in decision making; interpersonal conflict with other employees; and insufficient support from coworkers (Maslach, 1982).

We're at a good point in our discussion for you to complete the Work Stress Profile questionnaire in Figure 6.3. It was developed by Professor Philip Rice (1992) to identify individuals' various workplace stressors. Instructions for scoring your distress profile are included. The survey addresses three domains of work distress—interpersonal relations, physical demands, and level of mental interest—and these are totaled for an overall distress score.

The interpersonal scale measures distress related to personal relationships, or lack thereof, at work. Stressors in this category can emanate from communication breakdowns with coworkers or supervisors, lack of appropriate job training or recognition processes, infrequent opportunities for personal choice in work assignments or work processes, and insufficient social support from colleagues or team members.

The second scale of the distress questionnaire estimates the physical demands of work that can wear on an individual day after day. This includes environmental stressors such as noise, crowded conditions, and incessant work demands; personal stressors such as feeling overworked or ineffective; and interpersonal stressors like insufficient team support. When evaluating your score for this scale, it's important to understand the critical relationship between the outside world and your inside world—the world of your own perceptions. Remember, the same work demands and interpersonal stressors can result in stress and increased productivity for some employees but lead to distress and burnout for others.

Work Stress Profile

The interest scale reflects one's personal reaction to the stressors of his or her workplace. A high score reflects a low level of personal interest, commitment, and involvement for your job. This may indicate a need to change jobs or perhaps alter the way you view your work situation. You might perceive it as an opportunity rather than a necessity. A job should be perceived as something you *get* to do, not something you *got* to do. Interacting more effectively with work associates, especially through more active listening, can readily turn job distress into productive stress. More suggestions for decreasing personal distress are offered in the next section of this chapter.

After obtaining your survey totals for each of the three scales and adding these subtotals for an overall distress score, you can compare your results with the values given in Figure 6.4. The numbers here were obtained from a sample of 275 school psychologists (Rice, 1992) whose job responsibilities might be very different than yours. It might be more useful to compare your results with others in your work culture.

I urge you not to take your score, or relative rankings with others, too seriously. Surveys like this are only imperfect estimates of your perceptions and feeling states at the time you respond to the questions. Answering the questions and deriving a personal score will surely increase your understanding of both job stress and distress from an environmental, interpersonal, and personal perspective, but don't get discouraged by a high distress score. Your distress state can be changed with strategies we'll soon discuss. There is much you can do on your own, for yourself.

This questionnaire was adapted from Rice, 1992, with permission from the author.

The following statements describe work conditions, job environments, or personal feelings that people encounter in their jobs. After reading each statement, circle the answer that best reflects the working conditions at your place of employment. If the statement is about a personal feeling, indicate the extent to which you have that feeling about your job. The scale markers ask you to judge, to the best of your knowledge, the approximate percentage of time the condition or feeling is true.

NEVER = not at all true of my work conditions or feelings
RARELY = the condition or feeling exists about 25% of the time
SOMETIMES = the condition or feeling exists about 50% of the time
OFTEN = the condition or feeling exists about 75% of the time
MOST TIMES = the condition or feeling is virtually always present

	NEVER	RARELY	SOMETIMES	OFTEN	MOST TIMES
1. Support personnel are incompetent or inefficient.	1	2	3	4	5
2. My job is not very well defined.	1	2	3	4	5
3. I am not sure about what is expected of me.	1	2	3	4	5
4. I am not sure what will be expected of me in the future.	1	2	3	4	5
5. I cannot seem to satisfy my superiors.	1	2	3	4	5
6. I seem to be able to talk with my superiors.	5	4	3	2	1
7. My superiors strike me as incompetent, yet I have to take orders from them.	1	2	3	4	5
8. My superiors seem to care about me as a person.	5	4	3	2	1
9. There is a feeling of trust, respect, & friendliness between me & my superiors.	5	4	3	2	1
10. There seems to be tension between management and operators.	1	2	3	4	5
11. I have a sense of individuality in carrying out my job duties.	5	4	3	2	1
12. I feel as though I can shape my own destiny in this job.	5	4	3	2	1
13. There are too many bosses in my area.	1	2	3	4	5
14. It appears that my boss has "retired on the job."	1	2	3	4	5
15. My superiors give me adequate feedback about my job performance.	5	4	3	2	1
16. My abilities are not appreciated by my superiors.	1	2	3	4	5

Figure 6.3. Work Distress Profile ***(continued on the next three pages)***

	NEVER	RARELY	SOMETIMES	OFTEN	MOST TIMES
17. There is little prospect of personal or professional growth in this job.	1	2	3	4	5
18. The level of participation in planning and decision making is satisfactory.	5	4	3	2	1
19. I feel that I am over-educated for this job.	5	4	3	2	1
20. I feel that my educational background is just right for this job.	1	2	3	4	5
21. I fear that I will be laid off or fired.	1	2	3	4	5
22. In-service training for my job is inadequate.	1	2	3	4	5
23. Most of my colleagues seem uninterested in me as a person.	1	2	3	4	5
24. I feel uneasy about going to work.	1	2	3	4	5
25. There is no release time for personal affairs or business.	1	2	3	4	5
26. There is obvious sex/race/age discrimination in this job.	1	2	3	4	5

NOTE: Complete the entire questionnaire first! Then add all the values you circled for questions 1-26 and enter here.

Total 1-26_______

	NEVER	RARELY	SOMETIMES	OFTEN	MOST TIMES
27. The physical work environment is crowded, noisy, or dreary.	1	2	3	4	5
28. Physical demands of the job are unreasonable (heavy lifting, extraordinary concentration required, etc.).	1	2	3	4	5
29. My work load is never-ending.	1	2	3	4	5
30. The pace of work is too fast.	1	2	3	4	5
31. My job seems to consist of responding to emergencies.	1	2	3	4	5
32. There is no time for relaxation, coffee breaks, or lunch breaks on the job.	1	2	3	4	5
33. Job deadlines are constant and unreasonable.	1	2	3	4	5
34. Job requirements are beyond the range of my ability.	1	2	3	4	5
35. At the end of the day, I am physically exhausted from work.	1	2	3	4	5
36. I can't even enjoy my leisure because of the toll my job takes on my energy.	1	2	3	4	5

Continues on the next page

Continued from previous page

	NEVER	RARELY	SOMETIMES	OFTEN	MOST TIMES
37. I have to take work home to keep up.	1	2	3	4	5
38. I have responsibility for too many people.	1	2	3	4	5
39. Support personnel are too few.	1	2	3	4	5
40. Support personnel are incompetent or inefficient.	1	2	3	4	5
41. I am not sure about what is expected of me.	1	2	3	4	5
42. I am not sure what will be expected of me in the future.	1	2	3	4	5
43. I leave work feeling burned out.	1	2	3	4	5
44. There is little prospect for personal or professional growth in this job.	1	2	3	4	5
45. In-service training for my job is inadequate.	1	2	3	4	5
46. There is little contact with colleagues on the job.	1	2	3	4	5
47. Most of my colleagues seem uninterested in me as a person.	1	2	3	4	5
48. I feel uneasy about going to work.	1	2	3	4	5

NOTE: Complete the entire questionnaire first! Then add all the values you circled for questions 27-48 and enter here.

Total 27-48_______

49. The complexity of my job is enough to keep me interested.	5	4	3	2	1
50. My job is very exciting.	5	4	3	2	1
51. My job is varied enough to prevent boredom.	5	4	3	2	1
52. I seem to have lost interest in my work.	1	2	3	4	5
53. I feel as though I can shape my own destiny in this job.	5	4	3	2	1
54. I leave work feeling burned out.	1	2	3	4	5
55. I would continue to work at my job even if I did not need the money.	5	4	3	2	1
56. I am trapped in this job.	1	2	3	4	5
57. If I had it to do all over again, I would still choose this job.	5	4	3	2	1

NOTE: Now go back and add the values for questions 1-26. Do the same for questions 27-48. Enter the values where indicated. Then add all the values circled for questions 49-57.

Total 49-57_______

Continues on the next page

Continued from previous page

Last, enter those sums for each of the following groups of questions and add them all together to get a cumulative total.

QUESTIONS:	1-26 Interpersonal		27-48 Physical Conditions		49-57 Job Interest		1-57
TOTALS:	______	+	______	+	______	=	______

	←LOW DISTRESS→	←MEDIUM DISTRESS→	←HIGH DISTRESS→
Interpersonal	..39.....43.....46	...51.....54.....57....62	...68.....75..
Physical	..35.....40.....44	...48.....52.....55....58	...62.....67..
Interest	..13.....15.....17	...18.....19.....21....23	...25.....27..
Total	..91....101...111	..117...123...134..141	..151...167..
Percentile	..10.....20.....30	...40.....50.....60....70	...80.....90..

Note: Some terminology has been modified with the author's permission.

Figure 6.4. Normative data from a sample of 275 school psychologists to compare individual scores from the Work Distress Profile (Rice, 1992). Note: Some terminology has been modified with the author's permission.

Let's review the key points about stress and distress. Actually, the flow schematic in Figure 6.5 says it all. First an environmental event is perceived and appraised as a stressor to be concerned about, or as a harmless or irrelevant stimulus. Richard Lazarus (1966, 1991) refers to this stage of the process as primary appraisal. According to Professor Lazarus, an event is perceived as a stressor if it involves *harm* or loss that has already occurred, a *threat* of some future danger, or a *challenge* to be overcome.

Harm is how we appraise the impact of an event. For example, if you oversleep and miss an important safety meeting, the damage is done. In contrast, *threat* is how we assess potential future harm from the event. Missing the safety meeting could lower your team's opinion of you and reduce your opportunity to get actively involved in a new safety process. *Challenge* is our appraisal of how well we can eventually profit from the damage done. You could view missing the safety meeting as an opportunity to learn from one-on-one discussions with coworkers. This could demonstrate your personal commitment to the safety process and allow you to collect diverse opinions.

In this case, you are perceiving the stressor as an opportunity to learn and show commitment. This evaluation occurs during the secondary appraisal stage (Lazarus &

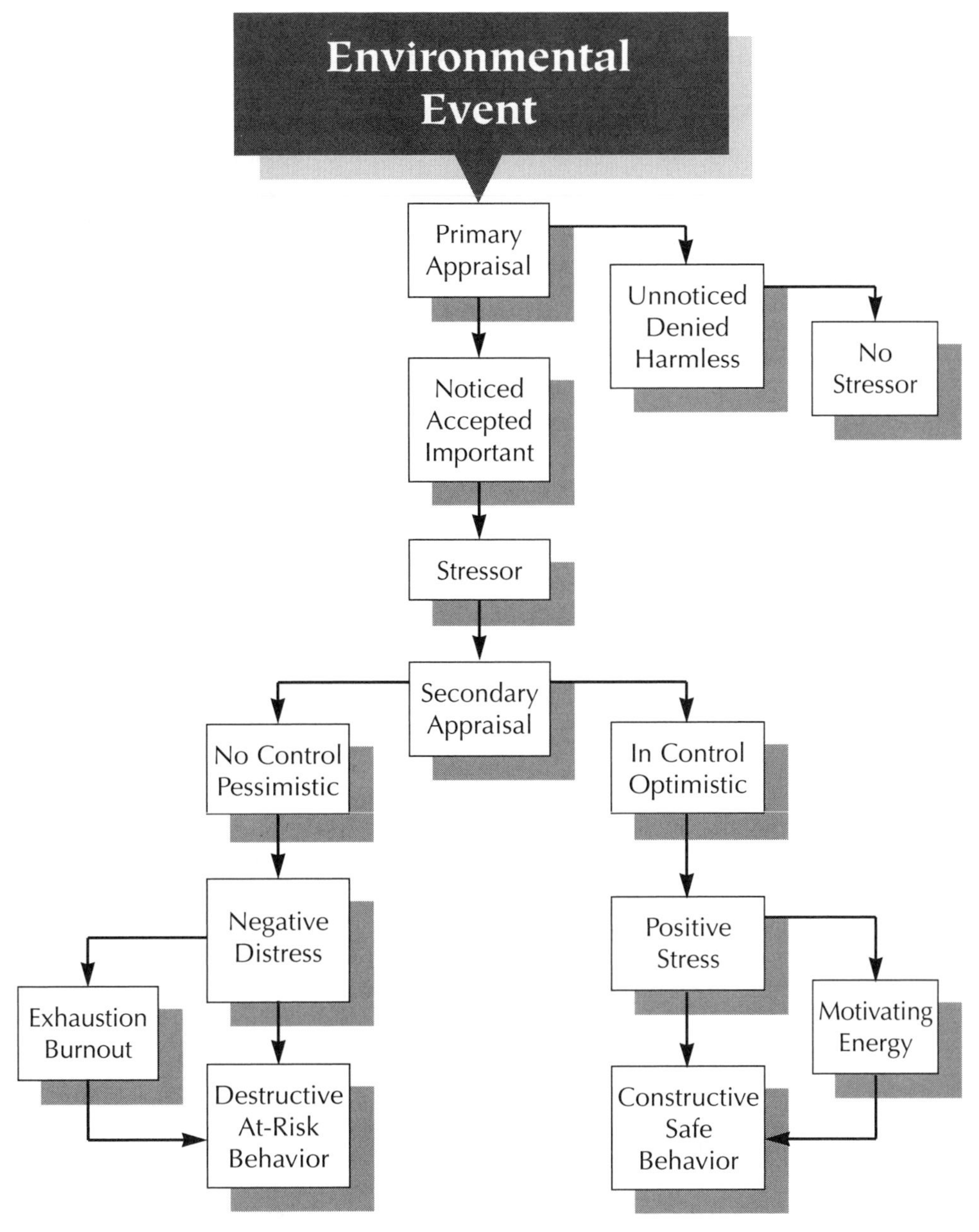

Figure 6.5. Through personal appraisal, people transform stressors into positive stress or negative distress.

Folkman, 1984), and the result can be positive and constructive. On the other hand, your appraisal could be downbeat—you see no recourse for missing the safety meeting, and so you do nothing about it. Now coworkers might think you don't care about the new safety process and withdraw their support. In turn, you might give up or actively resist participating. This outcome would be nonproductive, of course, and possibly destructive.

The secondary appraisal stage, as depicted in Figure 6.5, determines whether the stressor leads to positive stress and constructive behavior, or to negative distress and destructive behavior. The difference rests with the individual. Does he or she assess the stressful situation as controllable, and thus remain optimistic during attempts to cope with the stressor? Or are the stressors seen as uncontrollable and unmanageable? If so, a helpless or pessimistic attitude can prevail and lead to distress and destructive or at-risk behavior. Several personal, interpersonal and environmental factors influence whether this secondary appraisal leads to constructive or destructive behavior. This is the theme of the next section.

Coping with Stressors

Seek first to understand, says Stephen Covey (1989). This applies not only to relationships with people, but with stressors as well. Understanding the multiple causes of conflict, frustration, overload, boredom, and other potential stressors in our lives can sometimes lead to effective coping mechanisms. These include:

- Revising schedules to avoid hassles like traffic and shopping lines
- Refusing a request that will overload us
- Finding time to truly relax and recuperate from tension and fatigue
- Communicating effectively with others to clarify work duties, reduce conflict, gain support, or feel more comfortable about added job duties
- Getting reassigned to a task that better fits our present talents and aspirations

The fact is, though, it's often impossible to avoid sudden (acute) or continuous (chronic) stressors in our lives. We need to deal with these head-on. Believing you can handle the harm, threat, or challenge of stressors is the first step toward experiencing stress rather than distress, and acting constructively rather than destructively.

PERSON FACTORS

Certain personality characteristics make some people more resistant to distress. Individuals who believe they control their own destiny and generally expect the best from life are, in fact, more likely to gain control of their stressors and experience positive stress rather than distress, according to research (Bandura, 1982, 1986; Scheier & Carver, 1988, 1992). It's important to realize that these person factors— self-mastery and optimism—are not permanent inborn traits. They are states of mind or expectations derived from personal experience, and they can be nurtured. It's possible to give people experiences that increase feelings of being "in control," experiences that lead people to believe something good will come from their attempts to turn stress into constructive action.

LEARNING TO FEEL HELPLESS. When I help clients assess the safety climate of their workplaces, I often uncover an attitude among hourly workers, and some managers as well, that reflects an important psychological concept called "learned helplessness." For instance, when I ask workers what they do regularly to make their workplace safer, I often hear:

"Besides following the safety procedures there's not much I can do for safety around here."

"It really doesn't matter much what I do, whatever will be will be."

"There's not much I can do about reducing work injuries; if it's my time, it's my time."

This is learned helplessness. The concept was labeled more than 20 years ago by research psychologists studying the learning process of dogs (Maier & Seligman, 1976; Seligman, 1975). They measured the speed at which dogs learned to jump a low barrier separating two chambers in order to avoid receiving an electric shock through the grid floor. A tone or light would be activated, then a shock was applied to the grid floor of the chamber. At first the dogs didn't jump the hurdle until receiving the shock, but after a few trials the dogs learned to avoid the shock by jumping into the other chamber as soon as the warning signal was presented.

Some dogs experienced shocks regardless of their behavior before the regular shock-avoidance learning trials. These dogs did not learn to jump the barrier to escape the shock. Instead, they typically just laid down in the shock chamber and whimpered. The earlier bad experience with inescapable shocks had taught the dogs to be helpless. Dr. Martin Seligman and associates coined the term "learned helplessness" to describe this state. Their finding has been demonstrated in a variety of human experiments as well (Albert & Geller, 1978; Seligman & Garber, 1980).

Note how prior failures conditioned experimental subjects ranging from dogs to humans to feel helpless, in fact to be helpless. It's rather easy to assume that workers develop a "helpless" perspective regarding safety as a result of bad past experiences. If safety suggestions are ignored, or policies and procedures always come from management, workers might learn to feel helpless about safety. It's also true, however, that life experiences beyond the workplace can shape an attitude of learned helplessness. Certain individuals will come to work with a greater propensity to feel helpless in general, and this can carry over to feelings regarding occupational safety and health

LEARNED OPTIMISM. A bad experience does not necessarily lead to an attitude of learned helplessness. You probably know people who seem to derive strength or energy from their failures, and try even harder to succeed when given another chance. Similarly, Dr. Seligman and colleagues found that certain dogs resisted learned helplessness if they previously had success avoiding the electric shock. So it is that some people tend to give up in the face of a stressor, while others fight back.

What can be done to help those who feel helpless? How can we get them to commit to and participate in the proactive processes of injury prevention? The work climate can play a critical role here. This happens when employees are empowered to make a difference and perceive they are successful.

When workers believe through personal experience that their efforts can make a difference in safety, they develop an antidote for learned helplessness. This has been termed "learned optimism" (Scheier & Carver, 1992; Seligman, 1991). If the corporate climate empowers workers to take control and manage safety for themselves and their coworkers, they can legitimately attribute safety success to their own actions. This bolsters learned optimism and feelings of being in control.

FIT FOR STRESSORS

Fitness is another way to increase our sense of personal control and optimism. Being physically fit increases our body's ability to cope with the fight-or-flight syndrome discussed earlier. You probably know the basic guidelines for improving fitness, which include: stop smoking; reduce or eliminate alcohol consumption; exercise regularly, at least three times a week for about 30 minutes per session; eat balanced meals with decreased fat, salt, and sugar, and don't skip breakfast; and obtain enough sleep (usually 7 or 8 hours per 24 hour period for most people). Some of us find that following these guidelines over the long haul is easier said than done. We need support and encouragement to break a smoking or drinking habit, or to maintain a regular exercise routine.

Figure 6.6. Becoming a "mouse potato" by day and a "couch potato" by night can reduce our physical ability to cope with stressors.

Figure 6.6 illustrates the type of behavior that has come with the computer revolution. Low physical activity has become the way of work life for many of us. Often this inactivity spills over into home life. Survey research has shown that only one in five Americans exercises regularly and intensely enough to reduce the risk of stressor-induced heart disease (Dubbert, 1992). Figure 6.6 also depicts smoking behavior, considered to be the largest preventable cause of illness and premature death (before age 65) in the United States, accounting for approximately 125,000 deaths each year (American Cancer Society, 1989).

Also portrayed in Figure 6.6 is the positive influence of perceived control. Although the behavior is essentially the same at work (10 to 5) and at home (5 to 10), the individual is seemingly much happier at home. Why? Because at home he holds the remote control and therefore perceives more personal control.

Figure 6.7. Even the most obvious top-down situation allows for perceptions of bottom-up control.

But personal control is truly in the eyes of the beholder. Figure 6.7

depicts legitimate perceptions of control from the subjects of an experiment. These rodents are not usually considered "in control" of the situation, but in many ways they are. By simply changing our perspective, we can often perceive and accept more personal control at work. And this can turn negative distress into positive stress.

SOCIAL FACTORS

A support system of friends, family, and coworkers can do wonders in helping us reduce distress in our lives (Coyne & Downey, 1991; Janis, 1983; Lieberman, 1983). Social support can motivate us to do what it takes to stay physically fit. And the people around us can make a boring task bearable and even satisfying. Of course they can also turn a stimulating job into something dull and tedious. It works both ways. People can motivate us, or trigger conflict, frustration, hostility, a win/lose perspective, and distress. It's up to us to make the most of the people around us. We can learn from those who take effective control of stressful situations and expect the best. Or we can listen to the complaining, backstabbing, and cynicism of others, and fuel our own potential for distress.

It's obviously important to interact with those who can help us build resistance against distress and help us feel better about potential stressors. We can also set the right example, and be the kind of social support to others that we want for ourselves. The good feelings of personal control and optimism you experience from reaching out to help others can do wonders in helping you cope with your own stressors. And this actively caring stance builds your own support system, which you might need if your own stressors get too overwhelming to handle yourself.

The next section of this chapter introduces another means of reducing distress. It's a phenomenon that has particular implications for safety. In the aftermath of an injury or near hit, it can distort reports and investigations. The result: inappropriate or less-than-optimal suggestions for corrective action. This phenomenon of attributional bias can also create communication barriers between people, and limit the cooperative participation needed to achieve a Total Safety Culture.

Attributional Bias

Think back to the anecdote at the start of this chapter. I suggested that Judy's near-hit report was incomplete or biased. Specifically, Judy did not report the potential influence of her own distress on the incident. Rather, she focused on factors outside her immediate control—the poor bracket design for the wooden handle and the messy work area left by others. Giving up personal responsibility eliminated the incident as a stressor for her. She didn't have to deal with any guilt for almost hurting herself and damaging property. Her denial eased her distress, but biased the near-hit report. Psychologists refer to this as an attributional bias. By understanding when and how this phenomenon occurs, we can focus injury investigations on finding facts—not faults. This is a paradigm shift needed to achieve a Total Safety Culture.

THE FUNDAMENTAL ATTRIBUTION ERROR

Every day, we struggle to explain the actions of others. Why did she say that to me? Why did the job applicant refuse to answer that question? Why did Joe leave his work station in such a mess? Why did the secretary hang up on me? Why did Gayle take sick leave? Why does she allow her young children to ride in the bed of her pick-up truck? Why did the motorist pull a gun out of his glove compartment to shoot someone in the next car? Why were Nicole Brown-Simpson and Ronald Goldman murdered so brutally? In trying to answer questions like these, we point to external, environmental factors, such as equipment malfunctioning, excessive traffic, warm climate, and work demands; or to internal, person factors, such as personality, intelligence, attitude, or frustration.

Social psychologists have discovered a "fundamental attribution error" when systematically studying how people explain the behavior of others (Ross, 1977; Ross *et al*, 1977). When evaluating others, we tend to overestimate the influence of internal factors and underestimate external factors. We're more apt to judge the job applicant as rude or unaware (internal factors) than caught off-guard by a confusing or unclear question (external factor). Joe was sloppy or inconsiderate rather than overwhelmed by production demands. The injured employee was careless rather than distracted by a sudden environmental noise.

That's how we see things when we're judging others. It's different when we evaluate ourselves. The individuals performing the behaviors in the previous paragraph would say the causes were due more to external than internal factors.

Here's an example. My university students are quick to judge me as being an extrovert—outgoing and sociable. When I lecture in large classes of 600 to 800 students I'm animated and enthused, and they attribute my performance to internal personality traits. But I know better. I see myself in many different situations, and realize just how much my behavior changes depending on where I am. In many social settings I'm downright shy and reserved. I'm very sensitive to external influences.

THE SELF-SERVING BIAS

Students who flunk my university exams are quick to blame external factors, like tricky questions, wrong reading material assignments, and unfair grading. In contrast, students who do well are quite willing to give themselves most of the credit. It wasn't that I taught them well or that the exam questions were straightforward and fair; rather the student is intelligent, creative, motivated, and prepared. This real-world example, which I bet most readers can relate to, illustrates another type of attributional distortion, referred to as the self-serving bias (Harvey & Weary, 1984; Miller & Ross, 1975).

How does this bias affect incident or injury investigation? Think of Judy's near-hit experience. She protected her self-esteem by overestimating external causes, and underplaying internal factors. This aspect of the self-serving bias is illustrated clearly and humorously by the list of explanations for vehicle crashes given in Figure 6.8. These external and situational excuses were taken from actual insurance forms submitted by the drivers.

- The other car collided with mine without giving warning of its intentions.
- A pedestrian hit me and went under my car.
- The guy was all over the road. I had to swerve a number of times before I hit him.
- I had been shopping for plants all day, and was on my way home. As I reached an intersection, a hedge sprang up, obscuring my vision. I did not see the other car.
- As I approached the intersection, a stop sign suddenly appeared in a place where no stop sign had ever appeared before. I was unable to stop in time to avoid the accident.
- An invisible car came out of nowhere, struck my vehicle, and vanished.
- My car was legally parked as it backed into the other vehicle.
- The pedestrian had no idea which direction to go, so I ran over him.
- The telephone pole was approaching fast. I was attempting to swerve out of its path when it struck my front end.

Figure 6.8. People are reluctant to admit personal blame for their vehicle crashes. (Excerpted from the *Toronto Sun*, 1977)

People will obviously go to great lengths to shake blame for unintentional property damage or injury. This reduces negative stress or distress. No one wants to feel responsible for a workplace injury, especially if the company puts heavy emphasis on reducing "numbers," such as the plant's total recordable injury rate.

You can see how a focus on outcome statistics, perhaps supported with rewards for not having an injury, motivate people to cover up near hits and injuries whenever possible. It also motivates a self-serving bias during investigations.

Actually, there is good news here. By accentuating outside causes, victims remind us that behavior is indeed influenced by many external factors. And compared to internal factors, these are more readily corrected.

It's important for us to acknowledge how perceptions can be biased. Outsiders tend to blame the victim; victims look to extenuating circumstances. We should empathize with the self-serving bias of the victim because it will reduce the person's distress. It will shift attention to external factors that can be controlled more easily than internal factors related to a person's attitude, mood, or state of mind.

In Conclusion

In this chapter I explained the difference between stress and distress, and discussed some strategies for reducing distress or turning negative distress into positive stress. Stress and distress begin with a stressor, which can be a major life event or a minor irritation of everyday living. You can evaluate or appraise the stressor in a way that is constructive, resulting in safe behavior; or destructive, causing at-risk behavior. When people are physically fit, in control, optimistic, and able to rely on the social support of others, they are most likely to turn a stressor into energy to achieve success. This is positive stress.

When stressors are perceived as insurmountable and unavoidable, distress is likely. Without adequate support from others, this condition can lead to physical and mental exhaustion, at-risk behavior, and unintentional injury to oneself or others. We need to become aware of the potential stressors in our lives and in the lives of our coworkers. In addition, we need to develop personal and interpersonal strategies to prevent distress in ourselves and others.

Victims of a near hit or injury will likely feel stressed during their primary appraisal. If their secondary appraisal clarifies the incident as an uncontrollable failure, negative stress or distress is likely. But they could interpret it as an opportunity to collect facts, learn, and implement an action plan to prevent a recurrence. Now the victim is experiencing positive stress, and constructive behavior is likely.

The work culture, including policies, paradigms, and personnel, can have a dramatic impact on whether victims of near hits, injuries, or other adversities experience stress or distress. The fundamental attribution error, where we overestimate personal factors to explain others' behavior ("Judy broke the handle because she was tired, stressed-out, and careless") can provoke distress, and pinpoint the very aspects of an incident most difficult to define and control. Instead, a victim's natural tendency to reveal a self-serving bias when discussing the incident—by putting more emphasis on external, situational causes—should be supported by the work culture. This reduces the victim's distress and puts the focus on the observable factors, including behavior, most readily defined and influenced. I detail processes for doing just that in Part Three.

REFERENCES

Albert, M., & Geller, E. S. (1978). Perceived control as a mediator of learned helplessness. *American Journal of Psychology, 91*, 389-400.

American Cancer Society (1989). *Cancer facts and figures—1989.* Atlanta, Ga.: ACS.

American Heritage Dictionary (1991). Second College Edition, New York: Houghton Mifflin Company.

Bandura, A. (1982). Self-efficacy mechanism in human agency. *American Psychologist, 37*, 122-147.

———. (1986). *Social foundations of thought and action: A social cognitive theory.* Englewood Cliffs, NJ: Prentice-Hall.

Baron, R. A. (1995). *Psychology,* Third Edition (Chapter 13), Boston: Allyn and Bacon.

Covey, S. R. (1989). *The seven habits of highly effective people: Restoring the character ethic.* New York: Simon and Schuster.

Coyne, J. C., & Downey, G. (1991). Social factors and psychopathology: Stress, social support, and coping processes. *Annual Review of Psychology, 42*, 401-425.

Dubbert, P.M. (1992). Exercise in behavioral medicine. *Journal of Consulting and Clinical Psychology, 60*, 613-618.

Harvey, J. H., & Weary, G. (1984). Current issues in attribution theory. *Annual Review of Psychology, 35*, 427-459.

Holmes, T. H., & Masuda, M. (1974). Life change and illness susceptibility. In B. S. Dohrenwend & B. P. Dohrenwend (Eds.), *Stressful life events: Their nature and effects.* New York: Wiley.

Janis, I. L. (1983). The role of social support in adherence to stressful decisions. *American Psychologist,* 38, 143-160.

Jones, J. W. (1984). Cost evaluation for stress management. *EAP Digest,* p. 34.

Lazarus, R. S. (1966). *Psychological stress and the coping process.* New York: McGraw-Hill.

———. (1991). *Emotion and adaptation.* New York: Oxford University Press.

Lazarus, R. S., & Folkman, N. (1984). *Stress appraisal and coping.* New York: Springer.

Lieberman, M. A. (1983). The effects of social support on response to stress. In L. Goldbert & D. S. Breznitz (Eds.), *Handbook of stress management.* New York: Free Press.

Maier, S. F., & Seligman, M. E. P. (1976). Learned helplessness: Theory and evidence. *Journal of Experimental Psychology: General,* 105, 3-46.

Maslach, C. (1982). *Burnout: The cost of caring.* Englewood Cliffs, NJ: Prentice-Hall.

Miller, D. T., & Ross, M. (1975). Self-serving biases in the attribution of causality: Fact or fiction? *Psychological Bulletin,* 82, 213-225.

The New Merriam-Webster Dictionary (1989). Springfield, MA: Merriam-Webster Inc., Publishers.

Rice, P. L. (1992). *Stress & health,* second edition. Pacific Grove, CA: Brooks/Cole Publishing Company.

Ross, L. (1977). The intuitive psychologist and his shortcomings: Distortions in the attribution process. In L. Berkowitz (Ed.), *Advances in experimental social psychology* (Vol. 10), 173-220.

Ross, L. D., Amabile, T. M., & Steinmetz, J. L. (1977). Social roles, social control, and biases in social-perception processes. *Journal of Personality and Social Psychology,* 35, 485-494.

Scheier, M. F., & Carver, C. S. (1988). *Perspectives on personality.* Boston, MA: Allyn and Bacon.

———. (1992). Effects of optimism on psychological and physical well-being: Theoretical overview and empirical update. *Cognitive Therapy and Research,* 16, 201-229.

Seligman, M. E. P. (1975). *Helplessness: On depression development and death.* San Francisco, CA: Freeman.

———. (1991). *Learned optimism.* New York: Alfred A. Knopf, Inc.

Seligman, M. E. P., & Garber, J. (1980), (Eds.), *Human helplessness: Theory and application.* New York, NY: Academic Press.

Selye, H. (1976). *The stress of life,* Second Edition, New York, NY: McGraw-Hill.

———. (1974). *Stress without distress.* Philadelphia, PA: Lippincott.

Toronto Sun (1977, July 26th). Reports from insurance/accident forms. Toronto, Ontario, Canada.

Yerkes, R. M., & Dodson, J. D. (1908). The relation of strength of stimulus to rapidity of habit formation. *Journal of Comparative and Neurological Psychology,* 18, 459-482.

Part Three

Behavior-Based Psychology

Chapter 7
Basic Principles

To achieve a Total Safety Culture, we need to integrate behavior-based and person-based psychology and affect large-scale culture changes. The five chapters in Part Three explain principles and procedures founded on behavior-based research which can be applied successfully to change behaviors and attitudes throughout organizations and communities. This chapter describes the primary characteristics of the behavior-based approach to the prevention and treatment of human problems, and shows their special relevance to occupational safety. The three basic ways we learn are reviewed, and related to the development of safe versus at-risk behaviors and attitudes.

Specific safety techniques can be viewed as possible routes to reach a destination, in our case a Total Safety Culture. A particular route may be irrelevant or need to be modified substantially for a given work culture. The key is to begin with a complete and accurate map. In other words, it's most important to start with an understanding of the basic principles.

If you recall, our overall map or guiding principle is represented by the Safety Triad (Figure 2.2). Its reference points are the three primary determinants of safety performance—environment, person, and behavior factors. To achieve a Total Safety Culture, we need to understand and pay attention to each.

In Part Two, I addressed a number of person-based factors that can contribute to injuries, including cognitions, perceptions, and attributions. The BASIC ID acronym was introduced in Chapter 4 to express the complexity of human dynamics and the special challenges involved in preventing injuries. Behavior was the first dimension discussed, and it is implicated directly or indirectly in each of the other dimensions. Attitudes, sensations, imagery, and cognitions—the thinking, person side of the Safety Triad (Geller *et al,* 1989)—are each influenced by behavior. That's what is meant by the phrase, "Acting people into changing their thinking." When we change our behaviors, such as adopting a new strategy or paradigm, certain person factors change too.

The reverse is also true. Changes in attitudes, sensations, imagery, and cognitions can alter behaviors. But considerable research has shown that it's easier and more cost-effective to "act people into changing their thinking" than the reverse, especially in organizations and community settings (Glenwick & Jason, 1980, 1993; Goldstein & Krasner, 1987; Greene *et al,* 1987).

In Chapter 1, I justified a behavior-based approach to industrial health and safety by citing the article by Guastello (1993) that evaluated a variety of procedures. The two with the greatest impact on injury reduction, behavior-based and ergonomics, use principles and procedures from behavioral psychology. Actually, Guastello's review supports the power of behavior-based problem-solving. In the past 25 years, I have personally witnessed the large-scale effectiveness of this approach to:

- Treat agoraphobia (Brehony & Geller, 1981)
- Improve the teaching/learning potential in elementary schools (Geller, 1992c) and universities (Geller, 1972; Geller & Easley, 1986)
- Manage maximum security prisons efficiently and safely (Geller *et al,* 1977)
- Improve the impact of a community mental health center (Johnson & Geller, 1980)
- Control litter (Geller, 1980) and increase community recycling (Geller, 1980, 1981)
- Reduce excessive use of transportation energy (Reichel & Geller, 1981; Mayer & Geller, 1982)
- Prevent community crime (Geller *et al,* 1983; Schnelle *et al,* 1987)
- Improve sanitation during food preparation (Geller *et al,* 1980)
- Increase the use of vehicle safety belts in community and industrial settings (Geller, 1984, 1988, 1992a, 1993)
- Reduce alcohol abuse and the risk of alcohol-impaired driving (Geller, 1990; Geller & Lehman, 1985; Geller *et al,* 1991)
- Improve the effectiveness of child dental care (Kramer & Geller, 1987)
- Increase the immunization of children in Nigeria (Lehman & Geller, 1993)
- Protect the environment (Geller, 1987, 1992a,b; Geller *et al,* 1982)
- Increase safe driving practices among pizza deliverers (Ludwig & Geller, 1991)
- Increase the use of personal protective equipment (Streff *et al,* 1993)

Given these testimonials, let's examine the fundamental characteristics of the behavior-based approach.

Primacy of Behavior

Whether *treating* clinical problems (such as drug abuse, sexual dysfunction, depression, anxiety, pain, hypertension, and child or spouse abuse) or *preventing* any number of health, social, or environmental ills (from developing healthy and safe lifestyles to improving education and protecting the environment), overt behavior is the focus. Treatment or prevention is based on three basic questions:

1. What behaviors need to be increased or decreased to treat or prevent the problem?

2. What environmental conditions, including interpersonal relationships, are currently supporting the undesirable behaviors or inhibiting desirable behaviors?

3. What environmental or social conditions can be changed to decrease undesirable behaviors and increase desirable behaviors?

Thus, behavior change is both the outcome and the means. It is the desired outcome of treatment or prevention, and the means to solving the identified problem.

REDUCING AT-RISK BEHAVIORS

Heinrich's well-known Law of Safety implicates at-risk behavior as the root cause of most near hits and injuries (Heinrich *et al,* 1980). Over the past 20 years, various behavior-based research studies have verified this aspect of Heinrich's Law by systematically evaluating the impact of interventions designed to lower employees' at-risk behaviors. Feedback from behavioral observations was a common ingredient in most of the successful intervention processes, whether the feedback was delivered verbally, graphically by tables and charts, or through corrective action. (See, for exam-

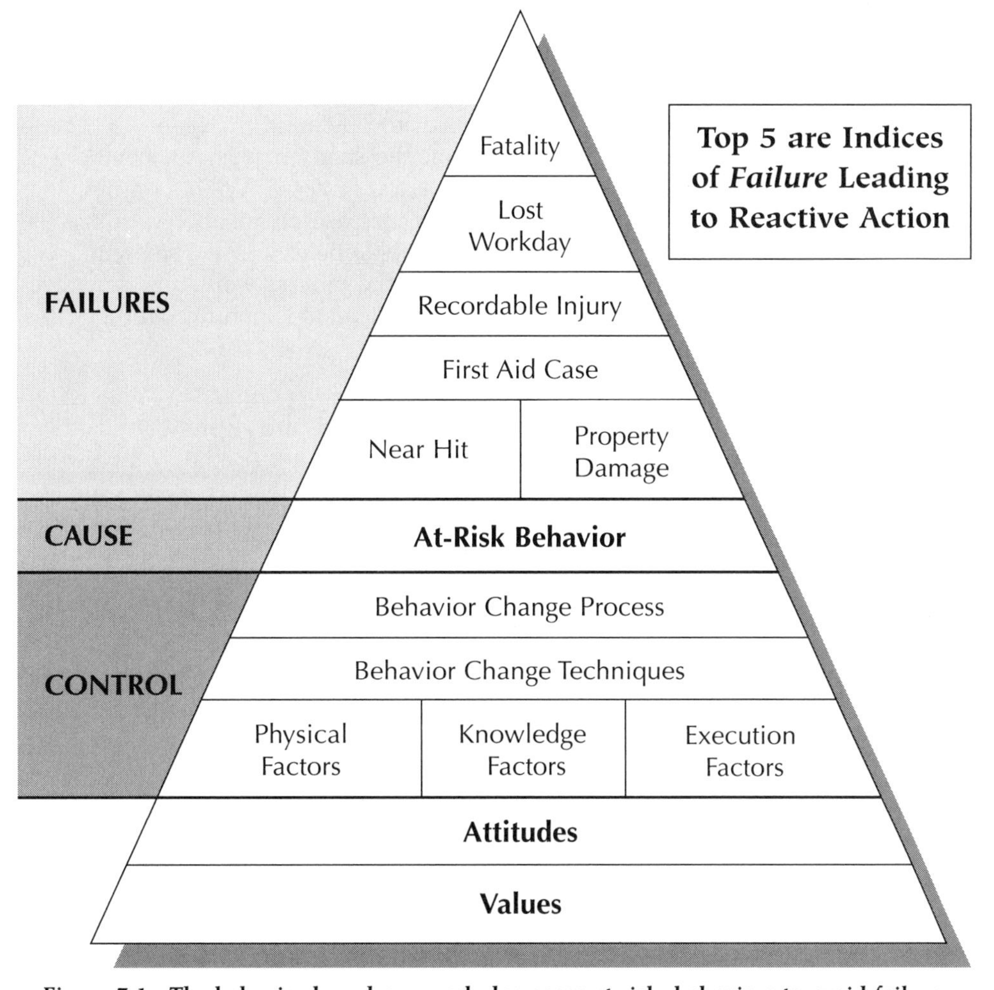

Figure 7.1. The behavior-based approach decreases at-risk behaviors to avoid failure.

ple, the comprehensive review by Dan Petersen, 1989, or individual research articles by Chhokar & Wallin, 1984; Geller *et al*, 1980; Komaki *et al*, 1980; and Sulzer-Azaroff & De Santamaria, 1980).

The behavior-based approach to reducing injuries is depicted in Figure 7.1. At-risk behaviors are presumed to be a major cause of a series of progressively more serious incidents, from a near hit to a fatality. According to Heinrich's Law, there are numerous risky acts for every near hit, and many more near hits than lost-time injuries. This is good news. But let's not forget that timing or luck is usually the only difference between a near hit and a serious injury.

Typically, behavior change techniques are applied to specific targets. It's necessary, of course, that participants know why targeted behaviors are undesirable and have the physical ability to avoid them. Education and engineering interventions are sometimes needed to satisfy the physical and knowledge factors of Figure 7.1. The execution factors represent the motivational aspect of the problem, and usually require the most attention. In other words, people usually know what at-risk behaviors to avoid and have the ability to do so, but their motivation might be lacking or misdirected. Behavior change techniques are used to align individual and group motivation with avoiding the undesired at-risk behavior.

Values and attitudes form the foundation of the pyramid in Figure 7.1. These obviously critical person factors need to support the safety process. Remember our discussion about risk compensation in Chapter 5, and Professor Wilde's warning that it's more important to reduce risk tolerance than increase compliance with specific safety rules (Wilde, 1994)? This happens when people believe in the safety process and help to make it work. Behavior helps to make the process work, and if involvement is voluntary and appropriately rewarded, it will lead to supportive attitudes and values to keep the process going.

Du Pont STOP. One popular behavior-based safety intervention is Du Pont's STOP (for Safety Training and Observation Program). Employees are given STOP cards to record the occurrence of at least one at-risk behavior or work condition each workday, along with their corrective action. At the end of the day the STOP cards are collected, compiled and recorded in a data log. Sometimes the data is transferred to a display chart or graph for feedback.

I have seen Du Pont STOP work well in some plants; in other plants I've noted substantial resistance. Why?

Interviews I've conducted with employees illustrate some important reminders for rolling out a behavior-based process. In some cases, employees felt like the program was not theirs, that it was forced on them by top management. I also talked to employees who didn't understand the rationale or underlying principles behind the program. There was also concern about its negativity. Behavior observation programs can't succeed if they are viewed as "gotcha" or "rat-on-your-buddy" campaigns. Employees will refuse to record the at-risk behaviors of their peers, or focus only on environmental conditions.

It should be noted that Du Pont has released an "Advanced STOP for Safety Auditing" program that the company says encourages the recording of safe work practices as well as unsafe acts. It's indeed important that observations offer positive reinforcement as well as punishment. Remember, in Chapter 3 we discussed the need to shift our orientation regarding safety from failure thinking to an achievement mindset. People have a more positive attitude when working to achieve rather than trying to avoid failure. This explains why employees might criticize and resist an intervention process that targets only failures.

The behavior-based approach illustrated in Figure 7.1 is failure oriented. It's also more reactive than proactive. The outcome measures are failures—fatalities, lost workdays, and the like—that require a fix. It's better to focus on increasing safe behaviors. This is being proactive; after all, behaviors are the root of the failures. Plus, by emphasizing safe behaviors, employees are more positive about the process and more willing to participate.

Increasing Safe Behaviors

Figure 7.2 illustrates a positive and proactive behavior-based model. I do not recommend this instead of the corrective action approach depicted in Figure 7.1. A complete behavior-based process should target both what's right and wrong about observed behavior. But again, more employees will participate with a positive attitude and remain committed over time if there is more recognition of achievements than correction of failures.

Monitoring achievement. The indices of achievement in Figure 7.2 are generally more difficult to record and track than those in Figure 7.1. Actually, the failure outcomes in Figure 7.1 are observed and recorded quite naturally. Except for near hits and first aid cases, the failures in Figure 7.1 have traditionally resulted in systematic investigation and formal reports. In contrast, the achievements in Figure 7.2 are somewhat difficult to define and record. In fact, it's impossible to obtain an objective record of the number of injuries prevented. A reasonable estimate of injuries prevented can be calculated, though, after you achieve a consistent decrease in injuries as a result of a proactive, behavior-based process.

It's possible to derive direct and objective definitions of the other success indices in Figure 7.2, and use these to estimate overall achievement. Involvement, for example, can be defined by recording participation in voluntary programs, and incidents of corrective action can be counted in a number of situations. You can chart the number of safety work orders turned in and completed, the number of safety audits completed and safety suggestions given, and the number of safety improvements occurring as a result of near-hit reports.

Throughout Part Three I'll be offering suggestions for monitoring achievements as I explain particular intervention strategies for teaching and motivating safe behavior. It's also possible to use surveys periodically and estimate successes from employee reactions to certain questions. A distress survey was presented in Chapter 6, for example, and a lower score on this survey would suggest improvement. In Part Four, I'll show you how to measure an individual's propensity to "actively care," or go beyond the call of duty for another person's safety. Increases in these measures indicate safety success.

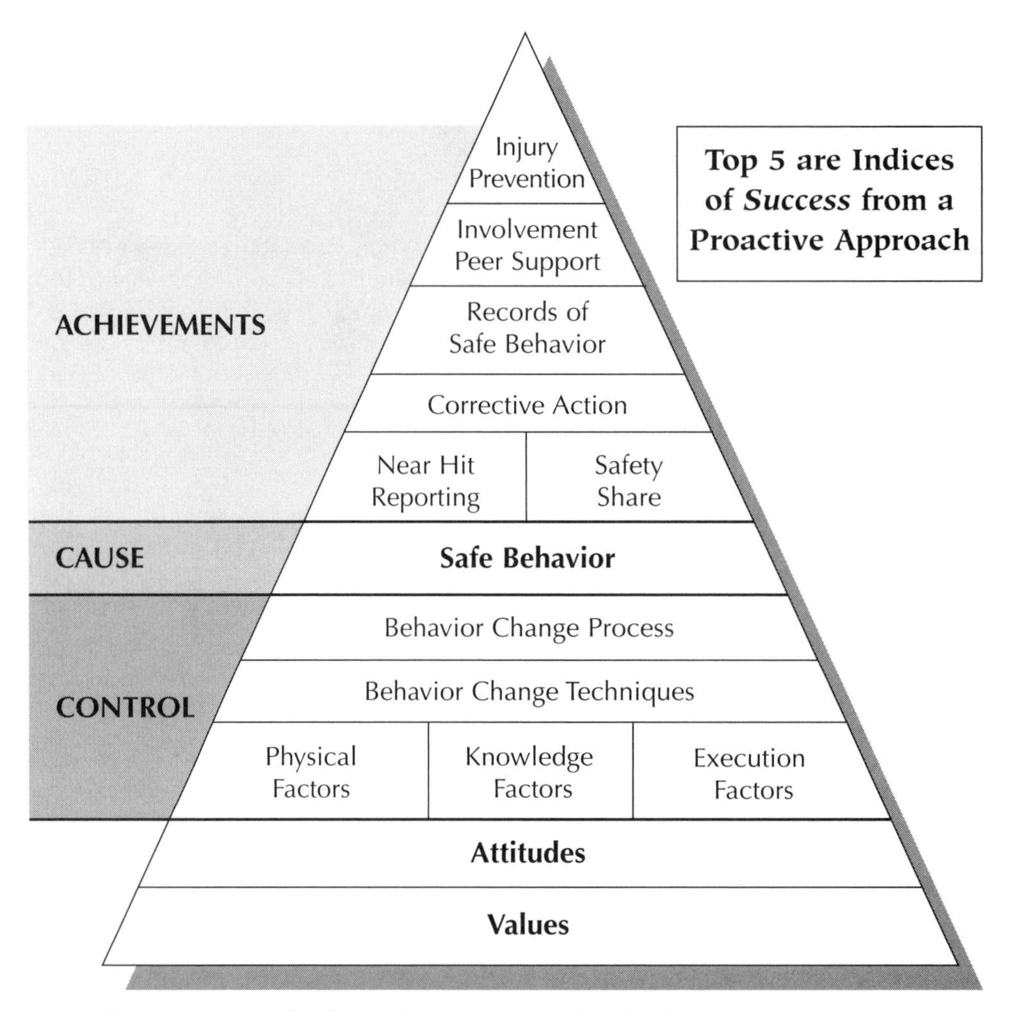

Figure 7.2. A total safety culture increases safe behaviors to achieve success.

SAFETY SHARE. The *safety share* noted in Figure 7.2 is a simple behavior-focused process that reflects my emphasis on achievement. At the start of group meetings, the leader asks participants to report something they have done for safety during the past week or since the last meeting. Because the safety share is used to open all kinds of meetings, safety is given special status and integrated into the overall business agenda. My experience is that people come to expect queries about their safety accomplishments, and go out of their way to have an impressive safety story to share.

This simple awareness booster—"What have you done for safety?"—helps teach an important lesson. Employees learn that safety is not only loss control, an attempt to avoid failure, but can be discussed in the same terms of achievement as productivity, quality, and profits. As a measurement tool, it's possible to count and monitor the number of safety shares offered per meeting as an estimate of proactive safety success in the work culture.

Direct Assessment and Evaluation

The roots of behavior-based interventions are in clinical psychology, because the focus on outward behaviors allows for an empirical assessment of therapeutic outcomes. Today this approach is the leading strategy for program evaluation, in part because of the research rigor of experimental behavior analysis (Skinner, 1938), and also because the focus is on behavior rather than the internal subjective concepts of the psychoanalytic, humanistic, and cognitive approaches to therapy. There is more solid research support for the validity of behavior-based approaches to solve diverse human problems than for all the other approaches combined, even though the behavior-based focus is one of the "youngest kids on the block."

BASELINE MEASURES. Typically, the impact of an intervention is evaluated in three stages. First, the behavior to be influenced is systematically assessed through direct observation in naturalistic settings, such as at home, school, or at work. This is done by "relevant" observers such as parents, spouses, teachers, supervisors, or co-workers. Often questionnaires are given to both those being observed and those doing the observing to obtain opinions, perceptions, and attitudes regarding the targeted problem and relevant environmental factors. Baseline information collected in this stage is used to set intervention goals and design ways to achieve them.

MONITORING DURING INTERVENTION. The targeted behavior is carefully monitored, sometimes by those responsible for the behavior, throughout the intervention process. Desired change often occurs as a result of feedback. The feedback needs to be a frequent and objective assessment of the target behavior and the circumstances where it occurs too frequently or not enough. Observing progress in changing a behavior is a powerful reward for all parties involved—those whose behavior is being changed, and those helping facilitate the change. Such a reward motivates continuous efforts from all involved. If feedback data indicates the intervention process is not working as expected, appropriate adjustments are made. Sometimes entirely different behavior-change techniques will be substituted.

FOLLOW-UP MEASURES. In a clinical sense, after the client is presumed "cured" and the intervention program is withdrawn, follow-up measurement occurs. This is how the long-term effectiveness of the intervention is assessed. Without an appropriate support system in the environmental settings where the problem behavior occurred or where a desirable behavior should occur more often, the client's problem is likely to resurface. Intermittent follow-up evaluations check for evidence of this support and indicate whether additional intervention is needed.

INTERVENTION BY MANAGERS AND PEERS. The remarkable success of the behavior-based approach to solve people's problems changed dramatically the role of the clinical psychologist. Therapists had been spending most of their days in the office working on psychotherapies with clients. But since behaviors are triggered by certain environmental circumstances, behavioral improvement requires changes in those settings. This means the therapist works with clients and potential support personnel where the problem exists.

Designing and refining an intervention process requires profound understanding of the problem's context, such as the environment and the people in it. Systematically, you must observe the target behavior in the field, and interview individuals close to the problem.

Here's another important point: The most cost-effective way to implement on-site intervention is to teach the natural managers of the setting—parents, teachers, supervisors, prison guards, peers—how to implement the process. After all, these people deal with the target behavior on an ongoing basis. So clinical psychologists using behavior-based techniques spend significant time in the field customizing environment-specific plans and teaching others how to execute and evaluate a behavior-change process.

Learning from Experience

A key assumption of behavior-based theory is that behavior (desirable and undesirable) is learned, and can be changed by providing people with new learning experiences. Diverse cultural, social, environmental, and biological factors interact to influence our readiness to learn behaviors, and they also support behaviors once they are learned. We don't understand exactly how the particulars work—diverse factors interact to influence behaviors for each individual. The basic ways we develop behavioral patterns have been researched, though, and it's possible to identify principles behind learning and maintaining human behavior.

Psychologists define learning as a change in behavior, or potential to behave in a certain way, resulting from direct and indirect experience. In other words, we learn from observing and experiencing events and behaviors in our environment (Bandura, 1986).

While the effects of learning are widespread and varied, it's generally believed that there are three basic models: *classical conditioning, operant conditioning,* and *observational learning.*

Laboratory methods have been developed to systematically study each type of learning. Although it's possible to find pure forms of each in real-world situations, it's likely that we learn simultaneously from different models as they overlap to influence what we do and how we do it.

CLASSICAL CONDITIONING

This form of learning became the subject of careful study in the early twentieth century with the seminal research of Ivan Pavlov (1849-1936), a Nobel Prize-winning physiologist from Russia. Pavlov (1927) did not set out to study classical conditioning; rather, his research focused on the process of digestion in dogs. He was interested in how reflex responses were influenced by stimulating a dog's digestive system with food. During his investigations he serendipitously found that his subjects began to salivate before actually tasting the food. They appeared to anticipate the food stimulation by salivating when they saw the food or heard the experimenters preparing it. Some dogs even salivated when seeing the empty food pan or the person who brought in the food. Through experiencing the relationship between certain stimuli and food, the dogs learned when to anticipate food. Pavlov recognized this as an important phenomena and shifted the focus of his research to address it.

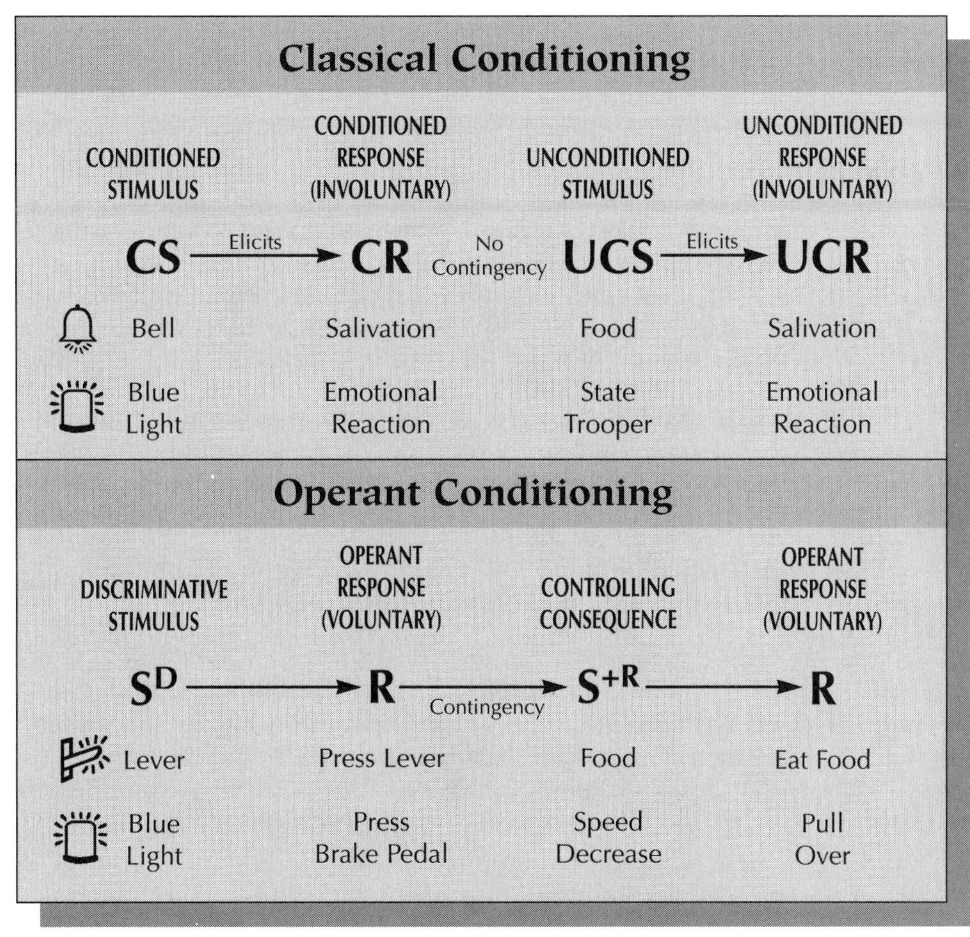

Figure 7.3. The stimulus-response relationships in classical and operant conditioning.

The top half of Figure 7.3 depicts the sequence of stimulus-response events occurring in classical conditioning. Actually, before learning occurs, the sequence includes only three events—conditioned stimulus (CS), unconditioned stimulus (UCS), and unconditioned response (UCR). The UCS elicits a UCR automatically, as in an autonomic reflex. That is, the food (UCS) elicited a salivation reflex (UCR) in Pavlov's dogs. In the same way, the smell of popcorn (UCS) might make your mouth water (UCR), a puff of air to your eye (UCS) would result in an automatic eyeblink (UCR), ingesting certain drugs (like Antabuse) would elicit a nausea reaction, and a state trooper writing you a speeding ticket is likely to influence an emotional reaction (like distress, nervousness, or anger).

If a particular stimulus (CS) consistently precedes the UCS on a number of occasions, the reflex (or involuntary response) will become elicited by the CS. This is classical conditioning, and occurred when Pavlov's "slobbering dogs" salivated when they heard the bell that preceded food delivery. Classical conditioning would also occur if your mouth watered when you heard the bell from the microwave oven tell you the popcorn is ready, if you blinked your eye following the illumination of a dim light that consistently preceded the air puff, if you felt nauseated after seeing and smelling the alcoholic beverage that previously accompanied the ingestion of Antabuse, and if you got nervous and upset after seeing a flashing blue light in your vehicle's rearview mirror.

OPERANT CONDITIONING

As shown in the lower portion of Figure 7.3, the flashing blue light on the police car might influence you to press your brake pedal and pull over. This is not an automatic reflex action, but is a voluntary behavior you'd perform in order to do what you consider appropriate at the time. In other words, you have learned (perhaps indirectly from watching or listening to others or from prior direct experience) to emit certain behaviors when you see a police car with a flashing blue light in your rearview mirror. Of course, you have also learned that certain driving behaviors (like pressing the brake pedal) will give a desired consequence (like slowing down the vehicle), and this will enable another behavior (like pulling over for an anticipated encounter with the police officer).

Selection by consequences. B. F. Skinner (1904-1990), the Harvard professor who pioneered the behavior-based approach to solving societal problems, studied this type of learning by systematically observing the behaviors of rats and pigeons in an experimental chamber referred to as a "Skinner Box" (much to Professor Skinner's disappointment). Skinner termed the learned behaviors in this situation *operants* because they were not involuntary and reflexive, as in classical conditioning, but instead they *operated* on the environment to obtain a certain consequence. A key principle demonstrated in the operant learning studies is that voluntary behavior is strengthened (increased) or weakened (decreased) by consequences (events immediately following behaviors).

The relationship between a response and its consequence is a contingency, and this relationship explains our motivation for doing most everything we choose to do. Thus, the hungry rat in the Skinner Box presses the lever to receive food and the vehicle driver pushes the brake pedal to slow down the vehicle. Indeed, we all select various responses to perform daily—like eating, walking, reading, working, writing, talking, and recreating—to receive the immediate consequences they provide us. Sometimes we emit the behavior to achieve a pleasant consequence, such as a reward. Other times we perform a particular act to avoid an unpleasant consequence—a punisher or penalty. And we usually stop performing behaviors that are followed by penalties.

Emotional reactions. As I discussed earlier in Chapter 3, we feel better when working for pleasant consequences than when working to avoid or escape unpleasant consequences. This can be explained by considering the classical conditioned emotions that naturally accompany the agents of reward versus penalty. How do you feel, for example, when a police officer flashes a blue light to signal you to pull over? When the instructor asks to see you after class? When you enter the emergency area of a hospital? When the dental assistant motions that "you're next"? When your boss leaves you a phone message to "see him immediately"? When you see your family at the airport after returning home from a long trip?

Your reactions to these situations depend, of course, on your past experiences. Police officers, teachers, doctors, dentists, and supervisors do not typically elicit negative emotional reactions in young children. However, through the association of certain cues with the consequences we experience in the situation, a negative emotional response or attitude can develop. And you don't have to experience these relationships directly. In fact, we undoubtedly learn more from observing and listening to others than from first-hand experience. This brings us to the third way in which we learn from experience—observational learning.

OBSERVATIONAL LEARNING

A large body of psychological research indicates that this form of learning is involved to some degree in almost everything we do (Bandura, 1977, 1986). Whenever you do something a particular way because you saw someone else do it that way, or because someone showed you to do it that way, or because characters on television or in a video game did it that way, you are experiencing observational learning. Whenever you attribute "peer pressure" as the cause of someone taking up an unhealthy habit (like smoking cigarettes or drinking excessive amounts of alcohol) or practicing an at-risk behavior (like driving at excessive speeds or adjusting equipment without locking out the energy source), you are referring to observational learning. When you remind someone to set an example for others, you are alluding to the critical influence of observational learning.

VICARIOUS CONSEQUENCES. As children we learned numerous behavioral patterns by watching our parents, teachers, and peers. When we saw our siblings or schoolmates receive rewards like special attention for certain behaviors, we were more likely to copy that behavior, for instance. This process is termed vicarious, or indirect, reinforcement. At the same time, when we

Figure 7.4. Children learn a lot from their parents through obvservational learning.

observed others getting punished for emitting certain behaviors, we learned to avoid these behaviors. This is referred to as vicarious punishment.

As adults we teach others by example. As illustrated in Figure 7.4, our children learn new behavior patterns, including verbal behaviors, by watching us and listening to us. In this way they learn what is expected of them in various situations. The prominent role of parental modeling in socializing children is illustrated by this humorous but true story of a father tucking his six-year-old daughter into bed:

> *When he came into her bedroom, his daughter requested "Daddy—would you tuck me in like you do mommy every night?" He said "Sure, honey," and pulled the covers up and underneath his daughter's chin. As he left the room, his daughter called after him, saying, "Wait, daddy—would you give me a good-night kiss like you do mommy every night?" Dad said, "Sure, honey," and he kissed his daughter on her cheek. As he was leaving the room again, his daughter called after him with one more request, "Daddy . . . Daddy, wait . . . would you whisper in my ear like you do mommy every night." "Sure, honey," he replied, and he leaned down and went "Bzzz, Bzzz, Bzzz" in his daughter's ear. Then she said, "Not tonight, daddy, I have a terrible headache."*

Our actions influence others to a greater extent than we realize. Without us being aware of our influence, children learn by watching us at home; and our co-workers are influenced by our practices at work. Not only does the performance of safe acts encourage similar behavior by observers, but verbal behavior can also be influential. If a supervisor is observed commending a worker for her safe behavior or reprimanding an employee for an at-risk practice, observers may increase their performance of similar safety behaviors (through vicarious reinforcement) or decrease the frequency of similar at-risk behavior (through vicarious punishment). Indeed, to make

Figure 7.5. Intentionally and unintentionally, we teach others through our examples.

safe behavior the norm—rather than the exception—we must always set an example both in our own work practices and in the verbal consequences we offer coworkers following their safe and at-risk behaviors. Figure 7.5 offers a memorable pictorial regarding the influence of example-setting on observation learning.

OBSERVATIONAL LEARNING AND TELEVISION. Behaviors by television performers can have a dramatic impact on viewers' behaviors. Research has shown, for example, that aggression can be learned through television viewing (Bandura *et al,* 1963; Baron & Richardson, 1994; Synder, 1991). When children and adults were exposed to certain ways of fighting they had not seen before, they later emulated these aggressive behaviors when frustrated, irritated, or angry. In addition to teaching new forms of aggression, movies and television programs often convey the message that aggression is an acceptable means of handling interpersonal conflict. In other words, violence on television gives us the impression that interpersonal aggression is much more common than it really is, and thus it reduces our tendency to hold back physical acts of aggression toward others (Berkowitz, 1984).

Given the potential for observational learning from television viewing, our comprehensive and systematic observations of violence and unsafe sex on television during the 1994-95 season was disappointing and alarming. In the fall of 1994, my students systematically coded 297 violent scenes from 152 prime-time episodes over a nine-week period. The Fox Television Network showed the most violence, with an average of almost three violent scenes per episode. The most commonly used weapon was the hand or fist (36.2%). A gun was used in 29.6% of the violent acts, and for only 21% of the scenes was a negative consequence indicated for initiating the violence. Furthermore, most of the negative consequences were legal and therefore delayed. For 14% of the scenes, the instigator of the violence received immediate positive consequences.

My students' coding of sexual behavior on prime-time television in 1994 revealed pervasive portrayal of irresponsible sexual behavior. As with violence, the Fox Television Network showed the most sexual behavior. Of the 81 scenes coded, 82.7% showed or clearly implied sexual intercourse. In only 2 of 81 scenes was there any discussion of contraception. A negative consequence for the irresponsible sexual behavior was rarely shown. A mere 7% of the characters showed guilt after the sexual act; only 4% appeared to have less respect for themselves; only 2% showed less respect for their partner.

LEARNING SAFETY FROM TELEVISION. Now consider the potential observational learning in showing television stars using versus not using vehicle safety belts. When seeing a television hero buckle up, some viewers, mostly children, learn how to put on a vehicle safety belt; others are reminded that they should buckle up on every trip; still others realize that safety belt use is an acceptable social norm. On the other hand, the frequent nonuse of safety belts on television teaches the attitude that certain types of individuals, perhaps macho males and attractive females, do not use safety belts.

As depicted in Figure 7.6, safety-belt use on television clearly increased across the three years, averaging 8% of 2094 driving scenes observed in 1984, 15% of 1478 driving scenes in 1985, and 22% of 927 driving scenes observed in 1986. Unfortunately, these percentages were significantly below the real-world averages.

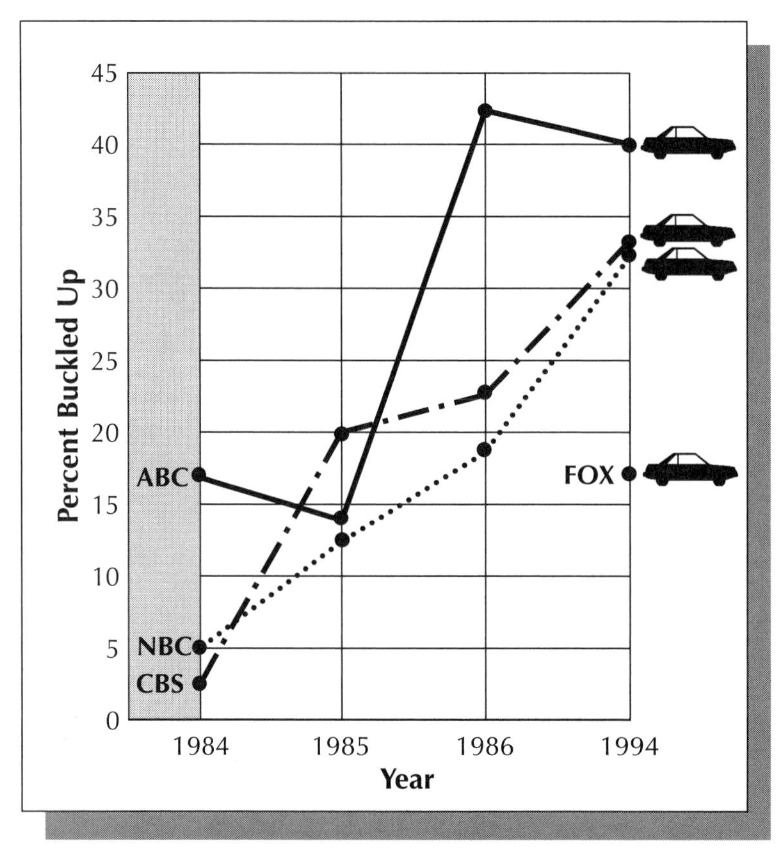

Figure 7.6. Percentage of driving scenes with a front-seat passenger using a safety belt as a function of network and year.

Our more recent television monitoring during the 1994-95 season showed an increase to an average of 29% safety belt use across 96 driving scenes observed on prime-time shows for ABC, CBS, NBC, and FOX. This level was well below the nationwide belt-use rate of 67% at the time (Novak, 1995). This is clearly irresponsible broadcasting and calls for social action. What can we do about such inappropriate observational learning generated by prime-time television shows?

In 1984, my students and I conducted a nationwide campaign to bring public attention to the nonuse of vehicle safety belts by television performers. We circulated a petition throughout the United States that described the detrimental learning effects of low safety-belt use on television. We received approximately 50,000 signatures from residents in 36 states. In addition, we distributed a list of 30 names and addresses of television stars along with instructions to write letters requesting safety-belt use by those who didn't buckle up, and to write thank-you notes to those who already buckle up on television.

As part of a creative writing assignment in elementary schools in Olympia, Washington, more than 800 third and fourth graders wrote a buckle-up request to Mr. T, a star on a popular action program at the time called "The A-Team." We believe it was no coincidence that Mr. T increased his use of safety belts from no belt use in

1984 to over 70 percent belt use in 1985, following the letter-writing campaign (Geller, 1988, 1989). Actually, Mr. T was the only A-Team member to buckle up during the 1985 season. In 1986 (the last year of this prime-time show), the entire A-Team was more likely to buckle up (39% of all driving scenes).

With graphs of the low use of safety belts on television from 1984 to 1986, I traveled to Hollywood and gave a special workshop to producers, writers, and actors on the need to buckle up on television and in the movies. The workshop was sponsored and marketed by the National Highway Traffic Safety Administration. The feedback graphs proved to be an influential means of convincing the large audience of a problem needing their attention.

My point here is that there are a number of things we can do to promote responsible broadcasting on television. If everyone contributes a "small win," the benefits can add up to a big difference. Considering the substantial influence of observational learning on behaviors and attitudes, and the millions of daily viewers of prime-time television shows, efforts to depict exemplary behavior among network stars—like safe driving practices—could potentially prevent millions of injuries and save thousands of lives. Television shows clearly influence our culture. Thus, to achieve a true societal Total Safety Culture, the behavior depicted on television needs to be consistent with such a vision.

Setting examples. The poem "Setting Examples" by Forest H. Kirkpatrick says it all. This poem is presented in Figure 7.7, and I recommend copying it and posting it for others to read. It's so easy to forget the dramatic influence we have on others by our own behaviors. Obviously, we need to take the slogan "walk the talk" very seriously. In fact, if we're not convinced a particular safeguard or protective behavior is necessary for us ("It's not going to happen to me"), we need to realize, at least, that our at-risk behaviors can endanger others.

The eye's a better teacher and more willing than the ear;
Fine counsel is confusing, but example's always clear;
And the best of all the preachers are the ones who live their creeds.
For to see the good in action is what everybody needs.

I can soon learn how to do it if you'll let me see it done;
I can watch your hands in action, but your tongue too fast may run;
And the lectures you deliver may be very wise and true.
But I'd rather get my lesson by watching what you do.

For I may not understand you and the high advice you give.
There's no misunderstanding how you act and how you live.

Figure 7.7. This poem, written by Forrest H. Kirkpatrick, illustrates the power of observational learning.

For example, I never get in my vehicle believing a crash will happen to me, so my rationale for buckling my combination lap and shoulder belt is to set the right example for others, whether they're in the car with me or not. Understanding the powerful influence of observational learning, we should feel obligated to set the safe example whenever someone could see us.

OVERLAPPING TYPES OF LEARNING

Laboratory methodologies have been able to study each type of learning separately, but the real world rarely offers such purity. In life, the usual situation includes simultaneous influence from more than one learning type. The operant learning situation, for example, is likely to include some classical (emotional) conditioning. As I indicated earlier, this is one reason rewarding consequences should be used more frequently than punishing consequences to motivate behavior change.

Remember, a rewarding situation (unconditioned stimulus) can elicit a positive emotional experience (unconditioned response), and a punitive situation (UCS) can elicit a negative emotional reaction (UCR). With sufficient pairing of rewarding or punishing consequences with environmental cues (such as a work setting, or particular people), the environmental setting (conditioned stimulus) can elicit a positive or negative emotional reaction or attitude (conditioned response). This can in turn facilitate (if it's positive) or inhibit (if it's negative) ongoing performance.

Figure 7.8 depicts a situation in which all three learning types occur at the same time. As discussed earlier and diagrammed in Figure 7.3, the blue flashing light of the police car signals drivers to press the brake pedal of their car and pull over. In this case, the blue light is considered a *discriminative stimulus* because it tells people when to respond in order to receive or avoid a consequence. Actually, drivers would apply their brakes to avoid punitive consequences, so this situation illustrates an avoidance contingency where drivers respond to avoid failure.

The flashing blue light might also serve as a conditioned stimulus eliciting a negative emotional reaction. This is an example of classical conditioning occurring simultaneously with operant learning. Our negative emotional reaction to the blue light might have been strengthened by prior observational learning. As a child, we might have seen one of our parents pulled over by a state trooper and subsequently observed a negative emotional reaction from our parent. The children in Figure 7.8 are not showing the same emotional reaction of the driver. Eventually, they will probably do so as a result of observational learning.

Figure 7.8. Three types of learning occur in some situations.

Later, their direct experience as drivers will strengthen this negative emotional response to a flashing blue light on a police car.

In Conclusion

In this chapter I have reviewed the basic principles underlying a behavior-based approach to the prevention and treatment of human problems. The variety of successful applications of this approach was discussed, based on my personal experiences. The behavior-based principles—the primacy of behavior, direct assessment and evaluation, intervention by managers and peers, and three types of learning—were explained with particular reference to reducing personal injury.

Since at-risk behaviors contribute to most if not all injuries, a Total Safety Culture requires a decrease in at-risk behaviors. Organizations have attempted to do this by targeting at-risk acts, exclusive of safe acts, and using corrective feedback, reprimands, or disciplinary action to motivate behavior change. This approach is useful, but less proactive and less apt to be widely accepted than a behavior-based approach that emphasizes recognition of safe behaviors. It will be easier to get employees involved in safety achievement if credit is given for doing the right thing more often than reprimands for doing wrong.

The three types of learning are relevant for understanding safety-related behaviors and attitudes. Most of our safe and at-risk behaviors are learned operant behaviors, performed in particular settings to gain positive consequences or avoid negative consequences. Classical conditioning often occurs at the same time to link positive or negative emotional reactions with the stimulus cues surrounding the experience of receiving consequences. These cues include the people who deliver the rewards or penalties. We often learn what to do and what not to do by watching others receive recognition or correction for their operant behaviors. This is observational learning, an ongoing process that should motivate us to try to set the safe example at all times.

REFERENCES

Bandura, A. (1986). *Social foundations of thought and action: A social cognitive theory.* Englewood Cliffs, NJ: Prentice-Hall.

———. (1977). *Social learning theory.* Englewood Cliffs, NJ: Prentice-Hall.

Bandura, A., Ross, D., & Ross, S. A. (1963). Imitation of film-mediated aggressive models. *Journal of Abnormal and Social Psychology, 66,* 3-11.

Baron, R. A., & Richardson, D. (1994). *Human aggression,* 2nd ed. New York: Plenum Publishers.

Berkowitz, L. (1984). Some effects of thoughts on anti- and pro-social influences of media events: A cognitive-neoassociation analysis. *Psychological Bulletin, 95,* 410-427.

Brehony, K. A., & Geller, E. S. (1981). Agoraphobia: A behavioral perspective and critical appraisal of research. In M. Hersen, R. M. Eisler, & P. M. Miller (Eds.), *Progress in behavior modification,* Vol. 8 (pp. 1-66). New York: Academic Press.

Chhokar, J. S., & Wallin, J. A. (1984). A field study of the effects of feedback frequency on performance. *Journal of Applied Psychology, 69,* 524-530.

Geller, E. S. (1993). Applications of behavioral science for road safety. In D. Glenwick & L. Jason (Eds.), *Promoting health and mental health: Behavioral approaches to prevention.* New York: Springer.

———. (1992a). *Applications of behavior analysis to prevent injury from vehicle crashes.* Monograph published by the Cambridge Center for Behavioral Studies, Cambridge, MA.

———. (1992b). Solving environmental problems: A behavior change perspective. In S. Staub & P. Green (Eds.). *In our hands: Psychology, peace, and social responsibility,* (pp. 248-268). New York: New York University Press.

———, ed. (1992c). *The educational crisis: Issues, perspectives, solutions.* Monograph Number 7. Lawrence, KS: Society for the Experimental Analysis of Behavior, Inc.

———. (1990). Preventing injuries and deaths from vehicle crashes: Encouraging belts and discouraging booze. In J. Edwards, R. S. Tindale, L. Heath, & E. J. Posavac (Eds.), *Social influence processes and prevention* (pp. 249-277). New York: Plenum Publishers.

———. (1989). Using television to promote safety belt use. In R. E. Rice & C. K. Atkin (Eds.), *Public communication campaigns,* 2nd ed. (pp. 201-203). Newberry Park, CA: SAGE Publications, Inc.

———. (1988). A behavioral science approach to transportation safety. *Bulletin of the New York Academy of Medicine,* 64, 632-661.

Geller, E. S. (1987). Environmental psychology and applied behavior analysis: From strange bedfellows to a productive marriage. In D. Stokols & I. Altman, (Eds.), *Handbook of environmental psychology,* Vol. I (pp. 361-388). New York: John Wiley & Sons.

———. (1984). Motivating safety belt use with incentives: A critical review of the past and a look to the future. In *Advances in belt restraint systems: Design, performance, and usage* (No. 141), Warrendale, PA: Society of Automotive Engineers, Inc.

———. (1981). Waste reduction and resource recovery: Strategies for energy conservation. In A. Baum & J. Singer (Eds.), *Advances in environmental psychology,* Vol.III (pp. 115-154). New Jersey: Lawrence Erlbaum Associates.

———. (1980). Applications of behavior analysis to litter control. In D. Glenwick & L. Jason (Eds.). *Behavioral community psychology: Progress and prospects* (pp. 254-283). New York: Praeger Press.

———. (1980). Saving environmental resources through waste reduction and recycling: How the behavioral community psychologist can help. In G. L. Martin & J. G. Osborne (Eds.), *Helping in the community: Behavior applications* (pp. 55-102). New York: Plenum Publishers.

———. (1972). A training program in behavior modification: Design, outcome, and implication. *JSAS Catalog of Selected Documents in Psychology,* 2, 29. (Ms. No. 95).

Geller, E. S., & Easley, A. T. (1986). Applied behavior analysis in the college classroom: Some ideas for educators. In L. Bishop (Ed.), *The art of teaching; Seven perspectives.* The Academy of Teaching Excellence, Virginia Tech, Blacksburg, VA.

Geller, E. S., Eason, S., Phillips, J., & Pierson, M. (1980). Interventions to improve sanitation during food preparation. *Journal of Organizational Behavior Management,* 2, 229-240.

Geller, E. S., Johnson, D. F., Hamlin, P. H., & Kennedy, T. D. (1977). Behavior modification in a prison: Issues, problems, and compromises. *Criminal Justice and Behavior,* 4, 11-43.

Geller, E. S., Koltuniak, T. A., & Shilling, J. S. (1983). Response avoidance prompting: A cost-effective strategy for theft deterrence. *Behavioral Counseling and Community Interventions,* 3, 29-42.

Geller, E. S., & Lehman, G. R. (1988). Drinking-driving intervention strategies: A person-situation-behavior framework. In J. R. Snortum, F. E. Zimring, & M. D. Laurence (Eds.), *Social control of the drinking driver* (pp. 297-320). Chicago and London: The University of Chicago Press.

Geller, E. S., Lehman, G. R., & Kalsher, M. J. (1989). *Behavior analysis training for occupational safety.* Newport, VA: Make-A-Difference, Inc.

Geller, E. S., Sleet, D., Elder, J., & Hovell, M. (1991). Behavior change approaches to deterring alcohol-impaired driving. In W. B. Ward & F. M. Lewis (Eds.), *Advances in health education and promotion,* Vol.III, Philadelphia, PA: Jessica Kingsley Publishers.

Geller, E. S., Winett, R. A., & Everett, P. B. (1982). *Preserving the environment; New strategies for behavior change.* New York: Pergamon Press.

Glenwick, D., & Jason, L. (1993) (Eds.). *Promoting health and mental health: Behavioral approaches to prevention.* New York: Springer.

Glenwick, D., & Jason, L. (1980) (Eds.). *Behavioral community psychology.* New York: Praeger Publishers.

Goldstein, A. P., & Krasner, L. (1987). *Modern applied psychology.* New York: Pergamon Press.

Greene, B. F., Winett, R. A., Van Houten, R., Geller, E. S., & Iwata, B. A. (1987) (Eds.). *Behavior analysis in the community: Readings from the Journal of Applied Behavior Analysis.* University of Kansas, Lawrence, KS.

Guastello, S. J. (1993). Do we really know how well our occupational accident prevention programs work? *Safety Science,* 16, 445-463.

Heinrich, H. W., Peterson, D., & Roos, N. (1980). *Industrial accident prevention: A safety management approach,* 5th Edition. New York: McGraw-Hill.

Johnson, R. P., & Geller, E. S. (1980). Community mental health center programs. In D. Glenwick & L. Jason (Eds.). *Behavioral community psychology: Progress and prospects* (pp. 147-174). New York: Praeger Press.

Komaki, J., Heinzmann, A. T., & Lawson, L. (1980). Effect of training and feedback: Component analysis of a behavioral safety program. *Journal of Applied Psychology,* 65(3), 261-270.

Kramer, K. D., & Geller, E. S. (1987). Community dental health promotion for children: Integrating applied behavior analysis and public health. *Education and Treatment of Children,* 10, 58-66.

Ludwig, T. D., & Geller, E. S. (1991). Improving the driving practices of pizza deliverers: Response generalization and moderating effects of driving history. *Journal of Applied Behavior Analysis,* 24, 31-44.

Mayer, J., & Geller, E. S. (1982-83). Motivating energy efficient travel: A cost-effective incentive strategy for encouraging usage of a community bike path. *Journal of Environmental Systems,* 12, 99-122.

Novak, L. (1995, May). National Highway Traffic Safety Administration, personal communication with E. S. Geller.

Pavlov, I. P. (1927). *Conditional Reflexes.* (G.V. Anrep, ed. and translator). London: Oxford University Press.

Petersen, D. (1989). *Safe behavior reinforcement.* New York: Aloray, Inc.

Reichel, D. A., & Geller, E. S. (1981). Applications of behavioral analysis to conserve transportation energy. In A. Baum & J. Singer (Eds.), *Advances in environmental psychology,* Vol. III (pp. 53-91). New Jersey: Lawrence Erlbaum Associates.

Schnelle, J. F., Geller, E. S., & Davis, M. A. (1987). Law enforcement and crime prevention. In E. K. Morris & C. J. Braukmann (Eds.), *Behavioral approaches to crime and delinquency: Application, research, and theory* (pp. 225-249). New York: Plenum Press.

Skinner, B. F. (1938). *The behavior of organisms.* New York: Appleton-Century-Crofts.

Snyder, S. (1991). Movies and juvenile delinquency: An overview. *Adolescence,* 26, 121-132.

Streff, F. M., Kalsher, M. J., & Geller, E. S. (1993). Developing efficient workplace safety programs: Observations of response covariation. *Journal of Organizational Behavior Management,* 13 (2), 3-15.

Sulzer-Azaroff, B., & De Santamaria, M. C. (1980). Industrial safety hazard reduction through performance feedback. *Journal of Applied Behavior Analysis,* 13, 287-295.

Wilde, G. J. S. (1994). *Target risk.* Toronto, Ontario, Canada; PDE Publications.

Chapter 8
Defining Critical Behaviors

The practical "how to" aspects of this book begin with this chapter. The overall process is called **DO IT***, each letter representing the four basic components of a behavior-based approach:* **D***efine target behaviors to influence;* **O***bserve these behaviors;* **I***ntervene to increase or decrease target behaviors; and* **T***est the impact of your intervention process. This chapter focuses on developing a critical behavior checklist for objective observing, intervening, and testing.*

Now the action begins. Up to this point I've been laying the groundwork (rationale and theory) for the intervention strategies described here and in the next three chapters. From this information you will learn how to develop action plans to increase safe behaviors, decrease at-risk behaviors, and achieve a Total Safety Culture.

Why did it take me so long to get here—to the implementation stage? Indeed, if you're looking for "quick-fix" tools to make a difference in safety you may have skipped or skimmed the first two parts of this text and started your careful reading here. I certainly appreciate that the pressures to get to the bottom line quickly are tremendous. But remember, there is no quick fix for safety. The behavior-based approach that is the heart of this book is the most efficient and effective route to achieving a Total Safety Culture. It is a never-ending continuous improvement process, one that requires ongoing and comprehensive involvement from the people protected by the process. In industry, these are the operators or line workers.

Long-term employee participation requires understanding and belief in the principles behind the process. Employees must also perceive that they "own" the procedures that make the process work. For this to happen it's necessary to teach the principles and rationale first (as done in this book), and then work with the process participants to develop specific procedures. This creates the perception of ownership and leads to long-term involvement.

As we begin here to define principles and guidelines for action plans, it's important to keep one thing in mind—you need to start with the conviction that there is rarely a generic best way to implement a process involving human interaction. For a behavior-based safety process to succeed in your setting, you will need to work out the procedural details with the people whose involvement is necessary. The process needs to be customized to fit your culture.

The DO IT Process

For the past several years I have taught applications of the behavior-based approach to industrial safety with the acronym depicted in Figure 8.1. The process is continuous and involves the following four steps:

D= Define critical target behavior(s) to increase or decrease

O= Observe the target behavior(s) during a preintervention baseline period to set behavior-change goals, and perhaps to understand natural environmental or social factors influencing the target behavior(s)

I = Intervene to change the target behavior(s) in desired directions

T = Test the impact of the intervention procedure by continuing to observe and record the target behavior(s) during the intervention program

From data recorded in the last item, you can evaluate the impact of your intervention and make an informed decision whether to continue it, implement another strategy, or define another behavior for the DO IT process.

This chapter focuses on the first two steps, defining and observing target behaviors. But before we get into those specifics, I want to briefly outline the DO IT process

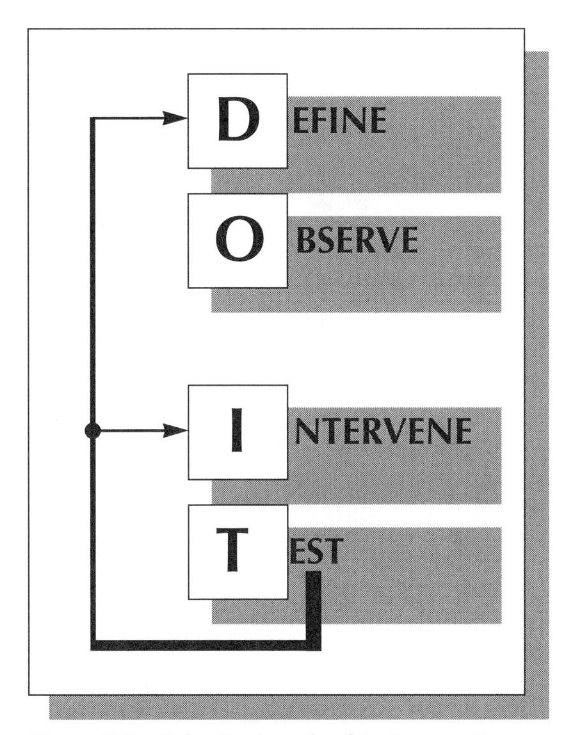

Figure 8.1. Behavior-based safety is a continuous four-step process.

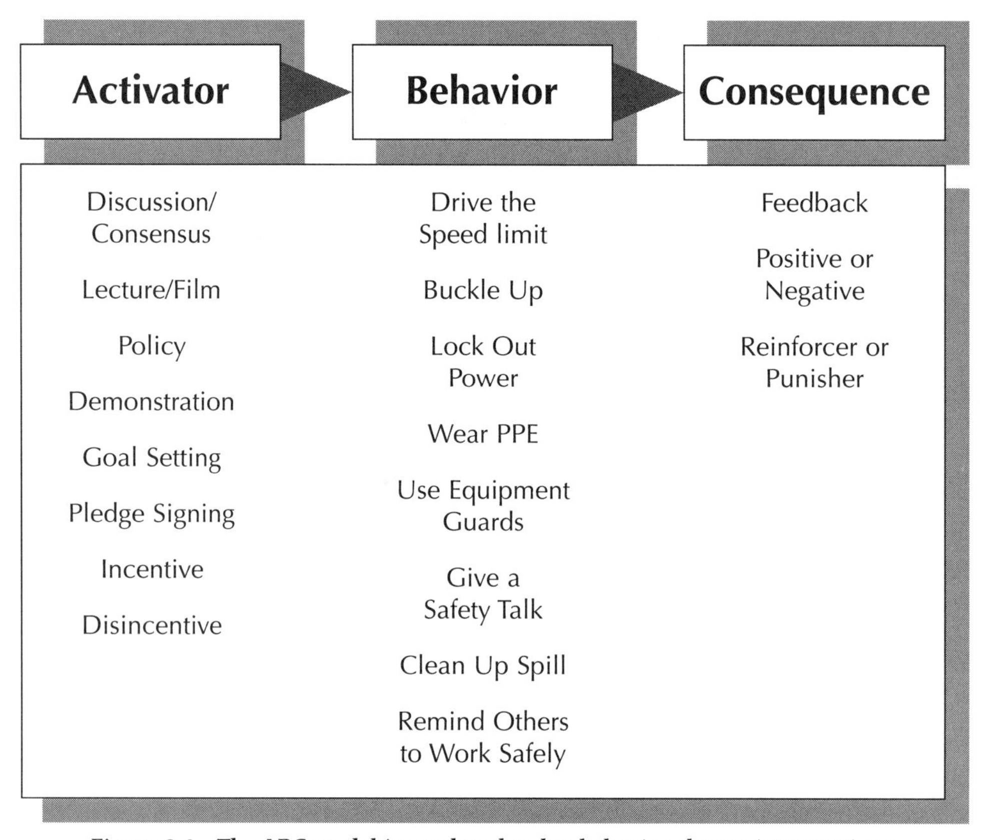

Figure 8.2. The ABC model is used to develop behavior change interventions.

to make an important point: What I'm explaining to you is all easier said than done. Remember, there is no quick fix for safety.

To begin, just what are clear and concise definitions of target behaviors? This is the first step in the DO IT process. There is so much to choose from: Using equipment safely, lifting correctly, locking out power appropriately, and looking out for the safety of others, to name just a scant few. The outcome of behaviors, such as wearing personal protective equipment, working in a clean and organized environment, and using a vehicle safety belt can also be targeted. If two or more people independently obtain the same frequency recordings when observing the defined behavior or behavioral outcome during the same time period, you have a definition sufficient for an effective DO IT process. Baseline observations of the target behavior should be made and recorded before implementing an intervention program. More details on this aspect of the process are given later in this chapter.

What about the intervention step? This phase of DO IT involves one or more behavior-change techniques, based on the simple ABC model depicted in Figure 8.2. Activators direct behavior (as when a ringing telephone or doorbell signals the need for certain behaviors from residents), and consequences motivate behavior (residents answer or do not answer the telephone or door depending on current motives or expectations developed from prior experiences). Activators direct behavior and consequences motivate behavior. The activators listed in Figure 8.2 are discussed in the next chapter. Conse-quences are discussed in Chapter 10.

The DO IT process is based on operant learning, which we discussed in Chapter 7. Figure 7.7, for example, shows the flashing blue light on the police car as a discriminative stimulus (or activator) that signals (or directs) the motorist to perform certain driving behaviors learned from past experience. We will respond to this activator if it is supported by a consequence. For example, if we believe the police officer will give us a ticket if we don't stop, we will pull over. Remember, the power of an activator to influence behavior is determined by the type of consequence(s) it signals (Skinner, 1953, 1969).

Let's talk a little more about consequences. The strongest consequences are soon, sizable, and certain. In other words, we work diligently for immediate, probable, and large positive reinforcers or rewards, and we work frantically to escape or avoid soon, certain, and sizable penalties. This helps explain why safety is a struggle in many workplaces. You see, safe behaviors are usually not reinforced by soon, sizable, and certain consequences. In fact, safe behaviors are often punished by soon and certain negative consequences, including inconvenience, discomfort, and slower goal attainment. Also, the consequences that motivate safety professionals to promote safe work practices—reduced injuries and associated costs—are delayed, negative, and uncertain (actually improbable) from an individual perspective.

Several illustrations make these points. The man in Figure 8.3 will surely comply with the "Don't Walk" activator, and stay on the curb until the light changes. In this case, negative consequences from stepping off the curb at the wrong time would occur very soon. Given the observable traffic flow, a negative consequence is almost certain to happen, and the consequence could be quite sizable, perhaps fatal. But most safety situations in the workplace are not like this one.

Take, for example, the man in Figure 8.4. He is complying with the activator, but there are no obvious (soon, certain, and sizable) consequences supporting this safe behavior. He might not see a reason for the hard hat, and without supportive consequences from peers his compliance could be temporary.

Research backs up these examples. When subjects in operant learning experiments no longer receive a consequence for making the desired response, they eventually stop performing the behavior. This is referred to as extinction, and it occurs in classical conditioning, too. When Pavlov stopped giving his dogs food (the unconditioned stimulus) following the bell (the conditioned stimulus), the dogs

Figure 8.3. Compliance with some activators is supported by natural consequences.

Figure 8.4. Compliance with some activators is not supported by natural consequences.

eventually stopped salivating (conditioned response) to the bell.

Competing with the lack of soon, certain, and sizable positive consequences for safe behaviors are soon, certain (and sometimes sizable) positive consequences for at-risk behaviors. Taking risks avoids the discomfort and inconvenience of most safe behaviors, and it often allows people to achieve their production and quality goals faster and easier. Supervisors sometimes activate and reward at-risk behaviors—unintentionally, of course—to achieve more production. Because activators and consequences are naturally available throughout our everyday existence to support at-risk behaviors in lieu of safe behaviors, safety can be considered a continuous fight with human nature (as discussed earlier in Part Two).

The DO IT process is a tool to use in this struggle with human nature. Developing and maintaining safe work practices often requires intervention strategies to keep people safe—strategies involving activators, consequences, or both. But we're getting ahead of ourselves. First we need to define critical behaviors as targets for our intervention. Let's see how this is done.

Defining Target Behaviors

The DO IT process begins by defining critical behaviors to work on. These become the targets of our intervention strategies. Some target behaviors might be safe behaviors you want to see happen more often, like lifting with knees bent, cleaning a work area, putting on personal protective equipment, or replacing safety guards on machinery. Other target behaviors may be at-risk behaviors that need to be decreased in frequency, such as misusing a tool, overriding a safety switch, placing obstacles in an area designated for traffic flow, stacking materials incorrectly, and so on. A DO IT process can define desirable behaviors to be encouraged or undesirable behaviors to be changed. What the process focuses on in your workplace depends on a review of your safety records, job hazard analyses, near-hit reports, audit findings, interviews with employees, and other useful information.

Critical behaviors to identify and target are:

- At-risk behaviors that have led to a substantial number of near hits or injuries in the past and safe behaviors that could have prevented these incidents
- At-risk behaviors that could potentially contribute to an injury (or fatality) and safe behaviors that could prevent such an incident

Determining which behaviors are critical is the first step of a DO IT process. A great deal can be discovered by examining the workplace and discussing with people how they have been performing their jobs. People already know a lot about the hazards of their work and the safe behaviors needed to avoid injury. They even know which safety policies are sometimes ignored to get the job done on time. They often know when a near hit occurred because an at-risk behavior or environmental hazard was overlooked. They also know which at-risk behaviors could lead to a serious injury (or fatality), and which safe behaviors could prevent a serious injury (or fatality).

Figure 8.5. Participants in behavior-based safety develop a specific way of communicating.

In addition to employee discussions, injury records and near-hit reports can be consulted to discover critical behaviors (both safe and at-risk). Job hazard analyses or standard operating procedures can also provide information relevant to selecting critical behaviors to target in a DO IT process. Obviously, the plant safety director or the person responsible for maintaining records for OSHA (Occupational Safety and Health Administration) can provide valuable assistance in selecting critical behaviors.

After selecting target behaviors, it's critical to define them in a way that gets everyone on the same page. All participants in the process must understand exactly what behaviors you intend to support, increase, or decrease. Defining target behaviors results in an objective standard for evaluating an intervention process.

There's another point to be made about the DO IT process: It involves translating educational concepts into operational (behavioral) terms. From education one learns general principles, procedures, or policy. Training is the process of translating knowledge from education into specific behaviors. Thus, when people describe critical behaviors in objective and observable terms, they are transferring knowledge into meaningful action plans. If done correctly, this can reduce biased subjectivity in various interpersonal matters. Such an influence of behavior-based psychology might even be reflected in changes in everyday language (as illustrated in Figure 8.5).

WHAT IS BEHAVIOR?

The key is to define behaviors correctly. Let's begin by stepping back a minute to consider: What is behavior? Behavior refers to *acts* or *actions* by individuals that can be *observed by others*. In other words, behavior is what a person does or says, as opposed to what he or she thinks, feels, or believes.

Yes, the act of saying words such as "I am tired" is a behavior because it can be observed or heard by others. However, this is not an observation of tired behavior. If the person's work activity slows down or amount of time on the job decreases, we might infer that the person is actually tired. On the other hand, a behavioral "slow down" could result from other internal causes, like worker apathy or lack of interest. The important point here is that feelings, attitudes, or motives should not be confused with behavior. They are *internal* aspects of the person that cannot be directly observed by others. It is risky to infer inner personal characteristics from external behaviors.

The test of a good behavioral definition is whether other persons using the definition can accurately observe if the target behavior is occurring. There are thousands of words in the English language that can be used to describe a person. From all these possibilities, the words used to describe behavior should be chosen for *clarity* to avoid being misinterpreted; *precision* to fit the specific behavior observed; *brevity* to keep it simple; and their *reference to observable activity*—they describe what was said or done.

OUTCOMES OF BEHAVIORS

Often it is easier to define and observe the outcomes of safe or at-risk behavior rather than the behavior itself. These outcomes can be temporary or permanent, but they are always observed after the behavior has occurred. For example, when observing a worker wearing safety glasses, a hard hat, or a vehicle safety belt, you are not actually observing a behavior, but rather you are observing the outcome of a pattern of safety behaviors (the behaviors required to put on the personal protective equipment). Likewise, a locked-out machine and a messy work area are both outcomes of behavior; one from safe behavior and one from at-risk behavior.

This distinction between direct observations of behavior versus behavioral outcomes is important. You see, evaluating an outcome cannot always be directly attributed to a single behavior or to any one individual. And the intervention to improve a behavioral outcome might be different than an intervention to improve behaviors observed directly. For example, direct guidance through instruction and demonstration (activators) might be the intervention of choice to teach the correct use of a respirator; verbal recognition (a consequence) would be more suitable to support the outcome of correctly wearing a respirator at the appropriate time and place.

PERSON-ACTION-SITUATION

Three elements comprise a complete description of behavior. The first is the *person* who is behaving. Secondly, what the person says or does constitutes an *action*. And last, but no less important, is the *situation* in which the person acts (when or where). The importance of safety-related behaviors depends on the situation in which the behavior occurs. Thus, target behaviors are often defined by environmental context: safety glasses and ear plugs are not needed in the personnel office, but are needed in other work areas.

Here's another important point: You can't study a behavior that doesn't happen. The nonoccurrence of a recommended safety action in a given situation is often defined as at-risk behavior. Although the lack of an appropriate safe action might put a person at risk, the absence of a behavior is not a behavior. It is important to define what is happening—the at-risk behavior(s) occurring in place of the safe behavior.

Such behavior can be observed and changed. Then it's possible to study the factors influencing this at-risk behavior and perhaps inhibiting safe behavior.

OBSERVABLE BEHAVIORS

A target behavior needs to be defined in observable terms so that multiple observers can watch one individual independently and obtain the same results regarding the occurrence or nonoccurrence of the target behavior. There should be no room for interpretation. "Isn't paying attention," "acting careless," or "lifting safely," for example, are not adequate descriptions of behavior, because observers would not agree consistently about whether the behavior occurred. In contrast, descriptions like "keeping hand on handrail," "moving knife away from body when cutting," and "using knees while lifting" are objective and specific enough to obtain reliable information from trained observers. In other words, if two observers watched for the occurrence of these behaviors, they would likely agree whether or not the behavior occurred.

The ultimate test for a behavioral description is to have two observers watch independently for the occurrence of the target behavior on a number of occasions, and then calculate the percentage of agreement between observers. More specifically, agreement occurs whenever the two observers report seeing or not seeing the target behavior at the same time. Disagreement occurs whenever one person reports seeing the behavior when the other person reports not seeing the behavior. Percentage of agreement is calculated by adding the number of agreements and disagreements and dividing the total into the number of agreements. The quotient is then multiplied by 100 to give percentage of agreement. If the result is 80 percent or higher, the behavioral definition is adequate and the observers have been adequately trained to use the definition in a DO IT process (Kazdin, 1994).

MULTIPLE BEHAVIORS

Let's look more closely at types of behaviors. Some workplace activities can be treated effectively as a single behavior. Examples include: "Looking left-right-left before crossing the road;" "Walking within the yellow safety lines;" "Honking the fork lift horn at the intersection;" "Returning tools to their proper location."

Some outcomes of behaviors also can be dealt with in singular terms: "Using ear plugs;" "Using a vehicle safety belt;" "Climbing a ladder that is properly tied off;" "Working on a scaffold with appropriate fall protection;" and "Repairing equipment that was locked-out correctly". With a proper definition, an observer could readily count occurrences of these safe behaviors (or outcomes) during a systematic audit.

But many safety activities are made up of more than one discrete behavior, and it may be important to treat these behaviors independently in a definition and an audit. "Bending knees while lifting," for example, is only one aspect of a safe lift. Thus, if safe lifting were the activity targeted in a DO IT process, it would be necessary to define the separate behaviors (or procedural steps) of a safe lift. This would include, at least, checking the load before lifting, asking for help in certain situations, lifting with the legs, holding the load close to one's body, lifting in a smooth motion, and moving feet when rotating (not twisting).

Each of the procedural steps in safe lifting requires a clear objective definition so two observers can determine reliably whether the behavior in a lifting sequence occurred. Observing reliably whether the load was held close and knees were bent would be relatively easy. However, defining a "smooth lift" so that observers could agree on 80 percent or more of the observations would be more difficult. And, for observers to reliably audit "asking for help," the "certain situations" calling for this response would need to be specified.

Role playing demonstrations are an important way to help define the behavioral steps of a procedure. A volunteer acts out the behaviors while observers attempt to determine whether each of the designated steps of the activity is "safe" or "at risk." Suppose, for example, your work group is interested in improving stair safety, and decided that the safety-related behaviors in this activity include: keeping one hand on the handrail, taking one step at a time, and walking rather than running. After defining these procedural steps more completely in a group session, the participants should go to a setting with stairs and observe people using the stairs. Observers should record independently whether each behavior of the activity is "safe" or "at risk." Then they can reconvene in a group meeting and compare notes.

Group discussions about practice observations might very well lead to changed or refined behavioral definitions. Additionally, it might be decided that some participants need additional education and training about the observation process. When observers can use behavioral definitions and agree on the "safe" versus "at-risk" occurrence of each behavior on 80 percent or more of the observation trials, you are ready for the next DO IT phase: Observation.

Observing Behavior

The acronym ***SOON*** depicted in Figure 8.6 reviews the key aspects of developing adequate definitions of critical behaviors to target for a DO IT process. You are ready for the observation phase when you have a checklist of critical behaviors with definitions that are:

Specific
Observable
Objective
Naturalistic

We have already considered most of the characteristics of behavioral definitions implied by these key words, and examples of behavioral checklists are provided later in this chapter, as well as in Chapter 11 on "safety coaching."

I have not yet explained one very important characteristic of "observable" behaviors. They are *quantifiable*. In other words, observers of a target behavior should be able to translate their experience into a form that can be counted and compared objectively with other observations. Some meaningful aspect (or property) of the target behavior needs to be recorded systematically so that changes (or improvements) in the behavior can be monitored over time. To do this, we have to consider various properties of behavior.

S PECIFIC
- Concise behavioral definition
- Unambiguous

O BSERVABLE
- Overt behaviors
- Countable & recordable

O BJECTIVE
- No interpretations nor attributions
- "What," not "Why"

N ATURALISTIC
- Normal interaction
- Real-world activity

Figure 8.6. Behavioral observations for the DO IT process should be "SOON."

PROPERTIES OF BEHAVIOR

One property of behavior is *intensity*. When a person says something, the sound can range from a low-intensity whisper to a high-intensity shout. Another property is *speed*. A fork lift, for example, can be operated at fast or at slower speeds.

Still another property of behavior is *duration*. Some behaviors last only a few seconds, like turning off the power to a machine, putting on safety glasses, or signing a "safety pledge." Others may continue for several hours, such as performing a series of responses at a particular work station or discussing safe work practices during a group meeting.

For our purposes, the property of *response frequency* is usually most important. A particular response may occur once in a given period of time, or it may occur several times. The rate of a behavior refers to its frequency of occurrence per unit of time.

Most safety-related behaviors can be considered in terms of frequency or rate. For example, the frequency or rate of operating a fork lift at an at-risk speed is a meaningful target for a DO IT process.

MEASURING BEHAVIOR

Certain safe and at-risk behaviors start and stop often during a work period. Consider lifting, smoking a cigarette, or praising a coworker. They are readily measurable in terms of rate. On the other hand, the opportunities for some safety-related behaviors, like locking out power to equipment, occur less often during the usual workday. Here, percentage of occurrence per opportunity might be a more meaningful property to measure than frequency or rate.

For example, situations that require locking-out power, stacking racks less than three high, and using cut-resistant gloves might vary considerably throughout the day. Thus, it is most meaningful to consider the number of occasions the target behavior actually occurs per total situations requiring that behavior. From these frequencies (total occurrences of safe behavior and total opportunities for the safe behavior) a percentage of safe behavior can be calculated and used to monitor safety achievement.

Some behaviors should continue throughout lengthy work sessions. Protective apparel such as safety glasses and ear plugs may need to be worn continuously. In these cases, it might be most appropriate to observe and record the duration, or the total amount of time the behavior occurs, rather than frequency. I'll return later to this issue of targeting and measuring the most appropriate properties of safety-related behaviors. For now, it's important to understand that the property targeted by a behavior-change intervention depends on specific situational factors and program objectives. Generally, the goal of the DO IT process is to increase the occurrence (frequency, rate, percentage, and/or duration) of safe behaviors, and decrease the occurrence of at-risk behaviors.

RECORDING OBSERVATIONS

Accurate and permanent records of observed behavior are essential for a job safety analysis, an injury investigation, and a successful DO IT process. Most existing records are in the form of written comments, and often do not offer an objective behavioral metric, or a measure of observable behavior. Before attempting to change a specific behavior, you should first observe and record a certain property of that behavior. By measuring the frequency, duration, or percentage of occurrence per opportunity of a particular behavior over a sufficient period of time you determine the extent to which that behavior needs to be changed.

Careful observation of response frequency, for example, helps answer several important questions:

How does the frequency of the target behavior vary among different individuals?

In what situations and at what times does the behavior occur most often?

When and where does the behavior occur least often?

How often does a person have an opportunity to make an appropriate safe behavior but does not make it?

What specific environmental changes occur before and after the target behavior occurs?

What environmental factors are supporting a particular at-risk behavior and/or inhibiting, perhaps demotivating, a particular safe behavior?

CALCULATING A BEHAVIOR RATE. Two basic requirements are necessary to record the rate of a behavior. First, a precise objective definition of the beginning and the end of the target behavior is required. Any observer should be able to count the number of times the behavior begins and ends within a given period. This results in a frequency measurement. Second, it's important to record the time you begin observing the target behavior and the time you stop. This gives you a record of the time interval during which the behaviors were counted, and enables you to convert response frequency into a response rate.

Response rate is calculated by dividing the frequency (for example, 45 occurrences of a behavior) by the length of the time interval (for example, 15 minutes). The response rate here is 3 responses per minute. Response rate is analogous to miles per hour. By translating frequencies into rates, comparisons between two measurements can be made even when the lengths of observation periods are different. Independent frequency records can only be compared if the length of the observation periods are the same; response rates are comparable regardless of the lengths of the different recording periods.

INTERVAL RECORDING. Some safety-related behaviors begin and end relatively infrequently during a workday, but once they occur they last for long durations. In this case, a frequency or rate measure would not be as informative as a record of the length of time the target behavior occurs. It might be more practical, though, to note periodically and systematically whether the target response is occurring *in a particular situation.* Instead of watching an individual and counting the start and end of a particular behavior during a given time period, an observer could intermittently look at the individual throughout the work period in a given environmental setting and note whether or not the target behavior is occurring. This measurement procedure is termed *interval recording,* in contrast to event recording where the occurrence of a discrete behavior is counted during an observation session and possibly converted to response rate. At work, interval recording is often the most practical approach to observing and recording critical behaviors. A checklist of critical behaviors is used and the observer merely watches an individual work for a set period of time and checks off "safe" or "at risk" for each behavior on the list. The number of "safe" and "at risk" checkmarks can be totaled and used to calculate the percentage of "safe" behaviors recorded in a particular interval. This is the approach recommended by Krause *et al,* (1990) and McSween (1995).

Several methods are available for objectively observing and recording safe versus at-risk behaviors in real-world settings. Different situations call for different procedures. I'll give a number of examples throughout this and subsequent chapters. I hope at least one of these methods will relate directly to the situation in which you want to apply the DO IT process.

A PERSONAL EXAMPLE

My daughter Krista asked me to drive her to the local Virginia Department of Motor Vehicles office to get her "learner's permit." She was 15 years old and thought she was ready to drive. Of course, I knew better, but how do you fight a culture that puts teenagers behind the wheel of motor vehicles before they are really ready for such a risky situation?

"Don't worry, Dad," my daughter said, "I've had driver's education in high school." Actually, that was part of my worry. She was educated about the concepts and rules regarding driving, but she hadn't been trained. She had not yet translated her education into operations or action plans.

In order to obtain a license to operate a motor vehicle in Virginia before the age of 18, teenagers with a learner's permit are required to take seven two-hour instructional periods of on-the-road experience with an approved driver-training school. For half of these sessions they must be the driver; the rest of the time they sit in the back seat and perhaps learn through observation. Thus, for seven one-hour sessions Krista drove around town with an instructor in the front seat and one or more students in the back, waiting for their turn at the wheel. This was an opportunity for my daughter to transfer her driver education knowledge to actual performance.

DRIVING ACTIVATORS AND CONSEQUENCES. On-the-job training obviously requires an appropriate mix of observation and feedback from an instructor. Practice does not make perfect. Only through appropriate feedback can people improve their performance. Some tasks give natural feedback to shape our behavior. When we turn a steering wheel in a particular direction we see immediately the consequence of our action, and our steering behavior is naturally shaped. The same is true of several other behaviors involved in driving a motor vehicle, from turning on lights, windshield wipers, and cruise-control switches to pushing gas and brake pedals.

However, many other aspects of driving are not followed naturally with feedback consequences, particularly those that can prevent injury from vehicle crashes. Although we get feedback to tell us our steering wheel, gas pedal, turn signal lever, and brakes work, we don't get natural feedback regarding our safe versus at-risk use of such control devices. Also, when we first learn to drive we don't readily recognize the activators (or discriminative stimuli) that should signal the use of various vehicle controls. This is commonly referred to as "judgment." From a behavior-based perspective, "good driving judgment" is recognizing environmental conditions (or activators) that signal certain vehicle-control behaviors, and then implementing the controls appropriately.

I wondered whether my daughter's driving instructor would give her appropriate and systematic feedback regarding her "driving judgment." Would he point out consistently the activators that require safe vehicle-control behaviors? Would he put emphasis on the positive, supporting my daughter's safe behaviors before criticizing her at-risk behaviors? Would he, or a student in the back seat, display negative emotional reactions in certain situations and teach Krista (through classical conditioning) to feel anxious or fearful in particular driving situations? Would some at-risk driving behaviors by Krista or the other student drivers be overlooked by the instructor, and lead to observational learning that some at-risk driving behaviors are acceptable?

DEVELOPING A DRIVING BEHAVIOR CHECKLIST. Even if the driving instruction is optimal, seven hours of such observation and feedback is certainly not sufficient to teach safe driving habits. I recognized a need for additional driving instruction for my daughter. We needed a DO IT process for driving. The first step was to define critical behaviors to target for observation and feedback. Through one-on-one discussion, my daughter and I derived a list of critical driving behaviors and then agreed on specific definitions for each item. My university students practiced using this critical behavior checklist (CBC) a few times with various drivers and refined the list and definitions as a result. The CBC we eventually used is depicted in Figure 8.7.

After refining the CBC and discussing the final behavioral definitions with Krista, I felt ready to implement the second stage of DO IT: **O**bservation. I asked my daughter to drive me to the university—about nine miles from home—to pick up some papers. I made it clear that I would be using the CBC on both parts of the roundtrip. When we returned home I totaled the safe and at-risk checkmarks and calculated the percentage of safe behaviors. Krista was quite anxious to learn the results, and I

Critical Behavior Checklist for Driving

Driver:	Date:	Day:
Observer 1:	Origin:	Start Time:
Observer 2:	Destination:	End Time:
Weather:		
Road Conditions:		

Behavior	**Safe**	**At-Risk**	**Comments**
Safety Belt Use			
Turn Signal Use			
Left turn			
Right turn			
Lane change			
Intersection Stop			
Stop sign			
Red light			
Yellow light			
No activator			
Speed Limits			
25 mph and under			
25 mph-35 mph			
35 mph-45 mph			
45 mph-55 mph			
55 mph-65 mph			
Passing			
Lane use			
Following Distance (2 sec)			
Totals			

$$\textbf{\% Safe} = \frac{\text{Total Safe Observations}}{\text{Total Safe + At-Risk Observations}} = \_\_\_\_\_\_\_\_$$

Figure 8.7. A critical behavior checklist (CBC) to improve driving safety.

looked forward to giving her objective behavioral feedback. I had good news. Her percentage of safe driving behaviors (percent safe) was 85 percent, and I considered this quite good for our first time.

I told Krista her percent safe score, and proceeded to show her the list of safe checkmarks (covering up the checks in the At-Risk column). Obviously, I wanted to make this a positive experience, and to do this, it was necessary to emphasize the behaviors I saw her do correctly. To my surprise, she did not seem impressed with her 85 percent safe score, and pushed me to tell her what she did wrong. "Get to the bottom line, Dad," she asserted, "Where did I screw up?" I continued an attempt to make the experience positive, saying "You did great, honey, look at the high number of safe behaviors." "But why wasn't my score 100 percent?" reacted Krista. "Where did I go wrong?"

This initial experience with the driving CBC was enlightening in two respects. It illustrated the unfortunate reality that the "bottom line" for many people is "where did I make a mistake?" My daughter, at age 15, had already learned that people evaluating her performance seem to be more interested in mistakes than successes. That obviously makes performance evaluation (or appraisal) an unpleasant experience for many people.

A second important outcome from this initial CBC experience was the realization that people can be unaware of their at-risk behaviors, and only through objective feedback can this be changed. My daughter did not readily accept my corrective feedback regarding her four at-risk behaviors. In fact, she vehemently denied that she did not always come to a complete stop. However, she was soon convinced of her error when I showed her my data sheet and my comment regarding the particular intersection where there was no traffic and she made only a rolling stop before turning right. I did remind her that she did use her turn signal at this and every intersection, and this was something to be proud of. She was developing an important safe habit, one often neglected by many drivers.

I really didn't appreciate the two lessons from this first application of the driving CBC until my daughter monitored *my* driving. That's right, Krista used the CBC in Figure 8.7 to evaluate my driving on several occasions. I found this reciprocal application of a CBC to be most useful in developing mutual trust and understanding between us. I found myself asking my daughter to explain my lower-than-perfect score, and arguing about one of the recorded "at-risk" behaviors. I, too, was defensive about being 100 percent safe. After all, I've been driving for 37 years and teaching and researching safety for more than 20 years, how could I not get a perfect driving score when I knew I was being observed?

From our experience with the CBC, my daughter and I learned the true value of an observation and feedback process. While using the checklist does translate education into training through systematic observation and feedback, the real value of the process is the active caring that occurs between people. In other words, we learned not to get too hung up on the actual numbers. After all, there is plenty of room for error in the numerical scores. Rather we learned to appreciate the fact that through this process people are caring for the safety and health of each other in a way that can truly make a difference. We also learned that even experienced people can perform at-risk behavior and not even realize it.

FROM UNCONSCIOUS TO CONSCIOUS. Figure 8.8 depicts the process we often go through when developing safe habits. Both Krista and I were unaware of some of our at-risk driving behaviors. For these behaviors we were "unconsciously incompetent." Through the CBC feedback process however, we became aware of our at-risk driving behaviors. But awareness did not necessarily result in 100 percent safe behavior scores. Several feedback sessions were needed before some safe driving behaviors

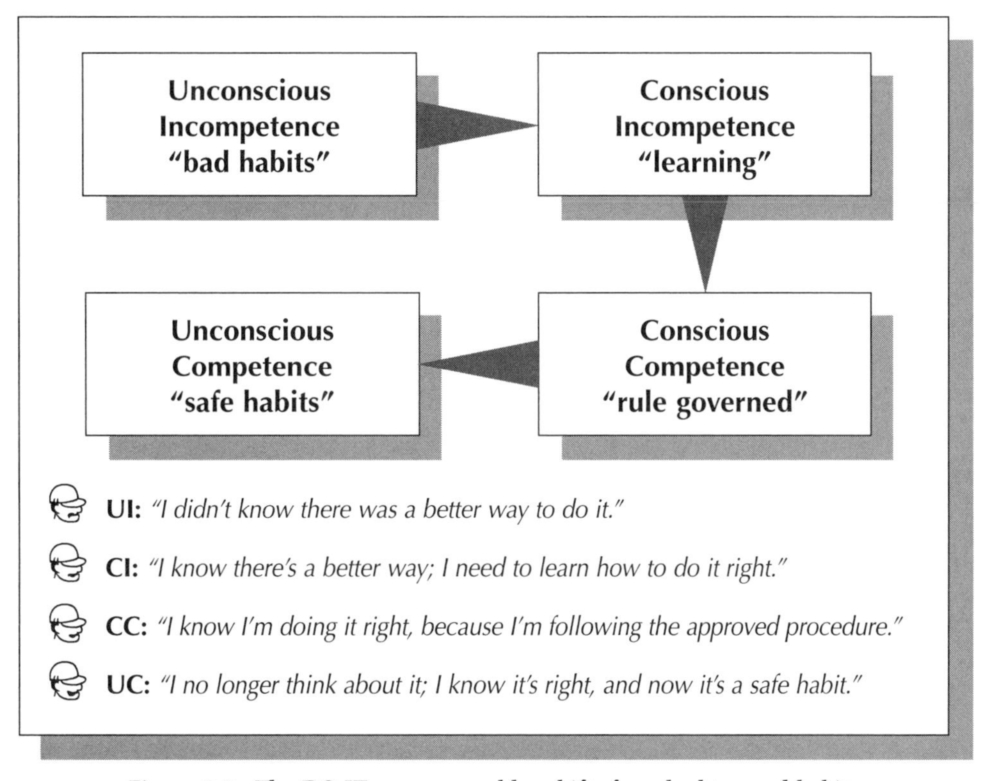

Figure 8.8. The DO IT process enables shifts from bad to good habits.

occurred regularly, and before some at-risk behaviors decreased markedly or extinguished completely. In other words, initial feedback made us "consciously incompetent" with regard to some driving behaviors. Continuous feedback and mutual support resulted in beneficial learning, as reflected in improved percent safe behavior scores on the CBC. Thus, for some driving behaviors we became "consciously competent."

Feedback made us aware of certain driving rules or the driving situation (activator) that calls for a particular safe behavior. Complying with these rules, developed with our CBC feedback process, is referred to as "rule governed" behavior (Malott, 1988, 1992). This stage involves thinking or talking to oneself to identify activators that require certain safe behaviors, and giving self-approval or self-feedback after performing the appropriate safe behavior. With continuous observation and feedback from both others and ourselves some safe behaviors become automatic or habitual. They reach the "unconscious competence" stage in Figure 8.8.

Some of my safe driving behaviors have progressed no further than the "conscious competence" stage. The behavior has not become a habit. I need to remind myself on every occasion to take the extra time or effort to set the safe example. These are the behaviors that benefited most from the CBC feedback process, because over time I had gotten careless about certain driving practices, especially stopping completely at intersections, maintaining a distance of two seconds behind vehicles in front, and staying in the right lane except to pass.

I am "unconsciously competent" about some safe driving practices, particularly safety-belt and turn-signal use, but these behaviors were not always habitual. With safety-belt use, I can recall going through each of the stages in Figure 8.8. When lap belts first appeared in vehicles, I barely noticed them. I even remained "unconsciously incompetent" to safety-belt use in 1974 when vehicles would not start unless the front-seat lap belts were buckled. Like numerous other drivers (observed by Geller *et al,* 1980), I merely buckled my front-seat lap belt and sat on it.

In the mid-'70s I learned well the statistics that justify the use of vehicle safety belts on every trip. In fact, I actually taught the value of using safety belts in my safety workshops at the time. However, the popular quote "Do as I say, not as I do" applied to me. Even though I knew the value of safety belt use, I still did not buckle up on every trip. I was "consciously incompetent" with regard to this safe behavior.

FROM INCOMPETENCE TO COMPETENCE. I started to buckle up consistently in the late '70s only after my students made my belt use the target of an informal DO IT process. My vehicle was visible from the large window in my research laboratory, and my students began observing whether I was buckled up when entering and leaving the faculty parking lot. After two weeks of collecting baseline data without me knowing it, they informed me of their little "experiment" by displaying a graph of weekly percentages of my safety-belt use.

I was appropriately embarrassed by the low percentages for the first two weeks of the "project." My students were holding me accountable for a behavior I should be performing. That was sufficient to change my behavior. From that day on I have always buckled up. For about a year I had to think about it each time. I was "consciously competent." Subsequently, safety belt use in my vehicle became a habit, and I moved to the optimal "unconscious competence" stage for this behavior. I bet many readers are now in this stage for safety-belt use, but can remember being at each of the earlier stages of habit formation.

At what stage of habit formation are you when you get in the back seat of someone else's vehicle, like a taxi cab? It's possible to be "unconsciously competent" in some situations, but be "consciously competent" or "consciously incompetent" in another situation for the same behavior. For example, wearing safety glasses, ear plugs, and steel-toed shoes might be a safe habit on the job, but what are the safe behaviors when mowing the lawn in your backyard? A DO IT process could increase our use of personal protective equipment at home as well as at work. First, we need to accept the fact that we can all be unconsciously or consciously incompetent with regard to some behaviors. Then we need to understand the necessity of behavioral feedback to improve our performance. Thirdly, we need others in our family or work team to observe us with a CBC and then share their findings as actively caring feedback.

Two Basic Approaches

The CBC examples described above illustrate two basic ways of implementing the Define and Observe stages of DO IT. The driving CBC I developed with my daughter illustrates the observation and feedback process recommended by a number of successful behavior-based safety consultants (Krause, 1995; Krause *et al,* 1990;

McSween, 1995). I refer to this approach as one-to-one safety coaching because it involves an observer using a CBC to provide instructive behavioral feedback to another person (Geller, 1995). This approach is reviewed briefly below; a more comprehensive description can be found in Chapter 11.

The second approach to the Define and Observe stages of DO IT involves a limited CBC (perhaps targeting only one behavior) and does not necessarily involve one-to-one coaching. This is the approach used in most of the published studies of the behavior-based approach to safety (for example, see reviews by Petersen, 1989; and Sulzer-Azaroff, 1982, 1987). This was the approach used by my students years ago when they observed, recorded, and graphed my safety-belt use as my vehicle entered and departed the faculty parking lot. In 1984, I taught this particular approach to plant safety leaders for 110 different Ford Motor Company plants (Geller, 1985, 1990). Vehicle safety belt use across all Ford plants increased from 8 percent to 54 percent, and this behavior change in 1984 alone saved the lives of at least eight employees and spared about 400 others from serious injury. Corporate cost savings were estimated at $10 million during the first year, and cumulated to $22 million by the end of 1985 (Gray, 1988).

Each of these approaches to the Define and Observe stages of DO IT are advantageous for different applications within the same culture. Thus, it's important to understand the basic procedures of each, and to consider their advantages and disadvantages. For some work settings, I have found it quite useful to start with the simpler approach of targeting only a few CBC behaviors. With immediate success, behaviors are then added until eventually a comprehensive CBC is developed, accepted, and used willingly throughout the culture. For instance, after Ford Motor Company obtained remarkable success with applications of the DO IT process to increase vehicle safety belt use, several Ford plants expanded the process to target numerous on-the-job work practices (Geller, 1990).

STARTING SMALL

This approach targets a limited number of critical behaviors, but does not require one-on-one observation. A work group defines a critical behavior or behavioral outcome to observe, as discussed earlier in this chapter. After defining their target so that two or more observers can reliably observe and record a particular property of the behavior, usually frequency of occurrence, the group members should give each other permission to observe this work practice among themselves. If some group members do not give permission, it's best not to argue with them. Simply exclude these individuals from the observations, and invite them to join the process whenever they feel ready[2]. They will likely participate eventually when they see that the DO IT process is not a "gotcha program" but an objective and effective way to care actively for the safety of others and build a Total Safety Culture.

2. Some applications of the DO IT process have worked well without this permission phase, as in numerous safety-belt promotion programs (Geller, 1985). However, obtaining permission first will help develop trust and increase opportunity to expand the list of critical behaviors to target.

Observer: **Date:**

TARGET BEHAVIOR	SAFE	AT RISK
load appropriate		
hold close		
use legs		
move feet; don't twist		
smooth motion; no jerks		

Comments (use back if necessary):

Figure 8.9. A sample behavior checklist to observe and record "safe lifting."

It often helps to develop a behavior checklist to use during observations. As discussed earlier, target behaviors like "safe lifting" and "safe use of stairs" include a few specific behaviors, either safe or at risk. Therefore, the CBC should list each behavior separately, and include columns for checking "safe" or "at risk." Figure 8.9 depicts a sample CBC for safe lifting. Through use of this CBC, a work group might revise the definitions, and possibly add a lifting-related behavior relevant to their work area.

Participants willing to be observed anonymously for the target behavior(s) use the CBC to maintain daily records of the safe and at-risk behaviors defined by the group. They do not approach another individual specifically to observe them. Rather they look for opportunities for the target behavior to occur. When they see a safe behavior opportunity (SBO), they take out their checklist and complete it. If the target behavior is "safe lifting," for example, observers keep on the lookout for an SBO for lifting. They might observe such an SBO from their work station or while walking through the plant. Of course, if they see an at-risk lifting behavior and are close enough to reduce the risk, they should put their CBC aside and intervene. Intervening to reduce risk must take precedence over recording an observation of at-risk behavior.

This process could be used to hold people accountable for numerous behaviors or behavioral outcomes. It's quite analogous to the standard environmental audit conducted throughout industrial complexes to survey equipment conditions, environmental hazards, and the availability of emergency supplies. Actually, most equipment and environmental audits reflect behaviors. An equipment guard in place, a tool appropriately sharpened, a work area neat and clean, and equipment power locked out properly are in that "safe" condition because of employees' behaviors. A behavior auditor might look for an SBO regarding any number of safe environmental conditions. When they see an opportunity for the safe target behavior to have occurred, they take out their CBC and record "safe" or "at risk" to indicate objectively whether the desired safe behavior(s) had occurred to make the condition safe.

OBSERVING MULTIPLE BEHAVIORS

As the list of targets on a CBC increases, it becomes more and more difficult to complete a checklist from a remote location. Auditing several critical behaviors usually puts observers in close contact with another person (the performer), resulting in a one-on-one coaching situation (Geller, 1995). The observer should seek permission from the performer before recording any observations, even though a work group might have agreed on the observation process in earlier education and training meetings. If the performer wishes not to be observed, the observer should leave with no argument and a friendly smile. This helps to build the trust needed to eventually reach 100 percent participation in the DO IT process.

Multiple behavior CBCs might be specific to a particular job or be generic in nature. The driving CBC I used with Krista was a job-specific checklist, only relevant for operating a motor vehicle. In contrast, a generic checklist is used to observe behaviors that may occur at various job sites. The CBC depicted in Figure 8.10 is generic because it is applicable for any job that requires the use of personal protective equipment (PPE). Since different PPE might be required on different jobs, certain PPE categories on the CBC may be irrelevant for some observations. For jobs requiring extra PPE, additional behaviors will be targeted on the CBC. Obviously, the observer needs to know PPE requirements before attempting to use a CBC like the one shown in Figure 8.10.

The CBC in Figure 8.10 includes a place for the observer's name, but the performer's name is not recorded. Also, this CBC was designed to conduct several one-on-one behavior audits over a period of time. Each time the observer performs an observation he or she places a checkmark in the left box (for total number of observations). If the performer was using all PPE required in the work area, a check would be placed in the right-hand box. From these entrees, the overall percentage of safe employees can be monitored.

The checkmarks in the individual behavior categories of the CBC in Figure 8.10 are totaled, and by dividing the total number of safe checks by the total safe and at-risk checks, the percentage of safe behaviors for each PPE category can be assessed. This enables valuable feedback regarding the relative use of various devices to protect employees. Such information might suggest a need to make certain PPE more comfortable or convenient to use. It might also suggest the need for special intervention, as discussed in the next three chapters. Chapter 11 also includes additional information on the design of CBCs for one-on-one behavior observation.

Critical Behavior Checklist for Personal Protective Equipment

Observation period (dates): ____________________

Observer: ______________________________

TOTAL NUMBER OF EMPLOYEES OBSERVED	NUMBER OF EMPLOYEES OBSERVED USING ALL REQUIRED PPE

PPE (FOR OBSERVED AREA)	SAFE OBSERVATION (PROPER USE OF PPE)	AT-RISK OBSERVATION (IMPROPER OR NO USE OF PPE)
Gloves		
Safety Glasses/Shield		
Safety Shoes		
Hard Hat		
Lifting Belt		
Hearing Protection		
TOTAL		

Figure 8.10. A generic critical behavior checklist (CBC) for monitoring the use of personal protective equipment (PPE).

In Conclusion

In this chapter we have gotten into the "nuts and bolts" of implementing a behavior-based safety process to develop a Total Safety Culture. The overall process is referred to as DO IT, each letter representing one of the four stages of behavior-based safety. This chapter focused on the first two stages: Define and Observe.

Defining critical behaviors to target for observation and intervention is not easy. A work team needs to consult a variety of sources, including the workers themselves, near-hit reports, injury records, job hazard analyses, and the plant safety director.

After selecting a list of behaviors critical to preventing injuries in their work area, the team needs to struggle through defining these behaviors so precisely that all observers agree on a particular property of each behavior at least 80 percent of the time. The behavioral property most often observed for industrial safety is frequency of occurrence, including a sampling of whether or not the target behavior occurred.

A critical behavior checklist (CBC) is used to observe and record the relative frequency (or percentage of opportunities) with which critical behaviors occur throughout a work setting. If the CBC contains only a few behaviors or behavioral outcomes (conditions resulting from behavior), it's possible to conduct observations without engaging in a one-on-one coaching session. This is often the best approach to use when first introducing behavior-based safety to a work culture. It's not as overwhelming or time-consuming as one-on-one coaching with a comprehensive CBC. Over time and through building trust, a short CBC can be readily expanded and lead to one-on-one safety coaching. Safety coaching is one very effective way to implement each stage of the DO IT process, and is detailed in Chapter 11. First, it's important to understand more about the third DO IT stage: Intervention. This is the topic of the next two chapters.

REFERENCES

Geller, E. S. (1985). *Corporate safety belt programs.* Blacksburg, VA: Virginia Polytechnic Institute and State University.

———. (1990). Performance management and occupational safety: Start with a safety belt program. *Journal of Organizational Behavior Management,* 11(1), 149-174.

———. (1995). Safety coaching: Key to achieving a Total Safety Culture. *Professional Safety,* 40(7), 16-22.

Geller, E. S., Casali, J. G., & Johnson, R. P. (1980). Seat-belt usage: A potential target for applied behavior analysis. *Journal of Applied Behavior Analysis,* 13, 94-100.

Gray, D. A. (1988, October). Introduction to invited address by E. S. Geller at the annual National Safety Council Congress and Exposition, Orlando, FL.

Kazdin, A. E. (1994). *Behavior modification in applied settings* (5th ed.) Pacific Grove, CA: Brooks/Cole Publishing Company.

Krause, T. R. (1995). *Employee-driven systems for safe behavior: Integrating behavioral and statistical methodologies.* New York: Van Nostrand Reinhold.

Krause, T. R., Hidley, J. H., & Hodson, S. J. (1990). *The behavior-based safety process: Managing involvement for an injury-free culture.* New York: Van Nostrand Reinhold.

Malott, R. W. (1992). A theory of rule-governed behavior and organizational behavior management. *Journal of Organizational Behavior Management,* 12(2), 45-65.

———. (1988). Rule-governed behavior and behavioral anthropology. *The Behavior Analyst,* 22, 181-203.

McSween, T. E. (1995). *The values-based safety process: Improving your safety culture with a behavioral approach.* New York: Van Nostrand Reinhold.

Petersen, D. (1989). *Safe behavior reinforcement.* Goshen, NY: Aloray, Inc.

Skinner, B. F. (1953). *Science and human behavior.* New York: Macmillan.

———. (1969). *Contingencies of reinforcement: A theoretical analysis.* New York: Appleton-Century-Crofts.

Sulzer-Azaroff, B. (1982). Behavioral approaches to occupational safety and health. In L. Frederiksen (Ed.), *Handbook of organizational behavior management.* New York: Wiley.

———. (1987). The modification of occupational safety behavior. The *Journal of Occupational Accidents,* 9, 177-197.

Chapter 9
Intervening with Activators

Intervention techniques to increase safe behaviors or decrease at-risk behaviors are either activators *or* consequences. *This chapter explains activators, with real-world examples showing how to develop effective strategies. Six principles for maximizing the impact of activators frame the discussion.*

Chapter 8 introduced the Activator-Behavior-Consequence (ABC) model as a framework for designing behavior-change interventions. Psychologists who use the behavior-based approach to solve human problems design activators (conditions or events preceding operant behavior) and consequences (conditions or events following operant behavior) to increase the probability that desired behaviors will occur and undesired behaviors will not. Activators precede and direct behavior. Consequences follow and motivate behavior. This chapter explains basic principles about activators to help you design interventions for increasing safe behavior and decreasing at-risk behavior. The next chapter focuses on the use of consequences to motivate safety achievement.

First, let me reiterate the need for safety interventions. As I've said before, maintaining our own safe behavior is not easy. It's usually one long fight with human nature, because in most situations activators and consequences naturally support risky behavior in lieu of safe behavior. At-risk behavior often allows for more immediate fun, comfort, and convenience than safe behavior, prompting the need for special intervention to direct and motivate safe behavior.

Activators are generally much easier and less expensive to use than consequences, so it's not suprising that they are employed much more often to promote safe behavior. Posters or signs are perhaps the most popular activators for safety:

Some bear only a general message—"Safety is a Condition of Employment"; others refer to a specific behavior—"Hard Hat Required in this Area"

Some signs request the occurrence of a behavior—"Walk," "Wear Ear Plugs in This Area"; others want you to avoid a certain behavior—"Don't Walk," "No Smoking Area"

Sometimes a relatively convenient response is requested—"Buckle Up," while other signs prompt relatively inconvenient behaviors—"Lock Out All Energy

Sources Before Repairing Equipment"

Some signs imply consequences—"Use Eye Protection: Don't be blinded by the light"; others do not—"Wear Safety Goggles"

We might be reminded of a general purpose—"Actively Care for a Total Safety Culture"; or challenged—"100 Percent Safe Behavior is Our Goal This Year"

Figure 9.1. Safety activators can be overwhelming and ineffective.

I've visited a number of work environments where all of these types of safety signs were displayed. In fact, I've seen situations that make the illustration in Figure 9.1 seem not very far fetched. Does this sort of overkill work to change behavior and reduce injuries? If you answered "no," then this time your common sense was correct, because you've been there and experienced the ineffectiveness of many safety signs.

Which signs would you eliminate from Figure 9.1? How would you change certain signs to increase their impact? What activator strategies would you use instead of the signs? This chapter will enable you to answer these questions—not on the basis of common sense, but from behavioral science research.

Let's move on to discuss six key principles for increasing the impact of activators. They are:

- Specify behavior
- Maintain salience with novelty
- Vary the message
- Involve the target audience
- Activate close to response opportunity
- Implicate consequences

Each of these principles will be illustrated with the help of some real-world examples in the following sections.

Principle #1: Specify Behavior

Behavioral research demonstrates that signs with general messages and no specification of a desired behavior to perform (or an undesirable behavior to avoid) have very little impact on actual behavior. But signs that refer to a specific behavior can be beneficial.

For example, my students and I conducted several field experiments in the 1970s on the behavioral effects of environmental protection messages. In one series of studies we gave incoming customers of grocery stores promotional fliers which included: 1) A general anti-litter message ("Please don't litter. Please dispose of properly"); 2) no environmental protection message; or 3) a specific behavioral request ("Please deposit in green trash can in rear of store"). Later we measured the impact of the different instructions.

Figure 9.2. Some activators are not specific enough.

Our findings were consistent over several weeks, across different stores, and with different research designs (Geller *et al*, 1976; Geller *et al*, 1977). There were three useful conclusions. The general anti-litter message was no more effective than no message (the control condition) in reducing littering or in getting fliers deposited in trash receptacles. In contrast, patrons receiving the fliers with the specific behavioral request were significantly less likely to litter the store, and 20 to 30 percent of these fliers were deposited in the "green trash receptacle." In addition, a message that gave a rationale for the behavioral request "Please help us recycle by depositing in green trash can in rear of store" was even more effective at directing the desired behavior.

Our research on the importance of response specificity in activator interventions has been replicated in other environmental protection research and in a few safety-belt promotion studies. For example, specific response messages reduced littering in a movie theater (Geller, 1975), increased the purchase of drinks in returnable bottles (Geller *et al*, 1973), directed occupants in public buildings to turn off room lights when leaving the room (Delprata, 1977; Winett, 1978), and reminded vehicle occupants to buckle up (Geller *et al*, 1985; Thyer & Geller, 1987). As you'll see, the activators in these studies had characteristics besides response specificity to help make them effective.

Figure 9.3. Some signs are too complex to be effective.

Figure 9.2 illustrates "explosively" the need to include sufficient response information with a behavioral request. But too much specificity can bury a message, as illustrated in Figure 9.3. Activators ought to specify a desired response, but not overwhelm with complexity, as I've seen in a number of industrial signs. Overly complex signs are easy to overlook—with time, they just blend into the woodwork. Keeping signs salient or noticeable is clearly a challenge.

Principle #2: Maintain Salience with Novelty

All of the field research demonstrating the impact of response-specific signs was relatively short term. None of the projects lasted more than a few months. The activators were salient because they were different, or novel. Customers rarely receive fliers when they enter grocery stores, so the fliers and their messages in our litter control and recycling research were quite novel and salient. Plus, most customers did not see the messages every day, since they rarely shopped more than once a week. Similarly, the subjects in the studies that showed effects of safety belt messages on dashboard stickers (Thyer & Geller, 1987) and on flash cards (Geller *et al,* 1985; Thyer *et al,* 1987) were exposed to the message only once, or on average less than once a day.

HABITUATION

It is perfectly natural for activators like sign messages to lose their impact over time. This process is called *habituation,* and it's considered by some psychologists to be the simplest form of learning (Carlson, 1993). Through habituation we learn not to respond to an event that occurs repeatedly. Habituation happens even among organisms with primitive nervous systems. For example, when you lightly tap the shell of a large snail it withdraws into its shell. After about 30 seconds the snail will extend its body from the shell and continue on its way. When you tap the shell again, the snail will withdraw again. However, this time the snail will stay inside its shell for a shorter duration. Your third tap will cause withdrawal again, but the withdrawal time will be even shorter. Each tap on the snail's shell results in successively shorter withdrawal time; eventually the snail will stop responding at all to your tap. The snail's behavior of withdrawal to the activator—shell tapping—will have habituated.

Habituation is perfectly consistent with an evolutionary perspective (Carlson, 1993). If there is no obvious consequence (good or bad) from responding to a stimulus, the organism, be it an employee or a snail, stops reacting to it. It is a waste of time and energy to continue responding to an activator that seems to be insignificant. What would a snail do in a rain storm if it did not learn to ignore shell taps that have no consequence?

Consider the distractions and distress you'd experience daily if you could not learn to ignore noises from voices, radios, traffic, and machinery. At first these environmental sounds might be quite noticeable and perhaps distracting, but through habituation they become insignificant background noise. They no longer divert attention nor interfere with ongoing performance.

I've heard of a much more dramatic illustration of habituation that I want you to only imagine. Please do not try this. If you were to take a frog and drop it in boiling water it would react immediately, leaping out to safety. But put a frog in cold water, slowly raise the heat over several hours, and the frog will not jump out but eventually cook in the boiling water. I confess I have not witnessed this myself nor read it in a scientific journal. But it does sound plausible, given the basic learning principle of habituation. Haven't you seen, for example, seemingly intolerable situations of noise, heat, or squalor which people seem to adjust to over time?

What is the relevance of habituation for safety? It's human nature to habituate to everyday activators in our environment that are not supported by consequences. And this is the case with most safety activators. Staying attentive to safety activators is a continuous fight with one aspect of human nature—habituation.

SAFETY-BELT REMINDERS. Safety-belt reminders are a lesson in how easy it is to ignore activators. How often do you notice the audible safety-belt reminder in your vehicle? I've found many people unable to describe the sound. Is it a continuous tone, a beeping signal, or a pleasant chime? How long does it last? Is it the same sound used for other warning signals in your vehicle?

Current safety-belt reminders in vehicles sold in the United States last from four to eight seconds, as mandated by the National Highway Traffic Safety Administration. Do they work? I suggest they are mostly ineffective (Geller, 1988), primarily due to lack of salience. The sound is usually the same for all warning signals in a vehicle. Not only do we habituate to this sound, but it must compete for our attention with the background noise initiating at the same time, from the roar of the engine, air conditioner or heater to music from a blaring radio. Even if safety-belt reminders were clearly audible, would they increase safety-belt use? Why should a four- to eight-second audible reminder activate people to buckle up?

Perhaps you never hear this reminder because you buckle up before turning on the ignition. According to our field research (von Buseck & Geller, 1984), about half of the drivers fasten their safety belt after turning on their ignition. Since the reminder starts upon ignition, there is no opportunity for these drivers to buckle up and avoid the reminder. When they buckle up, they are merely escaping the unpleasant activator. What if buckling up enabled a driver to avoid a reminder that was salient and somewhat unpleasant?

In 1987, General Motors Research Laboratories loaned me a 1984 Cadillac Seville to answer these important questions. This experimental vehicle was programmed to provide any of the following reminder systems: 1) A standard six-second buzzer or chime triggered by engine ignition; 2) a six-second buzzer or chime that initiated five seconds after ignition; 3) a voice reminder ("Please Fasten Your Safety Belt") that initiated five seconds after engine ignition and was followed by a "Thank You" if the driver buckled up; and 4) a second reminder option where the six-second buzzer, chime, or verbal prompt kicked in if the driver was not buckled when the vehicle made its first stop after exceeding ten miles per hour. This special vehicle had a portable computer in its trunk to record each instance of belt use by the driver.

My students and I studied the impact of these different reminder systems by having college students drive the experimental vehicle on a planned community course under the auspices of an energy conservation study. We asked subjects to stop and park the vehicle at six specific locations along this two-mile course and flip a toggle switch in the vehicle's trunk (presumably to record information on gasoline use). This gave each driver six opportunities to buckle up during a one-hour experimental session. Each subject returned periodically to participate in this so-called "energy conservation study." The number of days between sessions varied from one to five. Results were examined on an individual basis in order to study systematically the impact of a particular reminder system.

Our findings indicated that the more salient signals, especially the vocal reminder, increased safety belt use. A reminder signal had maximum impact if it could be avoided by buckling up (Berry & Geller, 1991; Geller, 1988). General Motors applied some of our findings in an innovative safety-belt reminder system for its line of Saturn vehicles. All Saturns have an airbag, automatic shoulder belt, and a manual lap belt. The innovative safety-belt reminder in the Saturn cannot be masked by other vehicle start-up noises because it sounds six seconds *after* these noises have initiated. Most importantly, this activator will not occur if the driver buckles the lap belt within that six-second window. Thus, Saturn drivers can buckle up after turning their ignition switch and avoid the safety-belt reminder.

I don't believe anyone has systematically evaluated whether the Saturn activator for lap-belt use is more effective than the standard system. There is plenty of evidence, however, that the lap belt offers optimal protection from vehicle ejection and fatalities (Evans, 1991). Plus, field observations have revealed a decrease in lap-belt use in vehicles with automatic shoulder belts (Williams *et al,* 1989). If any safety-belt reminder system can increase the use of vehicle lap belts, it will be the Saturn activator because it is based on theory and procedures developed from behavioral science research.

You can see how understanding some basic principles from behavioral science can improve the design of simple activators like safety-belt reminders. Such knowledge could also improve or alter public policy, as illustrated in the next example.

RADAR DETECTORS. A radar detector is a very effective activator to reduce speeding. Why? Because it's consistent with the behavioral science principles discussed here. The sound of this activator is distinct, so the user will not confuse this signal with other vehicle sounds. More importantly, this activator is linked to a particular negative consequence. If a driver is speeding and ignores this activator, a speeding ticket is possible, even likely. Finally, this activator is voluntarily purchased by the drivers who tend to speed. The purchaser is receptive to the information this activator provides.

I find it quite disappointing that this activator is outlawed in my home state of Virginia. If policymakers understood some basics of behavior-based safety and if they truly wanted to reduce excessive speeding, they would not only allow radar detectors but they would encourage their dissemination. Drivers who like to speed would purchase them, and readily slow down following this activator's distinct signal. Traffic enforcement agencies could saturate risky areas (metropolitan loops or bypasses) with radar devices and occupy every fifth or tenth device. Speeders would never know which signal was "real" and would thus reduce their speeding to avoid a negative consequence.

A basic behavioral science principle, supported by substantial research (Chance, 1994; Kimble, 1961), is that consequences occurring on an intermittent basis are much more effective at maintaining behavior change than consequences occurring on a continuous basis or after every response. When consequences are improbable, as is currently the case for receiving a speeding ticket, they can lose their influence entirely. The increased use of radar detectors and the strategic placement of staffed and unstaffed radar devices would make consequences for speeding more salient and immediate for those who most need control on their at-risk driving. I hope readers will teach policymakers and police officers these basic principles about activators and consequences whenever they have the opportunity.

WARNING BEEPERS: A COMMON WORK EXAMPLE

Figure 9.4. Some signals we rely on lose impact over time.

Figure 9.4 illustrates quite clearly the phenomenon of habituation and reduced activator salience with experience. I'll bet you can reflect on personal experiences quite similar to the one shown here.

Not only has the brick mason habituated to the familiar "beep" of the backing vehicle, but the driver is illustrating danger compensation (or risk homeostasis) as I discussed in Chapter 6. He is not looking over his shoulder to check for a potential collision victim. He assumes the warning beeper is sufficient to activate coworkers' avoidance behavior and prevent injury. So this particular activator has actually reduced the driver's perceived risk. This influences his at-risk behavior of looking forward instead of turning his head to check his blindspots. A key point: Understanding the basic learning phenomenon of habituation can prevent over-reliance on activators and support a need to work more defensively.

Principle #3: Vary the Message

What does habituation tell us about the design of safety activators? Essentially, we need to vary the message. When an activator changes it can become more salient and noticeable. The "safety share" discussed in Chapter 7 follows this principle. When participants in a group meeting are asked to share something they've done for safety since the last meeting, the examples will vary considerably. Similarly, group discussions of near hits and potential corrective actions will also vary dramatically. The messages from safety shares and near-hit discussions will also be salient because they are personal, genuine, and distinct.

CHANGEABLE SIGNS

Over the years I've noticed a variety of techniques for changing the message on safety signs. There are removable slats to place different messages. I'm sure most of you have seen computer-generated signs with an infinite variety of safety messages. Some plants even have video screens in break rooms, lunch rooms, visitor lounges, and hallways that display many kinds of safety messages, conveniently controlled by user-friendly computer software.

Who determines the content of these messages? I know who should—the target audience for these signs. The same people expected to follow the specific behavioral advice should have as much input as possible in defining message content. Many organizations can get suggestions for safety messages just by asking. But if employees are not accustomed to giving safety suggestions, they might need a positive consequence to motivate their input.

WORKER-DESIGNED SAFETY SLOGANS

This is what I'm talking about: In 1985, employees and visitors driving into the main parking lot for Ford World Headquarters in Dearborn, Michigan, passed a series of four signs arranged with sequential messages, like the old Burma Shave signs. The messages were rotated periodically from a pool of 55 employee entries in a limerick contest for safety-belt promotion. My three favorites are illustrated in Figure 9.5.

Note that the last sign in each series of safety-belt promotion messages at Ford World Headquarters includes the name and department of the author. This public recognition, with the author's permission, of course, provides a positive consequence

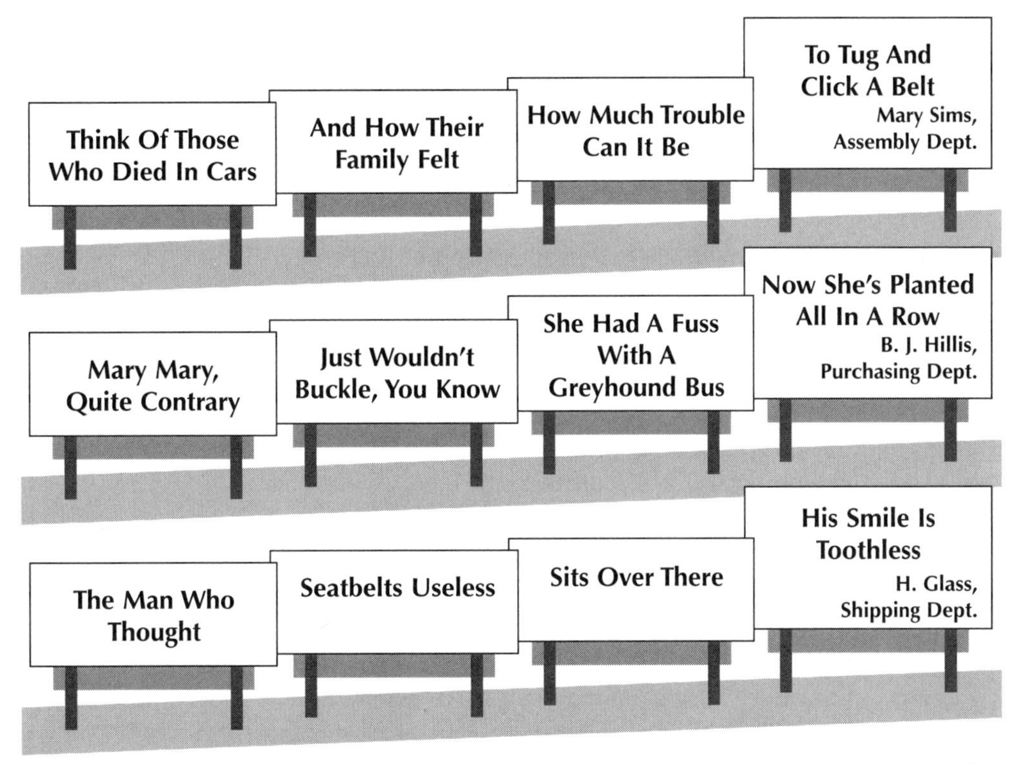

Figure 9.5. Sample buckle-up activators created by Ford employees and displayed at Ford World Headquarters in 1986.

or reward to the participant. It also reminds all sign viewers that many different people from various work areas are actively involved in safety. Through positive recognition and observational learning, including vicarious reinforcement, this simple technique promotes ownership and involvement in a safety process. This leads to the next principle: Involve the target audience.

Principle #4: Involve the Target Audience

This principle of activator design should be obvious by now. It's relevant for developing and implementing any behavior-change intervention. When people contribute to a safety effort, their ownership and commitment to safety increases. Of course, this principle works both ways. When individuals feel a greater sense of ownership and commitment, their involvement in safety achievement is more likely to continue. Thus, involvement feeds ownership and commitment, and vice versa.

Figure 9.6. Some activators imply ownership and increase caring behavior.

The simple activator in Figure 9.6 illustrates an ownership-involvement connection most of you can relate to. And it's a practical intervention strategy for many situations. The name plate in Figure 9.6 would not have to be as obtrusive in a real-world application to increase the perception of ownership. This would probably lead to a person taking greater care of the equipment, including more attention to safety-related matters.

This ownership-involvement principle is supported by litter control research that found much more littering and vandalism in public than private places (Ley & Cybriwsky, 1974; Newman, 1972). When public trash receptacles include the logos of nearby businesses, the merchants whose logos are displayed typically take care of the receptacle, and keep the surrounding area clean. This principle is also supported by the success of "Adopt-a-Highway" programs that have groups keep a certain roadway clear of litter and perhaps beautified with plants, shrubs, or flowers. Group ownership of a roadway typically leads to actively caring for its appearance.

I'd like to share three specific activator interventions that I have used to involve a target audience. All three made use of hand-held cards with safety messages.

SAFE BEHAVIOR PROMISE

First, in the mid-1980s, at a time when states did not have seat-belt laws and the use of vehicle safety belts in the United States was below 15 percent, companies which included General Motors; Ford; Corning Glass in Blacksburg, Virginia; Burroughs Welcome in Greenville, North Carolina; and the Reeves Brothers Curon

Plant in Cornelius, North Carolina, more than doubled the use of safety belts in company and private vehicles through "Buckle-Up Promise Cards" that employees were encouraged to sign (Geller & Lehman, 1991). Most of these cards were distributed after a lecture or group discussion about the value of using vehicle safety belts (Cope *et al,* 1986; Geller & Bigelow, 1984; Kello *et al,* 1988).

My students and I have also distributed "Buckle-Up Promise Cards" during church services (Talton, 1984), throughout a large university campus (Geller *et al,* 1989), and at the Norfolk Naval Base (Kalsher *et al,* 1989). In every case, a significant number of pledge-card signers increased their use of safety belts after their initial commitment behavior.

This simple activator approach also has had remarkable success in applications beyond safety-belt promotion. Streff *et al* (1994) found the technique successful at increasing the use of safety glasses. Dr. Richard Katzev and his colleagues at Reed College in Portland, Oregon, have used this activator to increase participation in community recycling programs (Katzev & Pardini, 1987-88; Wang & Katzev, 1990). Work teams at Logan Aluminum in Russellville, Kentucky, instituted a "Public Safety Declaration" that had employees sign a poster at the plant entrance specifying a safe-behavior commitment for the day—for example, "We wear hearing protection in all designated areas." Salience was maintained by changing the public commitment message weekly.

Figure 9.7 depicts a sample promise card for involving people in a commitment to perform a particular behavior. The target behavior to increase in frequency could be selected by a safety director, group leader, or through a group consensus discussion. This behavior is written on the promise card, perhaps by each individual in a group. Group members decide on the duration of the promise period, and write the end date on the card. Then each group member should be encouraged, not coerced,

Figure 9.7. A sample promise card for activating a behavioral commitment.

to sign and date their cards. I have found this group application of the safe behavior promise strengthens a sense of group cohesion or belongingness. Follow these procedural points for optimal results.

- Define the behavior specifically
- Involve the group in discussing the personal and group value of the behavior
- Make the commitment for a specified period of time that is challenging but not overwhelming
- Assure everyone that signing the card is only a personal commitment, not a company contract
- There should be no penalties (not even criticism) for breaking a promise
- Encourage everyone to sign the card, but do not use pressure tactics
- Each signer should keep the promise card in his or her possession, or post it in their work area as a reminder

The more involvement and personal choice solicited during the completion of this activator strategy, the better each individual feels about the process. Personal commitment to perform a specific behavior is activated as a result; and those involved in the process should feel obligated to fulfill the promise. Signing the card publicly in a group meeting also implicates social consequences to motivate compliance. That is, many participants will be motivated to keep their promise to avoid disapproval from a group member. Plus, when individuals keep their promise, recognition and approval from the group reinforces and supports maintenance of the targeted safe behavior.

"FLASH FOR LIFE"

The second activator intervention I want to relate dates back to 1984, when I developed the "Flash for Life." This is how it worked: A person displayed to vehicle occupants the front side of an 11″ × 14″ flash card that read, "Please Buckle-up—I Care." If a driver or passenger buckled up in response to the message, the "flasher" flipped the card over to display the bold words, "Thank You for Buckling Up." For the first evaluation of this behavior change intervention (Geller *et al*, 1985), the "flasher" was in the front seat of a stopped vehicle and the "flashee" was the driver of an adjacent, stopped vehicle. The flash card was shown to 1,087 unbuckled drivers, and of the 82 percent who looked at the card, 22 percent complied immediately with the buckle-up request.

My youngest daughter, Karly (3½ years old at the time), was the "flasher" for about 30 percent of the trials in the study (Fig. 9.8) On a few occasions we got a hand signal that was not used to indicate a right or left turn. Once Karly asked: "Daddy, what does that mean?" And I answered, "It means you're number one, honey—they're just using the wrong finger."

When hearing about this "Flash for Life" project, many of my colleagues expressed concern for my sanity. "Why do you waste your time?" some would say,

Figure 9.8. The "flashing" experience with my daughter in 1984. *Top,* Karly flashes a driver to buckle up while I look on. *Bottom,* Karly's activator works (the driver buckles up) so she flips over the flash card to give a consequence.

"Getting 22 percent to buckle up is not a big deal, and most of those who buckled up for your daughter only did it the one time. They probably won't buckle up the next day."

I had two answers to this sort of pessimism. First, achievement is built on "small wins." People need to break up big problems or challenges into small, achievable steps, and then work on each successive step, one at a time (Weick, 1984). We can't expect to solve a major safety problem like low use of personal protective equipment with one intervention technique, but we need to start somewhere. And if everyone contributed a "small win" for safety, the cumulative effects could be tremendous.

My second reply focused on the powerful influence of involvement. This intervention procedure enabled my young daughter to get involved in a safety project, even though she did not yet understand the concept of "safety." Every time she "flashed" another person to buckle up, her own commitment to practice the target behavior increased. I have never had to remind her to use her safety belt. Actually, she has reminded me to buckle up, and often monitors my driving speed. Now I never taught her about speed limits, but her early involvement in safety belt promotion generalized to caring about other safety-related behaviors. This was the real long-term benefit of involving Karly as a "Flash for Life" activator.

You see, people who actively care for safety by encouraging—or activating—others to practice safe behaviors strengthen their own personal safety commitment. When Karly was in fourth grade she won a speech contest for a talk on her "flashing" experiences years before. Her early involvement in safety led to this later role as a safety teacher, further strengthening her personal commitment to practice safe behaviors.

Bruce Thyer and colleagues (1987) demonstrated the benefits of another application of "Flash for Life" by posting college students at campus parking lot entrance/exit areas and asking them to "flash" vehicle occupants. Mean safety-belt use by vehicle drivers increased from 19.5 percent (n=629), during an initial one-week baseline, to 45.5 percent (n=635) during a subsequent week of daily flashing. The intervention was withdrawn during the third week, and average belt use decreased to 28.5 percent (n=634). When reinstating the intervention during the fourth week, the researchers observed a prominent increase in mean belt use to 51.5 percent (n=625).

A follow-up study (Berry *et al,* 1992) showed that this activator had a substantially greater impact when a person held the buckle-up sign, as opposed to the sign being attached to the stop sign by the exit. At a few industrial sites, notably the Hanford Nuclear site in Richland, Washington, employees have implemented this activator intervention in their parking lots. Vehicle occupants typically gave a smile or "thumbs-up" sign of approval when they saw their coworkers "flashing for safety." This rewarded the participants for their involvement, and increased the probability of their future participation in a safety project.

In another variation, Michael Roberts and his students (1990) disseminated to 10,000 school children vinyl folders with the "Flash for Life" messages on front and back. They observed children "flashing" throughout the community and *found higher rates of safety-belt use among children who received the flash card.* Again, this points out the power of involvement. I have personally distributed more than 3,000 "Flash for Life" cards nationwide, usually upon request by an individual who heard about the intervention procedure. In addition, a number of safety-belt groups in Ohio,

Tennessee, and Virginia have personalized the flash card for distribution and use throughout their states. I've heard numerous "small win" success stories from recipients of this "Flash for Life" activator.

THE AIRLINE LIFESAVER

And now, my third personal experience with an activator intervention, one that I've used since November, 1984, whenever boarding a commercial airplane (Geller, 1989). I hand the flight attendant a 3″ × 5″ Airline Lifesaver card like the one depicted in Figure 9.9. The card indicates that airlines have been the most effective pro-

Figure 9.9. The Airline Lifesaver Card.

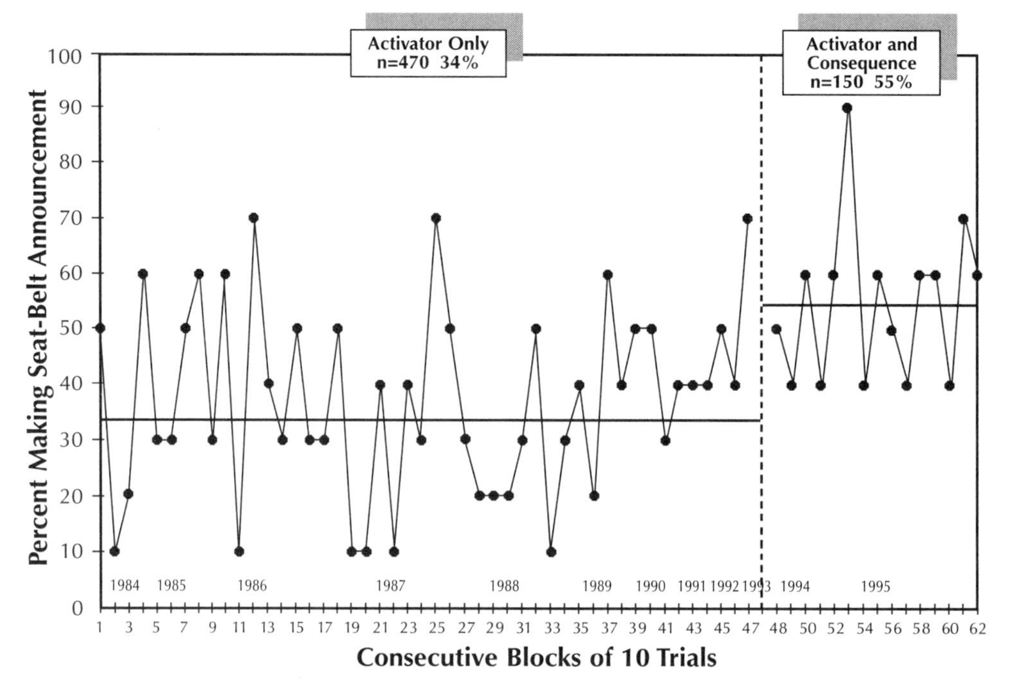

Figure 9.10. The percentage of compliance with the Airline Lifesaver announcement.

moters of seat-belt use, and requests that someone in the flight crew make an announcement near the flight's end to activate safety belt use in personal vehicles.

From November, 1984, to January, 1993, I distributed the Airline Lifesaver on 470 flights, and on 34 percent of these occasions the flight attendant gave a public buckle-up reminder. In the period from March, 1994, to February, 1995, I gave the Airline Lifesaver to 150 flight attendants and heard a buckle-up reminder on 55 percent of these flights. Figure 9.10 depicts a graph of the data collected over a decade of field observations.

Perhaps you're wondering why I separated the two time periods when reporting the results, and what could account for the significantly higher announcement percentages during the second time period? Well, I used different Airline Lifesaver cards during these periods. As shown in Figure 9.9, the cards distributed during the first phase merely requested the buckle-up announcement; cards used during the second phase offer prizes valued from $5 to $30 if the buckle-up reminder is given. Figure 9.11 illustrates the back of this new card. If the announcement is made, I give the attendant a postcard to mail to my office in order to redeem a reward. To date, of the attendants who received this reward opportunity, 65 percent stamped and mailed the postcard. And they received a prize.

As with the "Flash for Life" activator, many friends have laughed at the Airline Lifesaver, claiming I'm wasting my time. A common comment: "No one listens to the airline announcements anyway. Besides, do you really think that would be enough to motivate people to buckle up if they don't already?"

After you read the "buckle-up reminder," a passenger will give you a coupon redeemable for various prizes.

1 coupon ... $5.00*
2 coupons ... $15.00
3 coupons ... $30.00
(*Prize equivalent)

For more information call (703) 231-8145
Center for Applied Behavior Systems

Figure 9.11. The back of the new Airline Lifesaver card.

Ah, but consider this personal experience from the mid-1980s. I observed a woman approach the driver of an airport shuttle, asking her to "Please use your safety belt." The driver immediately buckled up. When I thanked the woman for making the buckle-up request, she replied that she normally wouldn't be so assertive but she had just heard a buckle-up reminder on her flight, "and if a stewardess can request safety-belt use, so can I."

Except for a few anecdotes like this one, it's impossible to assess the direct buckle-up influence of the Airline Lifesaver. But it's "safe" to assume that the beneficial, large-scale impact of this activator is a direct function of the number of individuals who deliver the reminder card to airline personnel. It's encouraging that several large corporations, including Ford Motor Company; Tennessee Valley Authority; and Air Products and Chemicals of Allentown, Pennsylvania, have distributed Airline Lifesaver cards to their employees for their own use during air travel. If the delivery of an Airline Lifesaver does not influence a single airline passenger to use a safety belt during ground transportation, at least the act of handing an Airline Lifesaver card to another person should increase the card deliverer's commitment to personal safety-belt use. Of course, the primary purpose of getting involved in a safety intervention is to prevent injury or improve a person's quality of life. Unfortunately, we rarely see these most important consequences. Thus, we need motivation: feedback, interpersonal approval, and self-talk. We tell ourselves that the safe behavior is "the right thing to do," and that someday an injury will be prevented. We can't count the number of injuries we prevent; we just need to "keep the faith."

On December 28, 1994, I received a special letter from Steven Boydston, assistant vice president of Alexander & Alexander of Texas, Inc., which helps me "keep the faith" that the Airline Lifesaver and other activator interventions for safety do make a difference. The motivating words in this letter are repeated in Figure 9.12. This success story is itself an activator for such proactive interventions as the Airline Lifesaver. It sure worked for me.

THANKS!!! On December 11, 1994 I was a passenger on Flt. 499 from Houston to San Francisco. At the end of the flight the pilot came on the speaker and said: "Now that the safest part of your journey is over and you are about to make the most dangerous part of your journey, please remember to use your seat belt...." I may have the words a little off, but I think you know the message. This was the second time I heard the message. The first was at the ASSE PDC in Dallas.

I am an obsessive seat belt user but when I got into the taxi, the seat belt was, as usual, buried in the seat. Usually when I find this in a cab I say, "the heck with it," but I can honestly say the pilot's message motivated me and I "dug out" the seat belt.

At over 70 mph the taxi hydroplaned and struck the guardrail. Thank God for the new barriers that prevent cars from being thrown back into traffic. Thank you and the pilot for the reminder. My wife and children are also grateful. I suffered a neck and shoulder injury—I think it is relatively minor. It could have been much worse. I should mention the driver was O. K. He was wearing his seat belt.

In safety it is seldom we can point to a particular event we can take credit for. This is one for you. Thanks again.

Figure 9.12. The encouraging letter I received from Steven Boydston on December 28, 1994.

Principle #5: Activate Close to Response Opportunity

Note that most of the effective activators discussed so far occurred at the time and place the target behavior should happen. The litter-control messages were on the fliers that needed disposal, the sign requesting lights to be turned off was below the light switch, the "Flash for Life" card was presented when people were in their vehicles and could readily buckle up, and I give airline attendants the Airline Lifesaver card when boarding the plane. Actually, I believe I would get more compliance with the request for a buckle-up announcement if I handed an attendant the announcement card at the end of the flight—closer to the opportunity to make the requested response. In fact, when I inquire about the lack of a buckle-up announcement while deplaning, some flight attendants tell me they forgot about the card.

POINT-OF-PURCHASE ACTIVATORS

In one study, my students and I systematically evaluated the impact of proximity between activator and response opportunity. We distributed handbills prompting the purchase of returnable drink containers at the entrance to a large grocery store or at the store location where drinks can be picked up for purchase (Geller *et al,* 1973; Geller *et al,* 1971). As predicted, customers purchased significantly more drinks in returnable than throwaway containers when prompted at the point of purchase. This "point-of-purchase advertising" is presumed to be an optimal form of product marketing (Tilman & Kirpatrick, 1972).

ACTIVATING WITH TELEVISION

You would think that product ad activators on television are less effective in directing behavior than promotions at store locations. Similarly, it's reasonable to predict that promoting vehicle safety belt use on television would be less effective than presenting buckle-up activators at road locations, as exemplified by the Flash-for-Life intervention. This assumption is supported by the classic and rigorous evaluation of safety-belt promotion in public service announcements on television by Robertson *et al* (1974).

In this study, six different safety belt messages were shown during the day and during prime time on one cable of a dual-cable television system. Residents in Cable System A (6,400 homes) received the safety belt messages 943 times over a nine-month period. Each viewer was exposed to the messages two to three times per week. The control residents in Cable System B (7,400 homes) did not receive any messages. In addition, for one month before and the nine months during the television activators, the use of safety belts by vehicle drivers was observed in a systematic rotating schedule from 14 different sites within the community. Vehicle license plate numbers were recorded and later matched with each owner's name and address from the files of the state Department of Motor Vehicles. The television viewers did not know they were in an experiment, and the field observers could not know the experimental condition of a particular vehicle observation.

Overall mean safety belt use among drivers was 8.4 percent for males and 11.3 percent for females for the intervention group, and 8.2 percent and 10.3 percent for the control group. It's easy to conclude that television public service announcements have no effect on whether a person buckles up (Robertson, 1976).

But consider that four of the six different television spots were based on a fear tactic, highlighting such negative consequences as disfigurement and disability. Research suggests that a fear-arousing approach is usually not desirable for safety messages (Leventhal *et al*, 1983; Winett, 1986). Anxiety elicited by a vivid portrayal of the disfiguring consequence of a vehicle crash can interfere with the viewer's attention and retention (Lazarus, 1980). It can cause viewers to "tune out" subsequent spots as soon as they appear (Geller, 1989). Consequently, many viewers may have missed the end of these public service activators, which demonstrated the problem's solution: using safety belts.

At least part of the ineffectiveness of activating with television is due to lack of proximity between the specific response message and the later opportunity to perform the target behavior. But this must be balanced with the great amount of exposure enabled by television. A one percent effect of a television ad could translate to thousands using their vehicle safety belt. It's also likely the naturalistic use of safety belts during actual television episodes, as discussed in Chapter 7, would have greater impact than a commercial activator or public service announcement (Geller, 1989; McGuire, 1984; Robertson, 1983). Still, if communities and corporations activated safety at the time and place for the desired behavior, the overall impact could be far greater than a television ad. And the cost could be minimal, as illustrated by the following field study.

BUCKLE-UP ROAD SIGNS

Over a two-year period my students evaluated the behavioral impact of buckle-up activators located along the road in my hometown of Newport—a small rural

Figure 9.13. The feedback sign located near the intersection of Highway 460 and Route 42 into Newport, Virginia.

community in southwest Virginia. We started collecting baseline data in March, 1993, by unobtrusively observing and recording the safety belt use of vehicle drivers and passengers from a parked vehicle near the intersection of a four-lane highway (Highway 460) and the two-lane road (Route 42) leading into Newport. Observations were taken of vehicles entering or leaving Route 42 to Newport, as well as vehicles continuing on Highway 460, during most weekdays from approximately 4:00 to 6:30 p.m., when the Newport traffic was heaviest.

After 13 weeks of baseline observation, the sign shown in Figure 9.13 was positioned approximately 7 feet from Route 42 and 300 feet from the intersection of Route 42 and Highway 460. The sign was 8′ × 4′ and the buckle-up message shown in Figure 9.13 was painted on both sides in black 8″ high letters against a white background. The sign could not be seen by occupants of vehicles continuing along Highway 460.

The identical message was posted on a 3′ × 6′ sign in front of the Newport Community Center, located about half a mile from the sign shown in Figure 9.13. The 5″ letters were black and removable. Every Monday the percentages were changed to reflect mean safety-belt use for males and females during the prior week.

Vehicle observations continued for 24 weeks, then the feedback sign was removed. After 21 weeks of observation during this withdrawal condition, the signs were reinstated, but with a different message. We wanted to see if safety-belt use could be activated with a sign that did not need to be changed weekly to reflect belt-use feedback. The new message was, "We Buckle Up in Newport to Set An Example for Our Children." Figure 9.14 shows how this message was displayed at the Newport Community Center. Note that this message specifies an actively caring (example setting) consequence for the requested behavior. We continued to record safety belt use

Figure 9.14. The second safety belt activator, as displayed at the Newport Recreation Center.

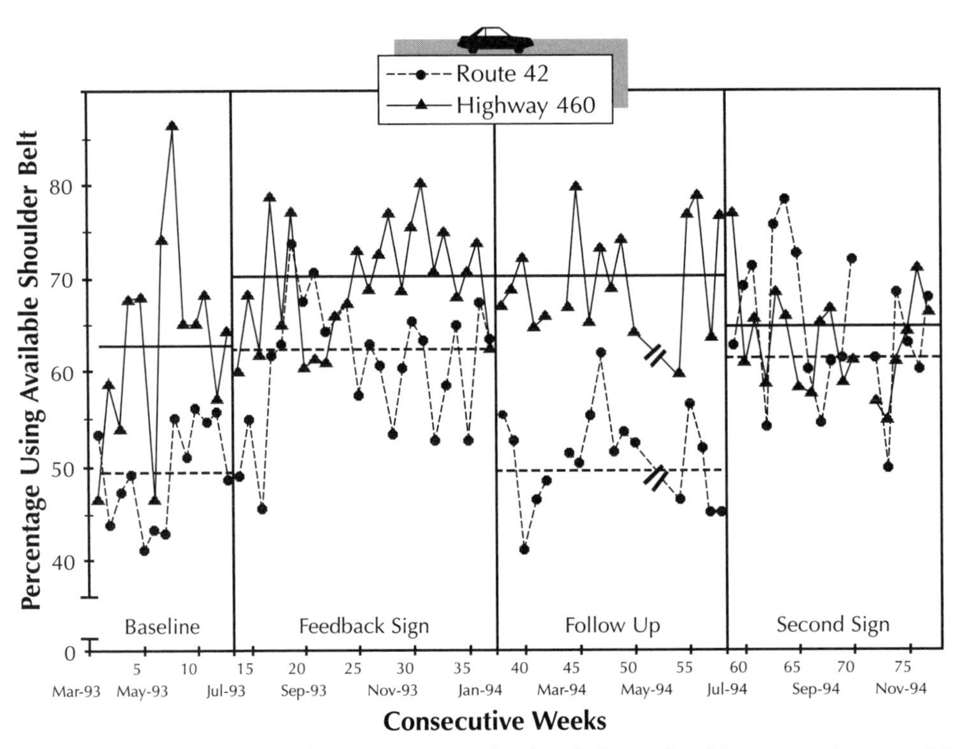

Figure 9.15. The mean weekly percentages of safety belt use for drivers entering or exiting Newport (Route 42) versus those continuing on Highway 460.

for 19 more weeks while this new buckle-up activator was in place at both community sign locations.

The results of our long-term field observations are depicted in Figure 9.15. The weekly percentages of drivers' safety belt use are graphed over the 77 weeks of the project. Percentages were calculated separately for vehicles entering or exiting Newport (the intervention group) versus those continuing on Highway 460 (the control group). Results show quite clearly that both signs increased safety belt use substantially. While mean safety-belt use in vehicles traveling on Highway 460 remained relatively stable, the mean safety belt use in vehicles entering or exiting the road on which the signs were placed fluctuated systematically with placement and removal of the buckle-up activators. The horizontal lines through the data points of the graph in Figure 9.15 depict mean driver safety belt use per phase and condition.

The overall impact of these activators was impressive, and suggests that large-scale increases in safety belt use would occur if communities and companies nationwide implemented this simple activator intervention. Other researchers have shown impressive effects of feedback signs to increase safety belt use at an industrial site (Grant, 1990) and to reduce vehicle speeds at various community locations (Van Houten & Nau, 1983; Ragnarsson & Björgvinsson, 1991). However, none of the other researchers evaluated sign effects for as long a period as our study. We showed relatively long-term benefits of the activator intervention with little habituation effects. In addition, our findings suggest that an activator message referring to actively caring

Figure 9.16. The second sign, thrown in my pond by vandals.

consequences can be as effective as a feedback sign that requires more effort to implement due to the need to collect behavioral data and post weekly feedback.

Finally, it's instructive to note that our second activator intervention ended abruptly when vandals carried the 70-pound sign about 100 yards and threw it in my pond, as shown in Figure 9.16. Consider this: We did not solicit community approval or involvement when developing or implementing this intervention. We just built the signs and put them in place. It's possible, even likely, that this apparent one-person decision to post a community sign irritated some residents. A few were so outraged that they reacted with countercontrol (Skinner, 1974), perhaps to gain a sense of freedom from our obvious attempt to control their behavior (Brehm, 1966).

We will return to this issue again in Chapter 14 when I discuss ways to increase perceptions of personal control and empowerment to boost involvement in efforts to achieve a Total Safety Culture.

Principle #6: Implicate Consequences

Field research shows the effective influence of salient activators implemented in close proximity to an opportunity to perform the specified target behavior. It's important to realize, though, that the target behaviors were all relatively convenient to perform. We're talking about depositing handbills in a particular receptacle, choosing certain products, using available safety glasses and safety belts. There is plenty of evidence that activators alone won't succeed when target behaviors require more than a little effort or inconvenience. My students and I, for example, could not activate water conservation behaviors (Geller *et al,* 1983) or the collection and delivery of recyclable newspapers (Geller *et al,* 1975; Witmer & Geller, 1976) without adding rewarding consequences. The same was true for a number of attempts to promote various energy conservation behaviors that involved more effort than flicking a light switch (Hayes & Cone, 1977; Heberlein, 1975; Palmer *et al,* 1978).

Many of the activator strategies illustrated in this chapter were explicitly or implicitly connected to consequences. Signing a promise card or public declaration, for example, implicates social approval versus disapproval for honoring versus disavowing a commitment. Consequences motivated employees to create safety slogans, and the most influential activators usually made reference to consequences. I have received more compliance with the Airline Lifesaver since offering rewards for making the buckle-up announcement. And the "Flash for Life" included a "Thank You" consequence if the "flashee" buckled up.

Vehicle buzzers designed to promote safety-belt use were improved by implicating consequences. When drivers of the Saturn buckle up within six seconds of turning the ignition key, they avoid receiving the audible reminder. Avoiding an annoying stimulus is a consequence that might motivate some people to buckle up. In a similar vein, the salient beep of a radar detector effectively motivates reduced vehicle speeds because it enables drivers to avoid a negative consequence—an encounter with a police officer.

INCENTIVES VERSUS DISINCENTIVES

Activators that signal the availability of a consequence are either incentives or disincentives. An *incentive* announces to an individual or group, in written or oral form, the availability of a reward. This pleasant consequence follows the occurrence of a certain behavior or an outcome of one or more behaviors. In contrast, a *disincentive* is an activator announcing or signaling the possibility of receiving a penalty. This unpleasant consequence is contingent on the occurrence of a particular undesirable behavior.

Research has shown quite convincingly that the impact of a legal mandate, for example, drunk driving or safety-belt use laws, varies directly with the amount of media promotion or disincentive (Ross, 1982). Similarly, the success of an incentive program depends on making the target population aware of the possible rewards. In other words, marketing positive or negative consequences with activators (incentives or disincentives) is critical for the motivating success of the consequence intervention.

The next chapter discusses how to design and apply consequences to motivate behavior. At this point, it's important to understand that *the power of an activator to motivate behavior depends on the consequence it signals.* Figure 9.17 illustrates this connection between activator and consequence. If a sign like the one shown in Figure 9.17 motivated a driver to attempt safer driving practices, it would work due to the potential consequences implied by the activator. Every time the driver got into the vehicle, she would be reminded of potential consequences for certain driving practices. Incidentally, do you perceive the sign on the vehicle in Figure 9.17 as an incentive or a disincentive?

Figure 9.17. The most powerful activators imply immediate consequences.

My guess is you perceived it as a *disincentive.* I'll bet you saw the sign as a threat to reduce at-risk driving, rather than an incentive to encourage safe driving. You wouldn't expect dad to get a phone call commending his daughter's driving. If any phone calls were made, they would be to criticize at-risk or discourteous driving. This is exactly how my daughter, Krista, perceived the activator in Figure 9.17. As the illustration suggests, she flatly refused to drive with such a sign on our car.

Much to Krista's chagrin, I actually painted and mounted the sign, but the only benefit from that effort was my daughter's increased motivation to work with the critical behavior checklist described in Chapter 8. "Let's lose this sign, dad," she asserted, "and focus on giving each other feedback with the checklist." I was happy to comply. This exercise simply reminded me that we're socialized to expect more negative consequences for our mistakes than positive consequences for our successes. The next chapter explores this unfortunate reality in greater detail.

SETTING GOALS FOR CONSEQUENCES

Let's talk about safety goals in the context of activators that imply consequences. Included among Dr. W. Edwards Deming's 14 points for quality transformation are "eliminate slogans, exhortations, and targets for the work force...eliminate work standards...management by objectives, and management by the numbers" (Deming, 1985, p 10.). Does this mean we should stop setting safety objectives and goals? Should we stop trying to activate safe behaviors with signs, slogans, and goal statements? Does this mean we should stop counting OSHA recordables and lost-time cases, and stop holding people accountable for their work injuries?

Answers to all of these questions are "yes," if you take Deming's points literally. However, my evaluation of Dr. Deming's scholarship and workshop presentations, and my personal communications with Dr. Deming in 1990 and 1991, have led me to believe that Deming meant we should eliminate goal setting, slogans, and work targets as they are currently implemented. Dr. Deming was not criticizing appropriate use of goal setting, management by objectives, and activators; rather he was lamenting the frequent incorrect use of these activator interventions. Substantial research evidence supports the use of objective goals and activators to improve behaviors if these behavior-change interventions are applied correctly (Latham & Yukl, 1975; Locke & Latham, 1990).

INCORRECT GOALS. Setting zero injuries as a safety goal (as illustrated in Figure 9.1) is a misuse of these principles and should in fact be eliminated. Holding people accountable for numbers or outcomes they do not believe they can control is a sure way to produce negative stress or distress. Some people won't be distressed because they won't take these outcome goals seriously. Experience has convinced them they cannot control the numbers, so they simply ignore the goal-setting exhortations. These individuals overcome the distress of unrealistic management objectives or goals by developing a sense, a perspective or attitude, of helplessness.

What does a goal of zero injuries mean anyway? Is this goal reached when no work injuries are recorded for a day, a month, six months, or a year? Does a work injury indicate failure to reach the goal for a month, six months, or a year? Does the average worker believe he or she can influence goal attainment, beyond simply avoiding personal injury?

SET SMART GOALS. I remember the techniques for setting effective goals with the acronym SMART, as illustrated in Figure 9.18. SMART goal setting defines what will happen when the goal is reached (the consequences), and tracks progress toward achieving the goal. Rewarding feedback from completing intermediate steps toward the ultimate goal is a consequence that motivates continued progress. Of course, it's critical that the people asked to work toward the goal "buy in" or believe in the goal, and believe they have the skills and resources to achieve it.

I recently talked with a group of hourly workers about setting safety goals, and each member of this safety steering committee was completely turned off to goal setting. Their past experiences with corporate safety goals created a barrier to listening about SMART goals. It was necessary for them to understand the difference between the right and wrong way to set safety goals, and to practice with feedback consequences SMART goal setting for safety.

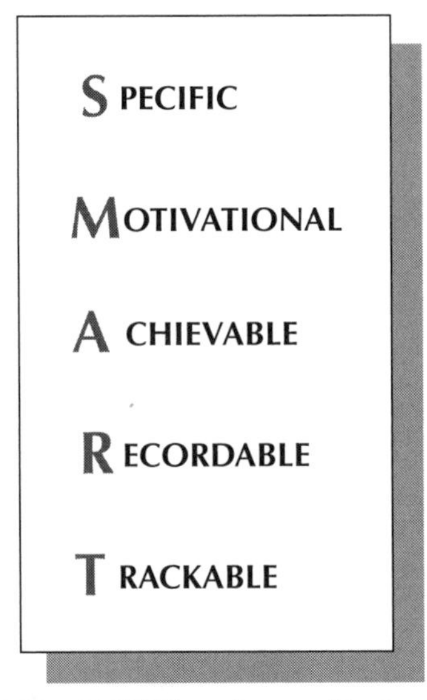

Figure 9.18. SMART goals are effective activators.

I've seen several corporate mission statements with the safety goal of zero injuries. As I've indicated earlier, this is obviously an example of incorrect goal setting. It's easy enough to track injuries, but employees' daily experiences lead them to believe that injuries are beyond their direct control. For one thing, injuries usually happen to someone else—what can they do about that? One injury in the workplace, perhaps resulting from another person's carelessness, ruins the goal of zero injuries. This leads to a perception of failure.

No one likes to feel like a failure. So people typically avoid situations where failure is frequent or prominent, or they at least attempt to discount their own possible contribution to the failure by blaming factors beyond their own control (as discussed in Chapter 6). This fosters the belief that injuries are beyond personal control, and creates the sense that safety goal setting is a waste of time. So do I make my feelings clear? "Zero injuries" should not be specified as a safety goal. Instead, zero injuries should be the aim or purpose of a safety mission—a mission that depends on various safety processes motivated in part by SMART goals.

FOCUS ON THE PROCESS. Safety goals should focus on process activities that can contribute to injury prevention. Workers need to discuss what they can do to reduce injuries, from reporting and investigating near hits to conducting safety audits of environmental conditions and work practices. The safety steering committee I mentioned earlier wanted to increase daily interpersonal communications regarding safety. They set a goal for their group to achieve 500 safety communications within the following month. To do this, they had to develop a system for tracking and recording "safety talk." They designed a wallet-sized "SMART Card" for recording their interactions with others about safety. One member of the group volunteered to tally and graph the daily card totals.

Another work group I consulted with recently set a goal of 300 behavioral observations of lifting. Employees had agreed to observe each other's lifting behaviors according to a critical behavior checklist they had developed. If each worker completed an average of one lifting observation per day, the group would reach their goal within the month. Each of these work groups reached their safety goals within the expected time period, and as a result they celebrated their "small win" at a group meeting, one with pizza and another with jelly-filled donuts. Perhaps subsequent group goals should focus on healthy diet choices!

These two examples illustrate the use of SMART goals, and depict safety as process-focused and achievement-oriented, rather than the standard and less effective outcome-focused and failure-oriented approach promoted by injury-based goals. More importantly, these goals were employee driven. Workers were motivated to initiate the safety process because it was their idea. They got involved in the process and owned it. And they stayed motivated because the SMART goals were like a roadmap telling them where they were going, when they would get there, and how to follow their progress along the way. This is much more effective than the typical top-down method for setting goals.

In Conclusion

In this chapter I have presented examples of intervention techniques called activators. They occur before desired or undesired behavior to direct potential performers. Based on rigorous behavioral science research and backed by real-world examples, six principles for maximizing effective activators were given:

- Specify behavior
- Maintain salience with novelty
- Vary the message
- Involve the target audience
- Activate close to response opportunity
- Implicate consequences

We're constantly bombarded with activators. At home we get telephone solicitations, junk mail, television commercials, and verbal requests from family members. At work it's phone mail, e-mail, memos, policy pronouncements, and verbal directions from supervisors and coworkers. On the road there's no escape from billboards, traffic signals, vehicle displays, radio ads, and verbal communication from people inside and outside our vehicles. As discussed in Chapter 5, we selectively attend to some of these activators, ignoring others. Only a portion of the activators we perceive actually influences our behavior. Understanding the six principles discussed in this chapter can help you predict which ones will influence behavior change.

Obviously, we don't need more activators in our lives. We certainly do need more effective activators to promote safety and health. It would be far better to make a few safety activators more powerful than to add more activators to a system already overloaded with information. We need to plan our safety activators carefully so the right safety directives receive the attention and ultimate action they deserve.

If you want an activator to motivate action, you need to imply consequences. The most powerful activators make the observer aware of consequences available following the performance of a target behavior. Consequences can be positive or negative, intrinsic or extrinsic to the task, and internal or external to the person. The next chapter will explain the preceding sentence, which is key to getting the most beneficial behavior from an intervention process.

REFERENCES

Berry, T. D., & Geller, E. S. (1991). A single-subject approach to evaluating vehicle safety belt reminders: Back to basics. *Journal of Applied Behavior Analysis*, 24, 13-22.

Berry, T. D., Geller, E. S., Calef, R. S., & Calef, R. A. (1992). Moderating effects of social assistance on verbal interventions to promote safety belt-use: An analysis of weak plys. *Environment and Behavior*, 24, 653-669.

Brehm, J. W. (1966). *A theory of psychological reactance*. New York: Academic Press.

Carlson, N. R. (1993). *Psychology: The science of behavior* (Fourth Edition). Needham Heights, MA: Allyn and Bacon.

Chance, P. (1994). *Learning and behavior* (Third Edition). Pacific Grove, CA: Brooks/Cole Publishing Company.

Cope, J. G., Grossnickle, W. F., & Geller, E. S. (1986). An evaluation of three corporate strategies for safety-belt use promotion. *Accident Analysis and Prevention,* 18, 243-251.

Delprata, D. J. (1977). Prompting electrical energy conservation in commercial users. *Environment and Behavior,* 9, 433-440.

Deming, W. E. (1985). Transformation of western style of management. *Interfaces,* 15(3), 6-11.

Evans, L. (1991). *Traffic safety and the driver.* New York: Van Nostrand Reinhold.

Geller, E. S. (1975). Increasing desired waste disposals with instructions. *Man-Environment Systems,* 5, 125-128.

———. (1988). A behavioral science approach to transportation safety. *Bulletin of the New York Academy of Medicine,* 65, 632-661.

———. (1989). The Airline Lifesaver: In pursuit of small wins. *Journal of Applied Behavior Analysis,* 22, 333-335.

———. (1989). Using television to promote safety belt use. In R. E. Rice & C. K. Atkin (Eds.). *Public communication campaigns* (Second Edition). Newbury Park, CA: SAGE Publications, Inc.

Geller, E. S., & Bigelow, B. E. (1984). Development of corporate incentive programs for motivating safety belt use: A review. *Traffic Safety Evaluation Research Review,* 3, 21-38.

Geller, E. S., Bruff, C. D., & Nimmer, J. G. (1985). "Flash for Life": Community-based prompting for safety belt promotion. *Journal of Applied Behavior Analysis,* 18, 145-149.

Geller, E. S., Chaffee, J. L., & Ingram, R. E. (1975). Promoting paper-recycling on a university campus. *Journal of Environmental Systems,* 5, 39-57.

Geller, E. S., Erickson, J. B., & Buttram, B. A. (1983). Attempts to promote residential water conservation with educational, behavioral, and engineering strategies. *Population and Environment,* 6, 96-112.

Geller, E. S., Farris, J. C., & Post, D. S. (1973). Promoting a consumer behavior for pollution control. *Journal of Applied Behavior Analysis,* 6, 367-376.

Geller, E. S., Kalsher, M. J., Rudd, J. R., & Lehman, G. R. (1989). Promoting safety-belt use on a university campus: An integration of commitment and incentive strategies. *Journal of Applied Social Psychology,* 19, 3-19.

Geller, E. S., & Lehman, G. R. (1991). The buckle-up promise card: A versatile intervention for large-scale behavior change. *Journal of Applied Behavior Analysis,* 24, 91-94.

Geller, E. S., Witmer, J. F., & Orebach, A. L. (1976). Instructions as a determinant of paper-disposal behaviors. *Environment and Behavior,* 8, 417-438.

Geller, E. S., Witmer, J. F., & Tuso, M. E. (1977). Environmental interventions for litter control. *Journal of Applied Psychology,* 62, 344-351.

Geller, E. S., Wylie, R. C., & Farris, J. C. (1971). *An attempt at applying prompting and reinforcement toward pollution control.* Proceedings of the 79th Annual Convention of the American Psychological Association, 6, 701-702.

Grant, B. A. (1990). Effectiveness of feedback and education in an employment-based seat belt program. *Health Education Research,* 5 (2), 197-205.

Hayes, S. C., & Cone, J. D. (1977). Reducing residential electrical use: Payments, information, and feedback. *Journal of Applied Behavior Analysis,* 14, 81-88.

Heberlein, T. A. (1975). Conservation information: The energy crisis and electricity consumption in an apartment complex. *Energy Systems and Policy,* 1, 105-117.

Kalsher, M. J., Geller, E. S., Clarke, S. W., & Lehman, G. R. (1989). Safety-belt promotion on a naval base: A comparison of incentives vs. disincentives. *Journal of Safety Research,* 20, 103-113.

Katzev, R., & Pardini, A. (1987-1988). The impact of commitment and token reinforcement procedures in promoting and maintaining recycling behavior. *Journal of Environmental Systems,* 17, 93-133.

Kello, J. E., Geller, E. S., Rice, J. C., & Bryant, S. L. (1988). Motivating auto safety-belt wearing in industrial settings: From awareness to behavior change. *Journal of Organizational Behavior Management,* 9, 7-21.

Kimble, G. A. (1961). *Hilgard and Marquis' conditioning and learning.* New York: Appleton-Century-Crofts.

Latham, G., & Yukl, G. (1975). A review of research on the application of goal-setting in organizations. *Academy of Management Journal,* 18, 824-845.

Lazarus, R. (1980). The stress and coping paradigms. In C. Eisdorfer, D. Cohen, A. Klienmen, & P. Maxim (Eds.), *Theoretical bases for psychopathology*. New York: Spectrum.

Ley, D., & Cybriwsky, R. (1974). Urban graffiti as territorial markers. *Annals of the Association of American Geographers*, 64, 491-505.

Leventhal, H., Shafer, M., & Panagis, D. (1983). The impact of communications on the self-regulation of health beliefs, decision, and behavior. *Health Education Quarterly*, 10, 3-29.

Locke, E., & Latham, G. (1990). *A theory of goal setting and task performance*. New Jersey: Prentice-Hall.

McGuire, W. J. (1984). Public communication as a strategy for inducing health promoting behavioral change. *Preventive Medicine*, 13, 289-319.

Newman, O. (1972). *Defensible space*. New York: Macmillan.

Palmer, M. H., Lloyd, M. E., & Lloyd, K. E. (1978). An experimental analysis of electricity conservation procedures. *Journal of Applied Behavior Analysis*, 10, 665-672.

Ragnarsson, R. S., & Björgvinsson, T. (1991). Effects of public posting on driving speed in Icelandic traffic. *Journal of Applied Behavior Analysis*, 24, 53-58.

Roberts, M. C., Alexander, K., & Knapp, L. (1990). Motivating children to use safety belts: A program combining rewards and "flash for life." *Journal of Community Psychology*, 18, 110-119.

Robertson, L. (1983). *Injuries: Causes, control strategies, and public policy*. Lexington, MA: Lexington Books.

———. (1976). The great seat belt campaign flop. *Journal of Communication*, 26, 41-46.

Robertson, L., Kelley, A., O'Neill, B., Wixom, C., Eisworth, R., & Haddon, W., Jr. (1974). A controlled study of the effect of television messages on safety belt use. *American Journal of Public Health*, 64, 1071-1080.

Ross, H. L. (1982). *Deterring the drinking driver: Legal policy and social control*. Lexington, MA: Lexington Books.

Skinner, B. F. (1974). *About behaviorism*. New York: Alfred A. Knoff.

Talton, A. (1984). *Increasing safety-belt usage through personal commitment: A church-based pledge card program*. Unpublished master's thesis. Blacksburg, VA: Virginia Polytechnic Institute and State University.

Thyer, B. A., & Geller, E. S. (1987). The "buckle-up" dashboard sticker: An effective environmental intervention for safety belt promotion. *Environment and Behavior*, 19, 484-494.

Thyer, B. A., Geller, E. S., Williams, M., & Purcell, S. (1987). Community-based "flashing" to increase safety belt use. *The Journal of Experimental Education*, 53, 155-159.

Tillman, R., & Kirkpatrick, C. A. (1972). *Promotion: Persuasive communication in marketing*. Homewood, IL: Richard D. Irwin, Inc.

Van Houten, R., & Nau, P. A. (1983). Feedback interventions and driving speed: A parametric and comparative analysis. *Journal of Applied Behavior Analysis*, 16, 253-281.

von Buseck, C. R., & Geller, E. S. (1984). *The vehicle safety belt reminder: Can refinements increase safety belt use?* Technical Research Report for General Motors Research Laboratories, Warren, MI.

Wang, T. H., & Katzev, R. (1990). Group commitment and resource conservation: Two field experiments on promoting recycling. *Journal of Applied Social Psychology*, 20, 265-275.

Weick, K. E. (1984). Small wins: Redefining the scale of social problems. *American Psychologist*, 39(1), 40-49.

Williams, A. F., Wells, J. D., Lund, A. K., & Teed, N. (1989). Observed use of seat belts in 1987 cars. *Accident Analysis and Prevention*, 19, 243-249.

Winett, R. A. (1986). *Information and behavior: Systems of influence*. Hillsdale, NJ: Erlbaum.

Winett, R. A. (1978). Prompting turning-out lights in unoccupied rooms. *Journal of Environmental Systems*, 6, 237-241.

Witmer, J. F., & Geller, E. S. (1976). Facilitating paper recycling: Effects of prompts, raffles, and contests. *Journal of Applied Behavior Analysis*, 9, 315-322.

Chapter 10

Intervening with Consequences

Consequences motivate behavior and related attitudes. There are various ways this happens. Consequences can be positive or negative, intrinsic (natural) or extrinsic (extra) to a task, and internal or external to a person. These characteristics need to be considered when designing and evaluating intervention programs. This chapter explains why, and provides principles and practical procedures for motivating people to work safely over the long term. I'll show you how to influence behavior and attitudes so that both are consistent with a Total Safety Culture.

People do what they do because of what happens to them when they do it.

This quote from Aubrey Daniels (1994, p. 25) represents a key principle of human motivation and behavior-based safety. Although supported by substantial research (Skinner, 1938), it actually runs counter to common sense.

Think about it: When people ask us why we did something, we're apt to say, "I wanted to do it," or "I was told to do it," or "I needed to do it." These explanations sound as if the cause of our behavior comes before we act. This perspective is supported by numerous "pop psychology" self-help books and audiotapes that say people motivate themselves with positive self-affirmations or optimistic thinking and enthusiastic intentions. In other words, behavior is caused by some external request, order, or signal; or by an internal force, drive, desire, or need.

Pop psychology often asserts that people cannot be motivated by others, only by themselves from within. This self-motivation is typically referred to as "intrinsic motivation," and is a prominent theme in recent popular books by W. Edwards Deming (1993), Steven R. Covey (1989), and Alfie Kohn (1993). It's also the theme of the classic best seller by Norman Vincent Peale (1952).

Indeed, Kohn reiterates throughout his book that any attempt to motivate people with extrinsic procedures—incentives, praise, recognition, grades, and penalties—will *detract* from intrinsic motivation and do more harm than good. Kohn concludes that interventions set up to motivate others, even achievement-oriented reward and recognition programs, are generally perceived as "controlling" and thus decrease "intrinsic" or self-motivation.

Fortunately there is much solid research in behavioral science to discredit Kohn's assertions (see, for example, reviews by Carr et al, 1995; and Flora, 1990). I say fortunately, because if all reward and recognition programs detracted from our intrinsic (or internal) motivation to perform in certain ways, many industry, school, and community motivational programs would be futile.

This chapter will explain the fallacy in Kohn's argument and show ways to maximize the impact of an extrinsic reward process. Again, the research-supported principle is that activators direct behavior, and consequences motivate behavior. But the type of consequence certainly influences the amount of motivation, as this chapter will explain.

Drs. Deming and Covey want people to act out of the knowledge that it's the right thing to do. Covey (1990) refers to this motivation as *principle-centered.* I certainly agree with the need for inner-directed, self-motivated behavior. When people consistently go out of their way for the safety of themselves and others, they are principle-centered. They have reached our ultimate purpose in safety. They hold safety as a value.

It seems Deming, Covey, Kohn, and others who have written about human motivation presume that people are already *principle-centered* for various activities, including safety. They give advice from the perspective that people are "willing workers," self-motivated to do the right thing. Throughout this book I've made the case that natural consequences often motivate people to do the wrong thing, like take risks, when it comes to safety. This is the basic distinction between the person-based, or humanistic, approach to safety and behavior-based safety.

Most people do not consistently avoid at-risk behavior. This calls for behavior-based safety (including the use of consequences) to bring people to the principle-centered, self-motivated stage. Recall the principle I've emphasized several times—people act themselves into new ways of thinking. In other words, people become principle-centered and self-directed through their routine actions. Behavior-based intervention is needed to make safe behavior the routine. Then principle-centered or value-based safety eventually follows.

The Power of Consequences

Popular author and humorist Robert Fulghum (1988) wrote *All I Really Need to Know I Learned in Kindergarten,* claiming he learned all the basic rules or norms for socially acceptable adult behavior as a young child. The list of rules in Figure 10.1 was excerpted from Mr. Fulghum's famous book. Rules like *share everything, play fair,*

- ❑ Share everything
- ❑ Play fair
- ❑ Don't hit others
- ❑ Put things back where you found them
- ❑ Clean up your own mess
- ❑ Don't take things that aren't yours
- ❑ Say you're sorry when you hurt someone
- ❑ Wash your hands before you eat
- ❑ Flush
- ❑ When you go out into the world, watch out for traffic, hold hands, and stick together
- ❑ And remember the Dick-and-Jane books and the first word you learned—the biggest word of all: **LOOK**

Figure 10.1. Basic rules of social life we learned well as children, but do not necessarily follow as adults. (Excerpted from Fulghum, 1988)

don't fight, and *clean up your own mess* were taught to most of us early on. These are clearly ideal edicts to live by. Perhaps you still recall a teacher or parent using these rules to try to shape your behavior. Did it work? Do you follow each of these basic norms regularly, for no other reason or consequence except your realization that it's the right thing to do?

Imagine what a better world we'd live in if everyone followed the simple rules listed in Figure 10.1 from a self-directed, principle-focused perspective. Alas, there are signs everywhere that this isn't so. Take the automatic flushers in public facilities like airports. They indicate that we've lost the ability to follow the simplest of these rules—"flush." And flushing the toilet is followed by a natural consequence that should increase future occurrences of this effortless response. Frankly, I like to control my own "flush," thank you, and I did not appreciate engineers taking that opportunity for personal control away from me.

The last two kindergarten rules in Figure 10.1 are directly relevant to safety, and in fact reflect basic themes of this text. As discussed previously, especially in Chapter 9, safety requires people to stick together in a spirit of shared belonging and interdependence. But sometimes we need activators to remind us of this critical rule, and consequences to keep us working together for safety.

"LOOK," Fulghum's last rule, is key to behavior-based safety and to achieving a Total Safety Culture. This implies the "defensive working style" employees need to adopt. In a Total Safety Culture, everyone looks for ways to improve safety by intervening to reduce at-risk behaviors and increase safe behaviors. In Chapter 9 we discussed ways to intervene with activators. Here we focus on the more powerful intervention approach—manipulating consequences.

CONSEQUENCES IN SCHOOL

Figure 10.2 reveals the power of consequences in school, the place where we heard most of the rules listed in Figure 10.1. Many students have difficulty staying focused on their studies. Everyone tells them to stick with the program, put up with uninteresting teachers, and do exactly as told. Why? Because if the student is diligent and patient, the hard work eventually pays off.

Figure 10.2. Students need consequences to keep them going.

Some students are able to hang in there for the distant consequence of attaining a college degree and/or getting a good job. Of course, it's necessary to remind them of these remote reasons. Sometimes this is done by emphasizing grades, claiming that high grades are necessary for a successful career. At any rate, academic behavior is typically motivated by consequences, the most sizable being distant and remote.

Many students, though, are lured away from their studies by more immediate and certain consequences for distracting behaviors. As a result, the principles of behavioral science discussed here for safety have been applied successfully to keep students on track. How? By making classroom activities more rewarding (Sulzer-Azaroff & Mayer, 1972, 1986, 1991).

Do any students get soon, certain, and positive consequences for their school-based behaviors? Who gets the letter-sweaters, awards banquets, newspaper recognition, and crowds of people cheering for their extra effort? Right—the athletes.

If my daughter, Karly, spent half the time working on academic-related tasks, even reading for pleasure, as she does on sports, I'd have no worry about her future. But what should I expect? She's been playing baseball and basketball since the fourth grade, and from the start she has received positive consequences for her performance, from trophies and plaques to encouraging words from coaches, peers, and family members. And guess who has been at almost every game, cheering her on? Dad, mom, grandma, and often her older sister.

What soon, certain, and valued consequences can keep Karly focused on improving her academic performance? Letters on a report card every six weeks cannot compete with the immediate ongoing rewards from her athletic performance. Academic activities are "boring," and sports activities are "fun." Homework is "work" to be avoided if possible. Sports conditioning is "work" also, but necessary to achieve those rewards of successful athletic performance. Actually, the soon, certain, and positive consequences available for any behavior can determine whether it's "boring" or "fun."

PEER INFLUENCE. Obviously, consequences from peers are powerful motivators. We work to achieve peer recognition or approval, and to avoid peer criticism or disapproval. Think about this: Do students receive peer support when they demonstrate extra effort in the classroom? In Karly's high school, students who ask questions and show special interest are often called "nerds." In college, I was in the fraternity with most of the school's sports heroes, and in order to fit in with the group I felt peer pressure to conceal my high grades.

Figure 10.3. Peer pressure can inhibit academic performance.

These days you're apt to see the names of honor students published in local newspapers. You might see a bumper sticker proudly asserting that someone in a family made the honor roll. Unfortunately, there has been a negative consequence to this sort of recognition, reflected in the peer pressure bumper sticker depicted in Figure 10.3. When I saw this on a pick-up truck in Blacksburg, Virginia, I was reminded once more of how hard it can be to see rewarding consequences for academic success or improvement.

We have no trouble in the United States finding a "dream team" for athletic activities, but too often we fail to motivate students to seek meaningful and admirable dreams through academic achievement. One root of the problem is misplaced consequences.

INTRINSIC VERSUS EXTRINSIC CONSEQUENCES

Most applied behavioral scientists view *intrinsic motivation* rather differently than the description used in pop psychology books (for example, Kohn, 1993). The behavior-based perspective is supported by research and our everyday experience. Plus, it's objective, practical, and useful for developing situations and programs to motivate behavior change.

Simply put, "intrinsic" does not mean "inside" people, where it cannot be observed, measured, and directly influenced (Horcones, 1987). Rather, "intrinsic" refers to the nature of the task in which an individual is engaged. Intrinsically motivated tasks, or behaviors, lead naturally to external consequences that support the behavior (rewarding feedback) or give information useful for improving the behavior (corrective feedback).

Most athletic performance, for example, includes natural or "intrinsic" consequences that give rewarding or correcting feedback. These consequences, intrinsic to the task, tell us immediately how well we have performed at swinging a golf club, shooting a basketball, or casting a fishing lure, for example. They motivate us to keep trying, sometimes after adjusting our behavior as a result of the natural feedback directly related, or intrinsic, to the task.

Figure 10.4. Some tasks are naturally motivating because of intrinsic consequences.

Take a look at the fisherman in Figure 10.4. Some psychologists would claim he is motivated from within, or self-motivated. They use the term "intrinsic motivation" to refer to this state (Deci, 1975; Deci & Ryan, 1995). In contrast, the behavioral scientist points to the external consequences naturally motivating the fisherman's behavior. These cause him to focus so completely on the task at hand that he's not aware of his wife's mounting anger—or he's ignoring her. He may also be unaware that his supply of fish is creating a potential hazard. In a similar way, safety can be compromised because of excessive motivation for production. Rewards intrinsic to production can cause this motivation.

Note that the "worker" in this picture does not receive a reward for every cast. In fact, he's on an intermittent reinforcement schedule. He catches a fish once in a while. This kind of reward schedule is most powerful in maintaining continuous behavior. Anyone who has gambled understands. Some say gambling is a disease, when in fact gambling is behavior maintained by intermittent rewarding consequences.

Some tasks do not provide intrinsic or natural feedback. In this case, it's necessary to add an extrinsic, or extra, consequence to support or redirect the behavior. Many, if not most, safety behaviors fall in this category.

In fact, many safety practices have intrinsic *negative* consequences, such as discomfort, inconvenience, and reduced pace, that naturally discourage their occurrence. Thus, there is often a need for extrinsic supportive consequences, like intermittent praise, recognition, novelties, and credits redeemable for prizes, to shape and maintain safe behaviors (Skinner, 1982). The intent is not to control people, but to help people control their own behavior by offering positive reasons for making the "safe" choice.

INTERNAL VERSUS EXTERNAL CONSEQUENCES

The intrinsic and extrinsic consequences discussed so far are external to the individual. In other words, they can be observed by another person. Behavioral scientists focus on these types of consequences to develop and evaluate motivational processes because they can be objective and scientific when dealing with external, observable aspects of people.

But behavioral scientists do not deny the existence of internal factors that motivate action. There is no doubt that we talk to ourselves before and after our behaviors, and this self-talk influences our performance. We often give ourselves internal verbal instructions, called intentions, before performing certain behaviors. After our activities we often evaluate our performance with internal consequences. In the process we might motivate ourselves to press on (with self-commendation) or to stop (with self-condemnation).

When it comes to safety and health, internal consequences to support the right behavior are terribly important. Remember, external and intrinsic (natural) consequences for safe behaviors are not readily available, and we can't expect to receive sufficient support (extra consequences) from others to sustain our proactive, safe, and healthy choices. So we need to talk to ourselves with sincere conviction to boost our intentions, and with genuine self-reinforcement after we do the right thing. When we receive special external consequences for our efforts, we need to savor these and use them later to bolster our self-reinforcement.

AN ILLUSTRATIVE STORY

A brief exchange I had with W. Edwards Deming at a seminar he conducted contrasts the behavioral perspective on intrinsic consequences with the humanistic and pop psychology view. Dr. Deming (1991) was describing how much he appreciated the special attention he received from a flight attendant—Debbie. Debbie helped him into a wheelchair at the arrival gate, pushed him a long distance across the airport to his ground transportation, and then helped get him into the limousine. Pleased with Debbie's actively caring behavior, Dr. Deming pulled a $5 bill from his pocket to give her. She quickly refused his offer, saying she was not allowed to accept gratuities from customers.

Dr. Deming told us he felt "so bad" about his attempt to reward Debbie. Later, he tried to find Debbie's last name so he could contact her and apologize for his "terrible mistake." He was so sure he had depreciated Debbie's "intrinsic" motivation by his attempt to give her an "extrinsic" reward. Dr. Deming used this story to explain the widespread pop psychology notion that motivation only comes from within a person, and that any attempts to increase it with extrinsic rewards will only decrease a person's "intrinsic" motivation.

I was disappointed in Dr. Deming's explanation of motivation, and was distressed that an audience of 600 or more might believe that any attempt to show appreciation for another person's performance with praise, some material reward, or award ceremonies would be done in vain, probably causing more harm than good. So I ventured timidly to a microphone to state a behavior-based perspective.[3]

I said that I was a behavioral scientist and university professor and would like to offer another perspective on his airline story. I began with the basic principle that behavior is motivated by consequences. Some consequences are natural or intrinsic to the task and others are sometimes added to the situation, like words of approval or

3. As anyone who has attended a Deming seminar will tell you, it was risky to voice a concern or question to Dr. Deming. Thus, my nervousness was quite rational.

money. Debbie's behaviors were motivated by intrinsic consequences occurring while she wheeled him to his destination, from observing sites along the way to enjoying conversation with a prominent scholar, teacher, and consultant. The 5 dollars was an extrinsic consequence which could add to or subtract from self-motivation depending upon personal interpretation.

If Debbie felt she deserved much more for her efforts, she might have been offended and thought less of her customer, but it would not have detracted from the ongoing intrinsic (natural) consequences that make her job enjoyable to her. On the other hand, she might interpret her job as quite boring or unsatisfying, meaning the intrinsic consequences are not enough to make her feel good about her work. In this case, any extrinsic consequence could help justify her behavior and make her feel better about her job.

Dr. Deming shook his head, saying, "Yes, thank you." My experiences at Deming's workshops led me to believe that such a reply from him represented sincere appreciation. Talk about consequences. I interpreted his extrinsic response as a reward and I felt good about my behavior—approaching the microphone. Plus, my self-motivation was increased further by kind words and approval I received from other workshop participants as I returned to my seat.

The bottom line is this: Our behavior is motivated by extrinsic or extra, as well as natural or intrinsic, consequences; our self-motivation is influenced by how these external consequences (intrinsic and extrinsic) are interpreted. Self-motivation can decrease if a motivational program is seen as an attempt to control behavior. Thus, it's important that praise, recognition, and other rewards are genuine expressions of appreciation. Individuals or groups being recognized must believe they truly earned this consequence through their own efforts. Rewards that we believe are genuine and earned by our own behaviors are likely to increase our inner drive; consequences perceived as not genuine, undeserved, or administered only to control our behavior could be counterproductive.

FOUR TYPES OF CONSEQUENCES

Figure 10.5 summarizes the different types of consequences. Relative to a task or activities, consequences can be natural (intrinsic) or extra (extrinsic). Natural consequences, produced by the target behavior, are usually immediate and certain. In contrast, extra consequences are added to the situation and are often delayed and may be uncertain. Extra consequences are necessary when the natural consequences are insufficient to motivate the desired behavior, as is often the case with safety-related activities (Sulzer-Azaroff, 1992).

Relative to the person performing the task, consequences can be considered external or internal. External consequences are observable by others, and thus can be studied objectively. Internal consequences are subjective and biased by the performer's perceptions. It's difficult to know objectively the exact nature of the internal consequences influencing an individual's performance. But we know from personal experience that internal consequences and evaluations accompany performance and dramatically influence motivation and subsequent performance.

Source of Consequence Relative to Task

Source of Consequence Relative to Person		Natural (immediate)	Extra (often delayed)
	Internal	Listening to Music Reading for Pleasure Watching Television	Reading for Homework Reading for Work Monitoring Instruments
	External	Playing Recreational Sports Doing Crossword puzzles Painting a Picture	Playing Professional Sports Doing Math Homework Working on Assembly Lines

Figure 10.5. Behavior is motivated by four different types of consequences.

I've eliminated the term "intrinsic" from this classification scheme, because of the different uses of this word. Note, however, that "natural" is synonymous with "intrinsic" from a behavior-based perspective (Skinner, 1957; Vaughan & Michael, 1982), while "internal" is the same as "intrinsic" from a humanistic (or person-based) perspective (Deci, 1975; Kohn, 1993).

Figure 10.5 classifies various activities according to the type of consequence relative to the task (natural versus extra) and the task performer (internal versus external). While these activities illustrate particular types of consequences available to motivate performance, the categorizations are neither mutually exclusive nor inclusive. Even the most straightforward task classifications, for example, can overlap with other categories, according to perceptions of the performer.

For example, if you play a musical instrument, complete a crossword puzzle (see Figure 10.6), plant a garden, or participate in recreational sports, natural and external consequences are immediately available. You've performed well, done a good job, or maybe you're not pleased with the results. Add to this the fact that you might compare your results to past results, or the accomplishments of others. This is adding a personal evaluation bias to the natural feedback—internalizing the external consequences. Now you've created internal consequences to accompany your activity.

Figure 10.6. Some tasks have natural rewarding consequences.

Let's take it a step further: Perhaps another person adds an extra consequence by commending or condemning your performance. This could dramatically influence your motivation. And what if you got paid for gardening, or playing the piano? Your motivation could be further influenced.

As we've discussed, some activities or behaviors are not readily motivated by certain types of consequences, thus requiring extra support. Figure 10.5 can be used to identify these tasks, and guide approaches for consequence intervention. Since safe behavior competes with at-risk behavior that is supported by external and natural consequences, it is usually necessary to support safe behavior with extra consequences. This leads us now to a discussion of two very popular safety topics: rewards and penalties.

Managing Consequences for Safety

At this point, I'm sure you appreciate the special message reflected in Figure 10.7. Submitting safety suggestions is an activity not typically followed by motivating consequences. In many work cultures the idea of safety suggestions has long since passed. The suggestion boxes are empty. Does this mean there are no more good suggestions? Is the work force not creative enough? You know the answer to both of these questions is a resounding "No."

Figure 10.7. Some tasks require supportive consequences.

Let me give you an example. I recently worked with safety leaders at a Toyota Motor manufacturing plant in Georgetown, Kentucky, whose 6,000 employees submitted more than 35,000 quality, production, or safety-related suggestions in 1994. And a greater number of suggestions were expected in 1995. Many employees in this culture are motivated internally to submit suggestions, but external consequences are in place to keep the process going. Employees receive timely feedback regarding the utility and feasibility of every suggestion, and if the suggestion is approved, they are empowered to implement it themselves. Also, the individual or team responsible receives 10 percent of the savings for the first year that result from the implemented suggestion. Such external, extra and meaningful—in this case economic—consequences motivate a large work force to make a difference.

FOUR BEHAVIOR-CONSEQUENCE CONTINGENCIES FOR SAFETY

A behavior-consequence contingency is a relationship between a target behavior to be influenced and a consequence that follows. Safety can be improved by managing—or manipulating—four distinct behavior-consequence relationships. Specifically, the probability of injury can be reduced by:

1. Increasing positive consequences of safe behavior
2. Decreasing negative consequences of safe behavior
3. Decreasing positive consequences of at-risk behavior
4. Increasing negative consequences of at-risk behavior.

The contingencies can involve natural or extra consequences. When PPE is made available that is more comfortable or convenient to use, a natural behavior-consequence contingency is put in place, decreasing the previous negative consequences of safe behavior—the possible feeling of discomfort and restricted movement that can come from wearing PPE.

Still, this contingency may not be sufficient to overpower the natural convenience and "get-the-job-done-quicker" contingency supporting the at-risk behavior of working without PPE. It might be advisable to add an incentive/reward contingency to increase PPE use, or implement a disincentive/penalty contingency to increase negative consequences of at-risk behavior.

AN ILLUSTRATIVE CASE STUDY. Last year I consulted with the managers and safety leaders of a large work group who were genuinely concerned about the work pace of their line employees. The probability of a cumulative trauma disorder, especially carpal tunnel syndrome, was certainly a direct function of the work pace (Silverstein *et al,* 1987). Their question was, "How can we reduce the work pace?" They essentially wanted my advice on an education or incentive program that would decrease the work pace and lessen the occurrence of cumulative trauma disorders (CTDs).

Before deriving a contingency to motivate behavior change, it's important to first examine the existing contingencies that support undesirable behavior. In this case, the behavior was a rapid work pace. The most obvious contingency supporting the at-risk behavior was the relationship between work pace and the workers' break time. As soon as employees finished their assignment they could go to the break area and remain until the next work period. According to supervisors, this contingency was necessary for the particular work process and the union contract.

Do you think I recommended an education program—which would be an activator—to reduce the work pace? Did I suggest positive consequences to motivate a slower pace? Or did I advise negative consequences for a rapid pace?

I'm sure you understand why the answer is "no" to each of these questions. I could not recommend a feasible extra consequence powerful enough to overcome the current negative consequence—less time in the break room—of a slower work pace and the ongoing positive consequence—more time in the break room—contingent on a fast work pace. I thought it necessary to alter the work-break reward to decrease positive support of the at-risk behavior. This sort of systems change was not possible at the time, and the probability of CTDs among these workers was not changed.

WHAT'S THE LESSON? Is there a lesson in my failure to make a difference? Perhaps the most important lesson here is that some behaviors cannot be changed by merely adding a consequence intervention to the situation. An existing behavior-consequence contingency might overpower the impact of a feasible intervention program. Actually, this is a frequent problem with efforts to improve safety. We should not expect activators or weak consequences to improve safety over the long term if natural and powerful behavior-consequence contingencies exist to support at-risk behavior. Sometimes it's necessary to change the existing contingencies first.

THE CASE AGAINST NEGATIVE CONSEQUENCES

To subdue influences supporting at-risk behavior, it's often tempting to use a punishment or penalty. All that's needed is a policy statement or some type of top-down mandate specifying a soon, certain, and sizable negative consequence following specific observable risky behaviors. Couldn't this contingency be powerful enough to override the many natural positive consequences for taking risks?

Yes, behavioral scientists have found negative consequences can permanently suppress behavior if the punishment is severe, certain, and immediate (Azrin & Holz, 1966). But before using "the stick," you should understand the practical limitations and undesirable side effects of using negative consequences to influence behavior.

B. F. Skinner (1953) deplored the fact that "the commonest technique of control in modern life is punishment" (p. 182). He protested against the human preoccupation with punishment until his death in 1990. Skinner's animal research with relatively mild punishment indicated that negative consequences merely suppress behaviors temporarily. Plus, the use of negative consequences to control behavior has four undesirable side effects: escape, aggression, apathy (Chance, 1994; Skinner, 1953; Sidman, 1989), and countercontrol (Skinner, 1974).

ESCAPE. Animals and people attempt to avoid situations with a predominance of negative consequences. Sometimes this means staying away from those who administer the punishment. Humans will often attempt to escape from negative consequences by simply "tuning out," or perhaps cheating or lying. Murray Sidman (1989) noted that the ultimate escape from excessive negative consequences is suicide. Indeed, it's not uncommon for an individual to commit suicide in order to escape control by aversive stimulation, which can include the intractable pain of an incurable disease, physical or psychological abuse from a family member, or perceived harassment by an employee or coworker.

Unpleasant attitudes or emotional feelings are produced when people work to escape or avoid negative consequences. As shown in Figure 10.8, negative consequences can influence behavior dramatically, but such situations are usually unpleasant for the "victim." Under fear arousal conditions, people will be motivated to do the right thing, but only when they have to. They feel controlled, and as discussed in Chapter 6, this can lead to distress and burnout. Obviously, this type of contingency and side effect is incompatible with a Total Safety Culture where people feel in control, and are ready and willing to go beyond the call of duty for another person's safety and health.

Figure 10.8. Fear of negative consequences is motivating.

AGGRESSION. Instead of escaping, people might choose to attack those perceived to be in charge. For example, murder in the workplace is rapidly increasing in the United States; and the most frequent cause appears to be reaction or frustration to control by negative means (Baron, 1993). But aggressive reaction to this kind of control might not be directed at the source (Oliver *et al*, 1974).

An employee frustrated by top-down aversive control at work might not assault his boss directly, but rather slow down production, sabotage a safety program, steal supplies, or vandalize industrial property. Or the employee might react with spousal abuse. Then the abused spouse might react by slapping a child. The child, in turn, might punch a younger sibling. And the younger sibling might punch a hole in a wall or kick the family pet—all as a result of perceived control by negative consequences.

APATHY. Apathy is a generalized suppression of behavior. In other words, the negative consequences not only suppress the target behavior but might also inhibit the occurrence of desirable behaviors. Regarding safety, this could mean a decrease in employee involvement. When people feel controlled by negative consequences, they are apt to simply resign themselves to doing only what's required. Going beyond the call of duty for a coworker's health or safety is out of the question.

COUNTERCONTROL. No one likes feeling controlled, and situations that influence these feelings in people do not encourage buy-in, commitment, and involvement. In fact, some people only follow top-down rules when they believe they can get caught, as typified by drivers slowing down when noticing a police car. Some people look for ways to beat the system they feel is controlling them, so you have vehicle drivers purchasing radar detectors. This is an example of "countercontrol" (Skinner, 1974), the fourth undesirable side effect of negative consequence contingencies.

I met an employee once who exerted countercontrol by wearing safety glasses without lenses; when wearing his "safety frames," he got attention and approval from certain coworkers. Perhaps these coworkers were rewarded vicariously when seeing him beat the system they perceived was controlling them also.

Figure 10.9 illustrates an example of countercontrol. Although the supervisors might view the behavior as "feedback," it is countercontrol if it occurred to regain con-

Figure 10.9. Countercontrol behavior is usually directed at those in charge of negative consequence contingencies.

trol or assert personal freedom. A perceived loss of control or freedom is most likely when a negative consequence contingency is implemented. Also, countercontrol behavior is typically directed at those in charge of the negative consequences.

DISCIPLINE AND INVOLVEMENT

Let's specifically discuss traditional discipline for safety—a form of top-down control with negative consequences. I've met many managers who include a "discipline session" as part of the corrective action for an injury report. The injured employee gets a negative lecture from a manager or supervisor whose safety record and personal performance appraisal were tarnished by the injury.

These "discipline sessions" are unpleasant for both parties, and certainly do not encourage personal commitment or buy-in to the safety mission of the company. Instead, the criticized and embarrassed employees are simply reminded of the top-down control aspects of corporate safety, usually resulting in increased commitment not to volunteer for safety programs, nor to encourage others to participate. In this case, the culture loses the involvement of invaluable safety participants.

Individuals who have been injured on the job have special insight into conditions and behaviors that can lead to an injury. If persuaded to discuss their injuries with others, they can be very influential in motivating safe work practices. Personal testimonies, especially by people known to the audience, have much greater impact than statistics summarizing the outcomes of a remote group (Sandman, 1991).

PROGRESSIVE DISCIPLINE. Whenever I teach behavior management principles and procedures, the question of how to deal with the repeat offender frequently comes up. Aren't there times when discipline is necessary? Doesn't an individual who breaks the rules after repeated warnings or confrontations deserve a penalty? Through progressive discipline these individuals receive top-down penalties, starting with a verbal warning, then written warnings, and eventually dismissal. In some cases, dismissal is indeed the best solution for uncooperative individuals who can be a divisive and dangerous factor in the workforce.

But keep in mind that progressive discipline can seriously damage your efforts to achieve a Total Safety Culture. Aggression or countercontrol moves from these individuals can not only decrease employee trust and involvement, but put employees at risk. The disgruntled worker might neglect a safety precaution. Also, escalating penalties for misbehavior can progressively turn off an employee who might have become a true supporter, even a leader, of the safety effort if initially approached differently. Thus, involvement and traditional discipline don't mix, even when the discipline is progressive.

ALTERNATIVES TO PROGRESSIVE DISCIPLINE. So what are alternatives for dealing with repeat safety offenders? One approach is to focus on the positive and give rewarding feedback for safe behavior. Then when corrective feedback is warranted it will be accepted if given appropriately. But what about the employee who seemingly does little to deserve rewarding feedback? Well, anyone can improve from their own baseline level, and improvement can be the target for rewarding feedback.

What about employees who commit serious safety violations which put themselves or, worse yet, *others* at risk, such as not complying with standard lock out/tag out procedures? Isn't it necessary to make a visible and immediate statement about such acts, and set an example that such infractions cannot be tolerated? Yes, indeed. But what kind of disciplinary action will have beneficial and lasting impact, without decreasing involvement and increasing countercontrol? I suggest that developing a "Total Safety Culture Empowerment Council" is the next step.

EMPLOYEE EMPOWERMENT COUNCIL. If a student at my university is caught cheating by an instructor or another student, his or her name is submitted to the "University Honor Council" along with details about the incident. Students volunteer to serve on the honor council, and a "Chief Justice" is elected by the entire student body. University faculty or staff only get involved in this discipline system when making a referral or when presenting evidence during the honor council's fact-finding mission.

After fact finding and deliberating, the honor council might dismiss the case, recommend a penalty for the alleged cheater, and/or suggest changes in the instructor's procedures or policies. In one case, an instructor was given advice on the use of different test forms and classroom seating arrangements. In another case, a professor was advised to eliminate his "closed book, take home" exams. The rationale behind the university honor council is that those most affected by cheating and those most capable of gathering and understanding the facts about alleged cheating should run the system. This disciplinary system is administered for and by the students it serves.

Given that employees typically have the most direct influence over their peers, and that top-down discipline usually decreases voluntary involvement in desirable safety processes, the idea of a TSC Empowerment Council seems logical and intuitively appealing. If a council of people representative of the entire work force serves the fact-finding and corrective-action functions of safety discipline, employee involvement would be enhanced rather than hurt by a discipline system. Such a council could offer the guidance, leadership, and counseling implied by the Latin roots of discipline: *disciplina,* meaning instruction or training, and *discipulus,* referring to a learner. *Disciple* is also derived from the same Latin roots.

PRACTICAL CONSIDERATIONS. If the side effects of negative consequences haven't convinced you to reduce their use in controlling behavior, then consider a few practical limitations. Research has shown that punishment contingencies can have a permanent impact on behavior *if* the negative consequences are soon, certain, and sizable (Azrin & Holz, 1966; Hineline, 1984; Kazdin & Wilson, 1978). That is, the negative consequence must be quite aversive and occur consistently after the undesirable behavior. In some controlled situations it is possible to administer swift and severe negative consequences for inappropriate behavior. And sometimes the target behavior is so maladaptive or dangerous—like a child throwing stones at another person or starting to cross a busy street—that risking undesirable side effects to stop the behavior immediately is warranted.

Most situations, though, do not allow for effective use of negative consequences. Plus, most undesirable behaviors can be decreased or eliminated by removing the positive consequences that support them. Consider, too, that punishment only temporarily suppresses the wrong behavior in most cases. Negative consequences provide no guidance or support for the right behavior.

Dos and Don'ts of Safety Rewards

Now let's look at the flip side of discipline: rewards. In one-on-one situations with children at home or in school, using positive consequences to increase desirable behavior is straightforward and easy. But using rewarding consequences effectively with adults in work settings is easier said than done, especially when it comes to safety. Throughout my 30+ years of professional experience in motivational psychology, I've seen more inappropriate reward programs in occupational safety than in any other area. This is unfortunate, because the effective use of extra positive consequences is often critically important to overcome the readily available influences supporting risky behavior.

By this point, I'm sure you understand the difference between an incentive and a reward. An incentive is an activator that promises a particular positive consequence (a reward) when a correct behavior occurs. Disincentives, on the other hand, are activators such as rules and policies that announce penalties for certain undesired behavior. Remember, the motivating power of incentives and disincentives depends on following through. Rules or policies that are not consistently and justly enforced with penalties for noncompliance are often disregarded. If promises of rewards are not fulfilled when the behaviorial criteria are reached, subsequent incentives might be ignored. This need for consistent delivery of consequences—whether positive or negative—makes it quite challenging to develop and manage an effective motivational program for safety.

DOING IT WRONG

Most incentive/reward programs for occupational safety do not specify behavior: Employees are rewarded for avoiding a work injury or achieving a certain number of "safe work days." So what behavior is motivated? Not reporting injuries.

If having an injury endangers your reward or, worse, the reward for an entire work group, there is pressure to avoid reporting that injury, if possible. Many of these non-behavioral, outcome-based incentive programs involve substantial peer pressure because they use a group-based contingency. That is, if anyone in the company or work group is injured, everyone loses their reward. Not surprisingly, I've seen coworkers cover for an injured employee in order to keep accumulating "safe days."

These incentive programs might decrease the numbers of reported injuries, at least over the short term, but corporate safety is obviously not improved. Indeed, such

Figure 10.10. The display of progress in an outcome-based safety incentive program at a large industrial complex of 1800 employees.

programs often create apathy or helplessness regarding safety achievement. Employees develop the perspective that they can't really control their injury record, but must cheat or beat the system to celebrate the "achievement" of an injury reduction goal.

Figure 10.10 shows how the result of an outcome-based incentive program was displayed to the 1800 employees of a large industrial complex. The man on the ladder (twice life size and named "I.M. Ready") climbed one step higher every day there was no lost-time injury. Whenever a lost-time injury occurred, I. M. Ready fell down the ladder and started his climb again. Every employee was promised a reward as soon as I. M. Ready reached the top of the ladder (30 days without a lost-time injury).

At first, this plan activated significant awareness, even enthusiasm, for safety. But no specific tools or processes were implemented to reduce the injury rate. Safety did not improve, and I. M. Ready did not reach the top of the ladder in 2½ years. Initial zeal for the program waned steadily. Eventually, people stopped looking at the display. The man on the ladder became a reminder of failure. Most of the employees were never personally responsible for the failure, yet they didn't know what to do to stop the failure. Many workers became convinced that they were not in control of safety at their facility, and developed a sense of learned helplessness (Seligman, 1975) about preventing lost-time injuries.

There is a happy ending to this story. The results-based incentive program was dropped, and a behavior-based model was implemented. I taught an incentive steering committee the guidelines presented below for doing it right, and the committee worked out the details. After about six months, I. M. Ready reached the top of the ladder and a plantwide celebration commemorated the achievement.

DOING IT RIGHT

Here are seven basic guidelines for establishing an effective incentive/reward program that manages the human element of industrial health and safety:

1. The behaviors required to achieve a safety reward should be specified and perceived as achievable by the participants.

2. Everyone who meets the behavioral criteria should be rewarded.

3. It's better for many participants to receive small rewards than for one person to receive a big reward.

4. The rewards should be displayed and represent safety achievement. Coffee mugs, hats, shirts, sweaters, blankets, or jackets with a safety message are preferable to rewards that will be hidden, used, or spent.

5. Contests should not reward one group at the expense of another.

6. Groups should not be penalized or lose their rewards for failure by an individual.

7. Progress toward achieving a safety reward should be systematically monitored and publicly posted for all participants.

Figure 10.11. Raffle drawings result in few "lucky" winners and many "unlucky" losers, and thus can have undesirable side effects.

Guideline 2 recommends against the popular lottery or raffle drawing. As illustrated in Figure 10.11, a lottery results in one "lucky" winner being selected and a large number of "unlucky" losers. The announcement of a raffle drawing might get many people excited, and if lottery tickets are dispensed for specific safe behaviors, there is some motivational benefit. Eventually, however, the valuable reward is received by a lucky few.

Also, I perceive a disadvantage in linking chance with safety. It's bad enough we use the word "accident" in the context of safety processes. The first definition of "accident" in my *New Merriam-Webster Dictionary* (1989) is "an event occurring by chance or unintentionally" (p. 23). We shouldn't add to this inference that injuries are chance occurrences.

I've worked with a number of safety directors who used a lottery incentive program and vowed they would never do it again. David Volk (1994) interviewed a number of safety directors who verified my observations. The big raffle prize, such as a snowmobile, pick-up truck or television set, was displayed in a prominent location. Everyone got excited—temporarily—about the possibility of winning. But their attention was directed at the big prize instead of the real purpose of the program: to keep everyone safe. *The material rewards in an incentive program should not be perceived as the major payoff.* Incentives are only reminders to do the right thing, and rewards serve as feedback and a token of appreciation for doing the right thing.

More important than external rewards is the way they are delivered. Rewards should not be perceived as a means of controlling behavior, but as a declaration of sincere gratitude for making a contribution. If many people receive this recognition, you have made deposits in the emotional bank accounts of potential actively caring participants in a Total Safety Culture. That's why it's better to reward many than few (Guideline 3).

When rewards include a safety logo or message (Guideline 4), they become activators for safety when displayed. Also, if the safety message or logo was designed by representatives from the target population, the reward takes on special meaning (as discussed previously in Chapter 9). Special items like these cannot be purchased anywhere, and from the perspective of internal consequences, they are more valuable than money.

As depicted in Figure 10.12, contests that pit one group against another can lead to an undesirable win-lose situation (Guideline 5). Safety needs to be perceived as win-win. This means developing a contract of sorts between each employee that makes everyone a stockholder in achieving a Total Safety Culture. Everyone in the organization is on the same team. Team performance within departments or work groups can be motivated by providing team rewards or bonuses for team achievement. Every team that meets the "bonus" criteria should be eligible for the reward. In other words, Guideline 2 should be applied when developing incentive/reward programs to motivate team performance.

Penalizing groups for individual failure (Guideline 6) reflects a problem I've seen with many outcome-based incentive/reward programs. The problem is typified in the I. M. Ready program. It's certainly easy to administer a contingency that simply withdraws reward potential from everyone whenever one person makes a mistake. But this can do more harm than good. As discussed earlier, it can promote unhealthy group pressure and develop feelings

Figure 10.12. Safety contests can motivate unhealthy competition.

of helplessness or lack of personal control. Displaying the results of such a program only precipitates these undesirable perceptions and expectations. On the other hand, when the incentive/reward program is behavior-based and perceived as equitable and fair, it's advantageous to display progress toward reaching individual, team, or company goals (Guideline 7). When people see their efforts transferred to a feedback chart, their motivation and sense of personal control is increased, or at least maintained.

Obviously, developing and administering an effective incentive/reward program for safety requires a lot of dedicated effort. There is no quick fix. But it is worth doing, if you take the time to do it right. As Aubrey Daniels wisely stated, "If you think this is easy, you are doing it wrong" (1994, p. 171). Let's examine an exemplary case study.

AN EXEMPLARY INCENTIVE/REWARD PROGRAM

In 1992, I consulted with the safety steering committee of a Hoechst Celanese company of about 2000 employees to develop a plant-wide incentive program that followed each of the guidelines given above. The steering committee, including four hourly and four salary employees, met several times to identify specific behavior-consequence contingencies. That is, they needed to decide what behaviors should earn what rewards. Their plan was essentially a "credit economy" where certain safe behaviors, which could be achieved by all employees, earned certain numbers of "credits."

At the end of the year, participants exchanged their "credits" for a choice of different prizes, all containing a special safety logo. The variety of behaviors earning credits included: attending monthly safety meetings; special participation in safety meetings; leading a safety meeting; writing, reviewing, and revising a job safety analysis; and conducting periodic audits of environmental and equipment conditions, and certain work practices. For a work group to receive credits for audit activities, the results of environmental and PPE observations had to be posted in the relevant work areas. Only one behavior was penalized by a loss of credits—the late reporting of an injury.

At the start of the new year, each participant received a "safety credit card" for tallying credit earnings each month. Some individual behaviors earned credits for the person's entire work group, thus promoting teamwork and group cohesion. It's noteworthy that this kind of incentive/reward program exemplifies a basic behavior-based principle for health and safety management—observation and feedback. Employees were systematically observed, and they received soon, certain, and positive feedback (a reward) after performing a target behavior. An incentive/reward program is only one of several methods to increase safe work practices with observation and feedback. In the next chapter I address feedback more specifically as an external and extra consequence to prevent injuries.

SAFETY THANK-YOU CARDS

I'd be remiss if I did not describe "Safety Thank-You Cards" in a discussion of exemplary incentive/reward approaches. Figure 10.13 depicts a card that was avail-

FRONT

C.C. Manufacturing

Thank You for ACTIVELY CARING

Date: ________

Please describe specifically the observed ACTIVELY CARING behavior: (see back for examples)

____

____

____

____

____

____

Observer's Code:

The first letter of the city where you were born	The first letter of your mother's maiden name	The number of the month you were born
____	____	____

Recipient's Code:

The first letter of the city where you were born	The first letter of your mother's maiden name	The number of the month you were born
____	____	____

______ ¢ **Thank You** Limit: 55 ¢

Observer's Name________________________

Recipient's Name________________________

BACK

Examples of ACTIVELY CARING Behaviors:

- Recognizing and correcting an unsafe condition.
- Reminding a coworker not to perform an unsafe act.
- Removing or cleaning unsafe objects or debris from a work area.
- Giving positive feedback to a coworker for working safely.
- Reporting a near miss.
- Making a task safer.
- Other

Hoechst Celanese

Elaine George Dave Salyer

Tom Tillman Jim Woods

Department 1490

Figure 10.13. Sample thank-you card used to recognize safe behavior.

able to all employees for distribution to coworkers whenever they observed them going out of their way for another person's safety (Roberts & Geller, 1995). The types of actively caring behaviors warranting recognition were listed on the back of the card, and involved such things as suggesting a safer way to perform a task, pointing out a potential hazard that might have been overlooked, or going beyond the call of duty to help another person avoid an at-risk behavior.

Over the years, I've seen a wide variety of thank-you cards designed by work teams and used successfully at a number of industrial sites, including Hoechst Celanese, Exxon Chemical, Ford, Phillip Morris, General Motors, Logan Aluminum, Westinghouse Hanford Company, Abbott Laboratories, Weyerhaeuser, and Kal Kan.

At some locations thank-you cards were used in a raffle drawing, exchangeable for food, drinks, or trinkets, or displayed on a plant bulletin board as a "safety honor roll." Sometimes the cards could be accumulated and exchanged for tee shirts, caps, or jackets with messages or logos signifying safety achievement. At several plants, the person who delivered a thank-you card returned a receipt naming the recognized

Figure 10.14. An actively caring thank-you card used at the Hoechst Celanese plant in Rock Hill, South Carolina.

employee and describing the behavior earning the consequence, thus creating objective information to define a "Safe Employee of the Month" (Geller, 1990).

At a few locations, the thank-you cards took on a special actively-caring meaning. Specifically, when deposited in a special collection container, each thank-you card was worth 25 cents toward corporate contributions to a local charity or to needy families in the community. The actively caring card used at the Hoechst Celanese plant in Rock Hill, South Carolina, is shown in Figure 10.14. The back of the card included a colorful peel-off symbol which the recognized employee could affix in any number of places as a personal reminder of the recognition. I was surprised but pleased to see a large number of these thank-you stickers on employees' hard hats. Obviously, this actively-caring thank-you approach to safety recognition has great potential as an inexpensive but powerful tool for motivating safe behavior.

In Conclusion

Writing this book was challenging, tedious, overwhelming, tiresome, sacrificing, and exhausting. Observers were apt to say I was self-directed and "intrinsically" motivated. Of course, I know better, and you do too.

Incidentally, I literally wrote the various drafts of this text. I have never learned to type, and therefore have never benefited from the technological magic of computer word processing. My colleagues explain that it's not necessary to be a skilled typist to reap the many "intrinsic" benefits of preparing a manuscript on a computer. "I type slowly with only one finger," some have said, "and still enjoy the wonderful benefits of high-tech computer word processing. I could never go back to preparing a manuscript by hand. You don't know what you're missing."

I'm sure you've noticed my disparate uses of "intrinsic" in the prior paragraphs, and you now understand the two meanings of this popular motivational term. Are my friends and colleagues so enthusiastic about computer-based word processing because of intrinsic (internal) motivation, or because of intrinsic (natural) consequences linked to their computer use? As a review of this chapter, I'm sure you see my point. Word processing on a computer allows for rapid "quick-fix" control of letters, words, sentences, and paragraphs. Computer users also can walk to a printer and obtain a typed "hard copy" of their document for study, revision, or dissemination. All these soon, certain, positive consequences are connected naturally to word-processing behavior. No wonder my friends and colleagues are motivated about computer word processing, and urge me to get on the high-tech "band wagon."

While I didn't reap the benefits of fast computer turnaround, there were plenty of external and natural consequences to keep me going. I experienced a sense of rewarding satisfaction (internal consequence) from seeing my thoughts and ideas take form. Almost daily I gave my writing to a secretary who processed my writing on computer disk (yes, I do realize this is a step I wouldn't need if I were computer literate). I got significant satisfaction (or the internalization of external consequences) from reading and refining the typed text. The next day my secretary delivered a refined version—another external consequence from my work.

Throughout the writing process I worked with a very talented illustrator who provided me with soon, certain, and positive consequences to feed my motivation. We talked about concepts I wanted to portray, and in a few days I saw his artwork. Sometimes I judged it ready. Other times we discussed revisions, and within a week or so I examined the fruits of our discussions.

Continuous feedback from others was invaluable as a motivator and necessary mechanism for continuous improvement. As soon as a chapter appeared close to my internal standard, I distributed copies to about ten colleagues and friends who had expressed interest in reading early drafts and offering feedback. The feedback I received from these earlier versions was valuable in refining this text and in motivating my progress. Feedback is obviously a powerful consequence intervention for improving and motivating behavior. In the next chapter, I'll discuss ways to maximize the beneficial impact of feedback consequences.

When we earn genuine appreciation and approval from others for what we do, we not only become self-motivated, we also maximize the chances that our activities will influence the behavior of others. In fact, this was my ultimate motivation for soliciting feedback on earlier drafts of this text and for painstakingly refining the presentations. Practice can only improve with feedback. We can only learn to communicate more effectively if we learn how we're coming across to others. If I communicate effectively and earn the approval and appreciation of readers for the principles and procedures presented in this text, injuries and fatalities could be reduced on a large scale. This would be an external and natural consequence of authoring this text—the remote but preeminent motivator for my writing behavior.

REFERENCES

Azrin, N. H., & Holz, W. C. (1966). Punishment. In W. K. Honig (Ed.), *Operant behavior: Areas of research and application*. New York: Appleton-Century-Crofts.

Baron, S. A. (1993). *Violence in the workplace: A prevention and management guide for businesses*. Ventura, CA: Pathfinders Publishing of California.

Carr, C., Mawhinney, T., Dickinson, A., & Pearlstein, R. (1995). Punished by rewards? A behavioral perspective. *Performance Improvement Quarterly*, 8(2), 125-140.

Chance, P. (1994). *Learning and behavior* (Third Edition). Pacific Grove, CA: Brooks/Cole Publishing Company.

Covey, S. R. (1989). *The seven habits of highly effective people*. New York: Simon and Schuster.

Daniels, A. C. (1994). *Bringing out the best in people*. New York: McGraw-Hill, Inc.

Deming, W. E. (1993). *The new economics for industry, government, education*. Cambridge, MA: Massachusetts Institute of Technology, Center for Advanced Engineering Study.

———. (1991, May). *Quality, productivity, and competitive position*. Four-day workshop presented in Cincinnati, Ohio by Quality Enhancement Seminars, Inc.

Deci, E. L. (1975). *Intrinsic motivation*. New York: Plenum.

Deci, E. L., & Ryan, R. M. (1985). *Intrinsic motivation and self-determination in human behavior*. New York: Plenum.

Fulgham, R. (1988). *All I really need to know I learned in kindergarten*. New York: Ivy Books.

Flora, S. R. (1990). Undermining intrinsic interest from the standpoint of a behaviorist. *The Psychological Record*, 40, 323-346.

Geller, E. S. (1990). Performance management and occupational safety: Start with a safety belt program. *Journal of Organizational Behavior Management*, 11(1), 149-174.

Hineline, P. N. (1984). Aversive control: A separate domain? *Journal of the Experimental Analysis of Behavior*, 42, 495-509.

Horcones. (1987). The concept of consequences in the analysis of behavior. *The Behavior Analyst*, 10, 291-294.

Kazdin, A. E., & Wilson, G. T. (1978). *Evaluation of behavior therapy: Issues, evidence, and research strategies*. Cambridge, MA: Ballinger.

Kohn, A. (1993). *Punished by rewards: The trouble with gold stars, incentive plans, A's, praise, and other bribes*. Boston: Houghton Mifflin.

New Merriam-Webster Dictionary (1989). Springfield, MA: Meriam-Webster, Inc., Publishers.

Oliver, S. D., West, R. C., & Sloane, H. N. (1974). Some effects on human behavior of aversive events. *Behavior Therapy*, 5, 481-493.

Peale, N. V. (1952). *The power of positive thinking*. New York: Prentice-Hall.

Roberts, D. S., & Geller, E. S. (1995). An "actively caring" model for occupational safety: A field test. *Applied & Preventive Psychology*, 4, 53-59.

Sandman, P. M. (1991). *Risk = Hazard + Outrage: A formula for effective risk communication.* Videotaped presentation for the American Industrial Hygiene Association. Environmental Communication Research Program, P.O. Box 231, Cook College, Rutgers University, New Brunswick, N.J.

Seligman, M. E. P. (1975). *Helplessness.* San Francisco, CA: W. H. Freeman.

Sidman, M. (1989). *Coercion and its fallout.* Boston, MA: Authors Cooperative.

Silverstein, B. A., Fire, L. J., & Armstrong, T. J. (1987). Occupational factors and carpal tunnel syndrome. *American Journal of Industrial Medicine*, 11, 343-358.

Skinner, B. F. (1938). *The behavior of organisms.* Acton, MA: Copley Publishing Group.

———. (1953). *Science and human behavior.* New York: Free Press.

———. (1957). *Verbal behavior.* New York: Appleton-Century-Crofts.

———. (1974). *About behaviorism.* New York: Alfred A. Knopf.

———. (1982). Contrived reinforcement. *The Behavior Analyst*, 5, 3-8.

Sulzer-Azaroff, B., & Mayer, G. R. (1972). *Behavior modification procedures for school personnel.* New York: Dryden Press.

——— & ———. (1986). *Achieving educational excellence using behavioral strategies.* New York: Holt, Rinehart & Winston, Inc.

——— & ———. (1991). *Behavior analysis for lasting change.* New York: Holt, Rinehart & Winston, Inc.

Sulzer-Azaroff, B. (1992). Is back to nature always best? In E. S. Geller (Ed.), *The educational crisis: Issues, perspectives, solutions* (pp. 68-69). Monograph No. 7, Society for the Experimental Analysis of Behavior, Inc.

Vaughan, M. E., & Michael, J. (1982). Automatic reinforcement: An important but ignored concept. *Behaviorism*, 10, 217-227.

Volk, D. (1994). Learn the do's and don'ts of safety incentives. *Safety & Health*, March, 54-57.

Chapter 11
Intervening as a Behavior-Change Agent

This chapter presents the principles and procedures of safety coaching—a key behavior-change process for safety improvement. The letters of **COACH** *represent the critical sequential steps of safety coaching:* **C***are,* **O***bserve,* **A***nalyze,* **C***ommunicate,* **H***elp. This coaching process is clearly relevant for improving behaviors in areas other than safety and in settings other than the workplace.*

The prior two chapters discussed guidelines for developing behavior change interventions. Chapter 9 focused on the use of activators to direct behavior change. Chapter 10 detailed the motivating role of consequences. Several examples employed both activators and consequences. This is applying the three-term contingency (activator-behavior-consequence), and is usually the most influential approach.

Large-scale behavior change is impossible without intervention agents—people willing and able to step in on behalf of another person's safety. In a Total Safety Culture, everyone feels responsible for safety, pursuing it daily. They go beyond the call of duty to identify at-risk conditions and behaviors, and intervene to correct them (Geller, 1994). This chapter is about becoming a behavior-focused change agent.

In simplest terms, this means observing and supporting safe behaviors or observing and correcting at-risk behaviors. It might involve designing and implementing a particular intervention process for a work team, organizational culture, or an entire community. Or it might mean merely engaging in behavior-focused communication between an observer (the intervention agent) and the person observed. This is safety coaching (Geller, 1995), and to be effective, certain principles and guidelines need to be followed.

ACTIVATORS

1. **Lecture:** Unidirectional oral communication concerning the rationale for specific behavior change.
2. **Demonstration:** Illustrating the desired behavior for target subjects(s).
3. **Policy:** A written document communicating the standards, norms, or rules for desired behavior in a given context.
4. **Written activator:** A written communication that attempts to prompt desired behavior.
5. **Commitment:** A written or oral pledge to perform a desired behavior.
6. **Discussion/consensus:** Bidirectional oral communication between the deliverers and receivers of an intervention program.
7. **Oral Activator:** An oral communication that urges desired behavior.
8. **Assigned individual goal:** One person decides for another person the level of desired behavior he or she should accomplish by a certain time.
9. **Individual goal:** An individual decides the level of desired behavior (the goal) that should be accomplished by a specific time.
10. **Individual competition:** An intervention promotes competition between individuals to see which person will accomplish the desired behavior first (or best).
11. **Individual incentive:** An announcement to an individual in written or oral form of the availability of a reward following one or more designated behaviors.
12. **Individual disincentive:** An announcement to an individual specifying the possibility of receiving a penalty following one or more undesired behaviors.
13. **Assigned group goal:** A group leader decides the level of desired behavior a group should accomplish by a certain time.
14. **Group goal:** Group members decide for themselves a level of group behavior they should accomplish by a certain time.
15. **Group competition:** An intervention promotes competition between specific groups to see which group will accomplish the desire behavior first (or best).
16. **Group incentive:** An announcement specifying the availability of a group reward following the occurrence of desired group behavior.
17. **Group disincentive:** An announcement specifying the possibility of receiving a group penalty following the occurrence of undesired group behavior.

CONSEQUENCES

18. **Individual feedback:** Presentation of either oral or written information concerning an individual's desired or undesired behavior.
19. **Individual reward:** Presentation of a pleasant item to an individual, or the withdrawal of an unpleasant item from an individual, for performing desired behavior.
20. **Individual penalty:** Presentation of an unpleasant item to an individual, or the withdrawal of a pleasant item from an individual, following undesired behavior.
21. **Group feedback:** Presentation of either oral or written information concerning a group's desired or undesired behavior.
22. **Group reward:** Presentation of a pleasant item to a group, or the withdrawal of an unpleasant item from a group or team, following desired group behavior.
23. **Group penalty:** Presentation of an unpleasant item to a group, or the withdrawal of a pleasant item from a group or team, following undesired group behavior.

Figure 11.1. Brief definitions of 23 different intervention techniques used to influence behavior. (Adapted from Geller, 1992, and Geller *et al*, 1990)

Selecting an Intervention Approach

The number of ways to intervene on behalf of safety by using activators, consequences, and their combination is almost limitless. A steering committee for an organization or community needs to design specific procedures for each aspect of the three-term contingency. What are the desired (or undesired) target behaviors? How will the target behavior(s) be activated? What consequences can be employed to motivate behavior change?

Figure 11.1 gives brief definitions of 23 different ways to use activators and consequences for improving safety-related behaviors. These come from the research literature on techniques to change behaviors at individual and group levels.

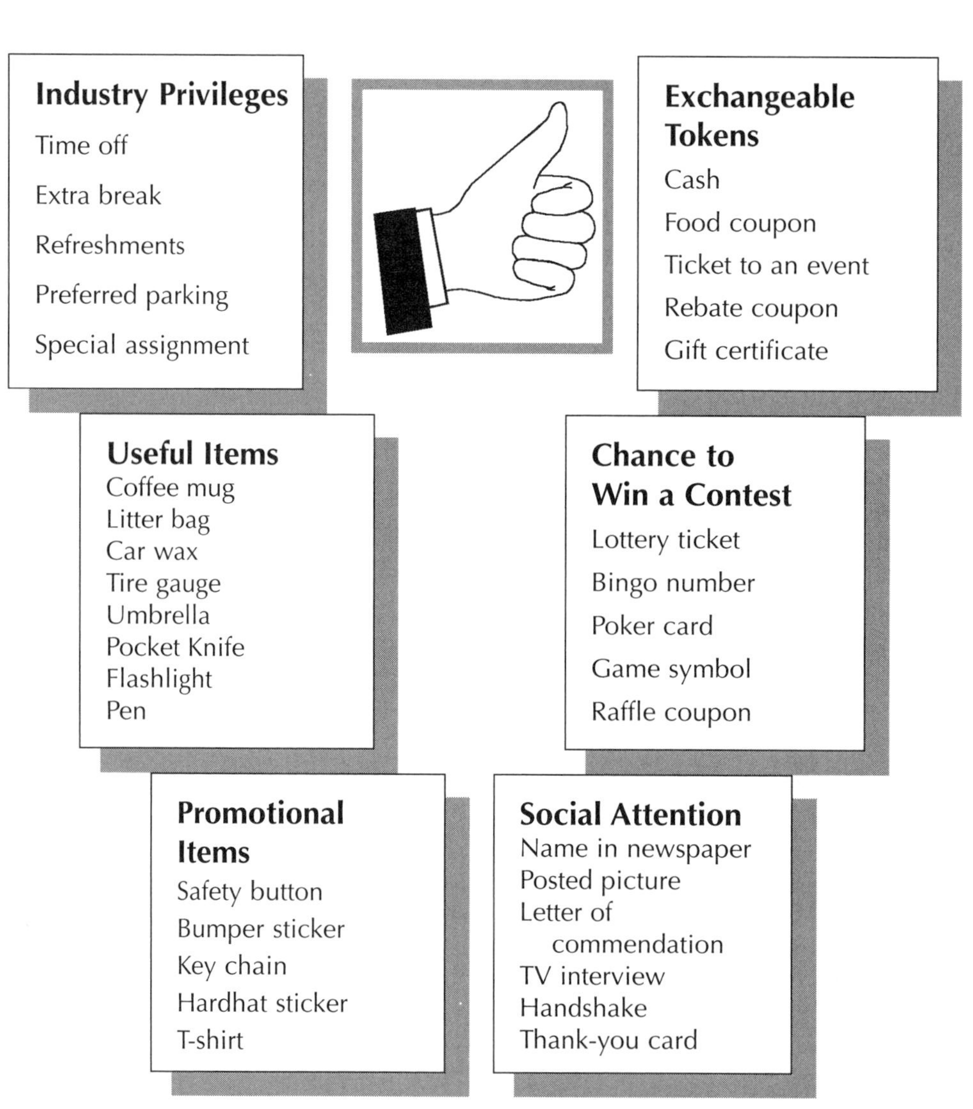

Figure 11.2. The variety of possible rewards available to motivate safe behaviors in organizational settings.

Representative sources include Cone and Hayes (1980), Glenwick and Jason (1980, 1993), Geller *et al* (1982), and most research articles published in the *Journal of Applied Behavior Analysis* from 1968 until the present (for example, Greene *et al*, 1987).

The first 17 approaches are activators, occurring before the target behavior is performed. They attempt to persuade or direct people, can be classified as passive or active, and target individuals or groups. The three basic types of consequence approaches—reward, penalty, and feedback—can be given to an individual or to a group. Therefore, Figure 11.1 defines six different consequence procedures.

As discussed earlier, most interventions consist of various techniques listed in Figure 11.1. Education programs to promote safety and health, for example, often include discussion/consensus building; demonstrations; lectures; and written activators, including signs, newsletters, memos, or verbal reminders. An incentive/reward program requires a variety of activators (incentives) to announce the availability of a reward for certain behaviors, and the consequence (reward) can be given to individuals or groups.

Also, as depicted in Figure 11.2, various items or events can be used for rewards, from special privileges and promotional items to special individual or team recognition. Note that receiving a reward for particular behavior is a form of feedback—information regarding the occurrence of desired behavior. A person can receive feedback, however, without acquiring any of the rewards listed in Figure 11.2. Safety coaching, for instance, always involves feedback. When the coaching process recognizes safety behavior, it's usually perceived as a social attention reward.

MULTIPLE INTERVENTION LEVELS

Interventions fluctuate widely in terms of cost, administrative effort, and participant involvement. Written activators like signs, memos, and newsletter messages are relatively effortless. Other activators like demonstrations, commitment techniques, and consensus-building discussions take considerable time and effort to design and deliver. And, as discussed in Chapter 10, implementing an incentive/reward process correctly requires continuous attention and periodic refinement from a team of intervention agents.

Some people benefit from the simplest and least expensive interventions, such as signs or policy statements specifying the correct behavior for a certain situation. But for a variety of reasons, many of which were considered in Part Two, some people do not alter their at-risk behavior after reading a simple safety message. They require a more intensive, intrusive and expensive intervention. Exposing them to more signs, posters, and memos is generally a waste of time.

By the same token, when people do the right thing following the least intrusive intervention technique, such as passive activators, it's a waste of time and effort to target them for a more compelling intervention. Instead, enroll these folks as agents of change (Katz & Lazarfeld, 1955). In other words, don't "preach to the choir"—enlist the "choir" to preach to others.

Figure 11.3 depicts a multiple intervention level (MIL) hierarchy (adapted from Geller, 1992). It summarizes the important points about different levels of interven-

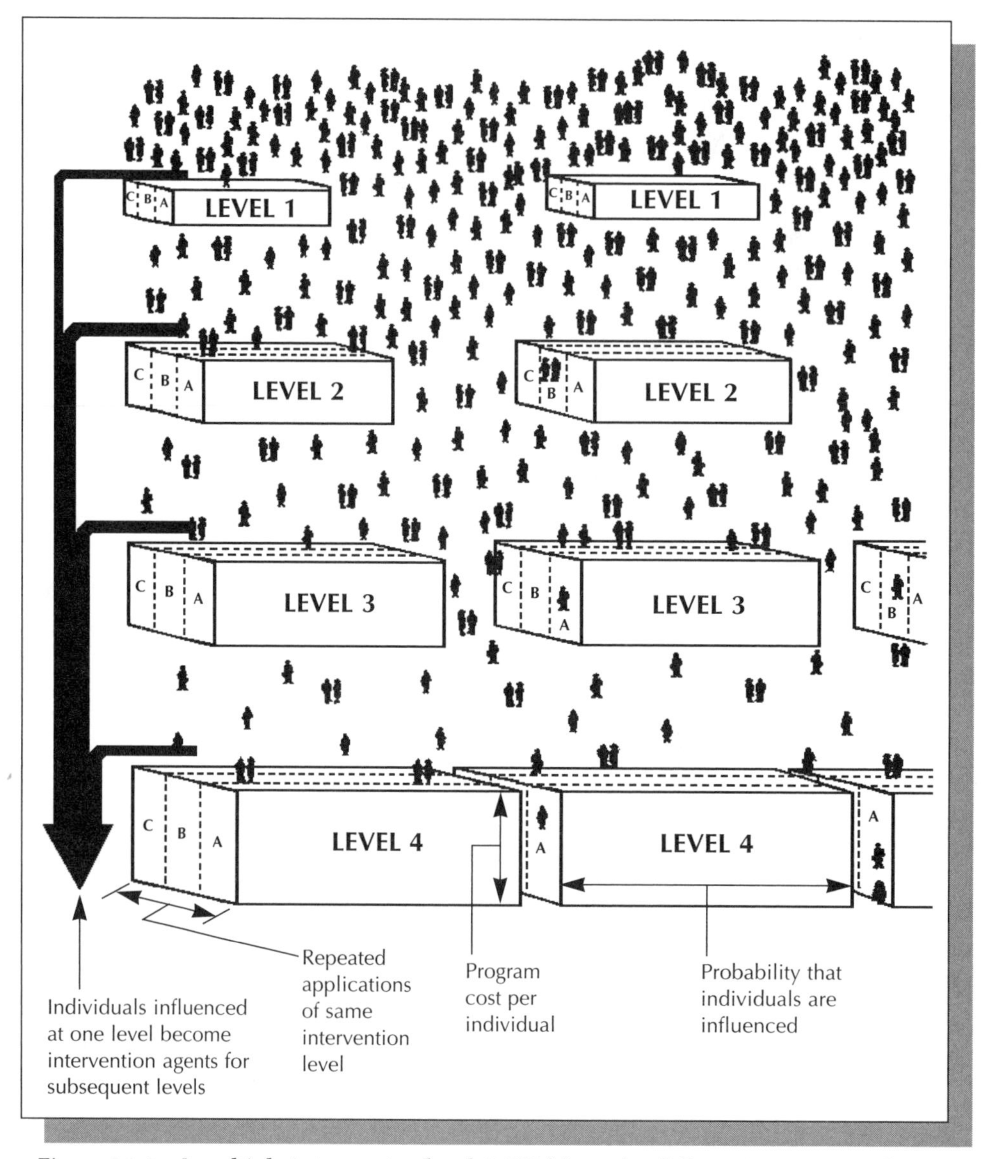

Figure 11.3. A multiple intervention level (MIL) hierarchy differentiates repeated interventions at the same impact level from those interventions progressively more influential at changing behavior. (Adapted from Geller, 1992)

tion impact and intrusiveness, and illustrates the need to multiply the number of intervention agents as the level of intensity and intrusiveness increases. At the top of the hierarchy (Level 1), the interventions are least intense and intrusive. They target the maximum number of people for the least cost per person. At this level, an intervention is designed to have maximum large-scale appeal, while minimizing contact between target individuals and intervention agents. Those showing the desired target behavior at a particular intervention level may continue to benefit from repeated exposure to similar interventions. However, individuals unaffected by initial exposures to a particular intervention level will "fall through the cracks." Repeated exposure to interventions at the same level will have no effect. These individuals require a more influential and costly intervention.

This hierarchy lists four intervention levels, but to date this number has not been empirically verified. Each level has height, length, and width, representing different characteristics of the intervention. The height of each intervention box represents the "financial investment per person" to participate in or experience that particular intervention. Notice how the investment per person increases as the levels increase. The length of each box represents the "probability" that an individual will be influenced to change his or her behavior as a result of experiencing that intervention. This probability increases as the level increases.

The width of each intervention level is marked with the letters "A," "B," "C," etc. This indicates repeated applications of the same intervention approach. The hierarchy predicts that repeating the same intervention over and over will typically not influence additional people. This is because those who were susceptible to changing their behavior at that level have already done so. The others have "fallen through the cracks." Let's look at some examples.

A Level 1 intervention might include relatively inexpensive signs, posters, or billboards with safety messages or slogans. Placed around the plant, people typically notice them when they're first put in place or when the message changes, but soon the activators are forgotten or ignored. A certain percentage of the plant population might change their behavior and perform more safely as a result, but others are not influenced by this level of intervention intrusiveness.

For these people a Level 2 intervention is in order. This might include weekly safety meetings for each work group, where employees talk about safety issues for an allotted time. Meetings require more participation and involvement from the employees. They give participants a greater sense of personal control over safety, and offer more opportunities for social support. This intervention should have a greater impact. Also, due to the more intense nature of Level 2 programs, the personal investment per participant has increased. This will likely add more converts to our safety task force, but for nonparticipants it will not be enough.

So these employees "fall through the cracks" again, to be faced with a Level 3 intervention. This might be an incentive/reward program or a commitment and goal-setting process. Of course these techniques require more time and effort from both intervention agents and participants.

Finally, note the bold arrow on the left of the diagram in Figure 11.3. It indicates that once individuals are influenced at any given level, they can participate in the next intervention level as a behavior-change agent, helping others improve their safety performance. Higher level interventions require more change agents. The highest level intervention for safety is one-on-one, the level for safety coaching.

INCREASING INTERVENTION IMPACT

Based on an extensive literature review and our own studies in safety-belt promotion, my students and I (Geller *et al,* 1990) proposed that the success of any intervention program is a direct function of:

- The amount of specific response information transmitted by the intervention
- The degree of external consequence control
- The target individual's perception of personal control or personal choice regarding the behavior change procedures
- The degree of group cohesion or social support promoted
- The amount of participant involvement facilitated by the intervention

Response information varies according to the amount of new behavioral knowledge transmitted by the intervention. As discussed in Chapter 9, this can be facilitated by increasing the salience of the information presented and the proximity between the time a behavioral request is made and the opportunity or ability to perform the desired response. *External control* is determined by the type of behavior-consequence contingency used; and as covered in Chapter 10, the nature of the consequence will influence the target individual's perception of personal control over the behavior change procedures.

Negative consequences and nongenuine or insincere positive consequences decrease *personal control*—and the long-term benefits of an intervention process. When people get to choose aspects of an intervention, such as which behaviors to focus on and what rewards to offer, their perceptions of personal control increase, and the intervention's impact is enhanced. The concept of personal control is discussed more completely in Chapters 13 and 14.

Social support is shaped by the amount of peer, family or friend encouragement resulting from the intervention process. Person factors, such as an individual's natural tendency to interact in group settings, and various group dynamics, such as degree of group cohesion or belongingness, also affect the amount of perceived social support associated with an intervention. The person-based aspects of perceived social support are discussed in more detail in Chapter 13.

Participant involvement in an intervention also depends on certain person factors, including an individual's degree of introversion versus extroversion (Eysenck, 1970; Eysenck & Eysenck, 1985) and perceived locus of control (Strickland, 1989). Extroverts, for example, usually participate more than introverts in interventions involving a high level of activity and social interaction. In addition, people with an internal locus of control typically prefer situations that allow them greater personal control rather than being at the mercy of others or chance factors. The reverse is true for individuals with an external locus of control (Phares, 1973).

More important than personal factors in determining involvement is the ratio of number of intervention agents to individuals in the target population. More intervention agents per target population usually promote greater participation. A one-to-one agent-to-target ratio is the highest level of intervention intensity and effectiveness, and occurs in safety coaching. This is the ultimate intervention for safety, and the remainder of this chapter details the ingredients of an effective safety coaching intervention process.

Intervening as a Safety Coach

Coaching is essentially a process of one-on-one observation and feedback. The coach systematically observes the behaviors of another person, and provides behavioral feedback on the basis of the observations. Safety coaches recognize and support the safe behaviors they observe, and offer constructive feedback to reduce the occurrence of any at-risk behaviors. This chapter specifies the steps of safety coaching, points out trainable skills needed to accomplish the process, and illustrates tools and support mechanisms to increase effectiveness.

ATHLETIC COACHING VERSUS SAFETY COACHING

The term "coach" is very familiar to us in an athletic context. In fact, winning coaches practice the basic observation and feedback processes needed for effective safety coaching. They follow most of the guidelines reviewed here. As illustrated in Figure 11.4, the most effective team coaches observe the ongoing behaviors of individual players, and record their observations in systematic fashion, using a team roster, behavioral checklist, or videotape. Football coaches, for instance, will spend hours and hours analyzing film. Then they deliver specific and constructive feedback to team members to support or increase desirable behavior and/or to decrease undesirable behavior. Sometimes the feedback is given in a group session, perhaps by critiquing videotapes of team competition. At other times, the feedback is given individually in a personal one-on-one communication. Usually, the one-to-one format has greater impact on individual performance.

Figure 11.4. Systematic observation and feedback are key to effective coaching.

DIFFERENTIAL ACCEPTANCE. The most effective athletic coaches communicate feedback so that team members learn from the exchange and increase their motivation to continuously improve. The same is true for safety coaches. The basic principles and procedures for effective coaching are the same whether communicating behavioral feedback to improve athletic or safety performance. It is usually more challenging, however, to coach for safety than for sports.

People are generally more accepting of information to improve their performance in sports. Indeed, they usually offer sincere words of appreciation for feedback to improve an athletic behavior, such as a golf swing, a tennis stroke, a batter's stance, a basketball maneuver, or a football blocking technique. But how often do you hear an individual offer genuine thanks for being corrected on a safety-related behavior? Safety coaching is often viewed as a personal confrontation. In fact, we are usually more willing to accept and appreciate advice regarding work production and quality than work safety.

CONSEQUENCES OF COACHING. When we adjust our behavior following constructive athletic coaching, it doesn't take very long to notice an improvement in our performance. We see a direct connection between the improvement and the coach's feedback. Sometimes we even see an increase in the individual or team scores as a result of individual or group feedback. Thus, the process of athletic coaching is often supported by consequences occurring naturally soon after the behaviors targeted by the coach. The value of athletic coaching becomes obvious.

Usually the value of safety coaching is not obvious, because on a day-to-day basis there is no clear connection between safety coaching and the ultimate purpose of coaching—reduced injuries. When people follow the advice of a safety coach they usually do not perceive an immediate difference, either in their own safety or the company's injury rate. People don't expect injuries to happen to them, and since their everyday experience supports this belief, they don't perceive a personal need for advice from a safety coach.

Changing the way we keep score for safety can increase acceptance and appreciation for safety coaching. While injury reduction is the ultimate purpose of coaching, the immediate goal is behavior change. Since at-risk behavior is involved in almost every workplace injury, noting an increase in safe work behavior or a decrease in at-risk behavior due to safety coaching should result in coaching being appreciated as a proactive, upstream approach to reducing injuries. This requires a behavior-based evaluation process, and tools for accomplishing this are covered here.

As I've emphasized throughout this text and have written in other articles (Geller, 1994, 1995, 1996), a behavior-based approach to safety treats safety as an *achievement-oriented* (rather than failure-oriented) *process* (not outcome), that is *fact finding* (not fault finding) and *proactive* (rather than reactive). This chapter illustrates coaching techniques that meet these criteria, and demonstrates the critical value of safety coaching for achieving a Total Safety Culture.

The Safety Coaching Process

As shown in Figure 11.5, the five letters of the word ***COACH*** can be used to remember the basic ingredients of the most effective coaching—whether coaching the members of a winning athletic team or the individuals in a work group striving for safe behaviors. This is my favorite instructional acronym, because it not only contains the components of an effective coaching process, it lists them in the sequence in which they should occur.

"C" FOR CARE

Caring is the basic underlying motivation for coaching. Safety coaches truly care about the health and safety of their coworkers, and they act on such caring. In other words, they "actively care" (Geller, 1991, 1994, 1996). When people realize from a safety coach's words and body language that he or she cares, they are more apt to lis-

C ARE
- Show that you care
- Set caring examples

O BSERVE
- Define target behaviors
- Record behavioral occurrences

A NALYZE
- Identify existing contingencies
- Identify potential contingencies

C OMMUNICATE
- Listen actively
- Speak persuasively

H ELP
- Recognize continuous improvement
- Teach and encourage the process

Figure 11.5. The five letters of COACH represent the basic ingredients of effective safety coaching.

ten to and accept the coach's advice. When people know you care, they care what you know.

OUR EMOTIONAL BANK ACCOUNTS. Stephen Covey (1989) explained the value of interdependence among people—exemplified by appropriate safety coaching—with the metaphor of an "emotional bank account." People develop an emotional bank account with others through personal interaction. Deposits are made when the holder of the account views a particular interaction to be positive, as when they feel recognized, appreciated, or listened to. Withdrawals from a person's emotional bank account occur whenever that individual feels criticized, humiliated, or less appreciated, usually as a result of personal interaction.

Sometimes it is necessary to offer constructive criticism, or even state extreme displeasure with another person's behavior. But if such negative discourse occurs on an "overdrawn or bankrupt account," this corrective feedback will have limited impact. In fact, continued withdrawals from an overdrawn account can lead to defensive or countercontrol reactions (Skinner, 1974). The person will simply ignore the communication or actually do things to discredit the source or undermine the process or system implicated in the communication. Thus, safety coaches need to demonstrate a caring attitude through their personal interactions with others. This maintains healthy emotional bank accounts—operating in the black. The woman in Figure 11.6 is requesting a deposit along with the withdrawal. Our emotional reaction to police officers depends on the proportion of deposits versus withdrawals we have experienced with them.

A shared responsibility. People are often unwilling to coach or to be coached for safety because they view safety from an individualistic perspective. To them, it's a matter of individual or personal responsibility. This is illustrated by the verbal expression or internal script, "If Molly and Mike want to put themselves at risk, that's their problem, not mine." For some people a change in personal attitude or perspective is needed in order to motivate coaching. People need to consider safety coaching a shared responsibility to prevent injuries throughout the entire work culture. This requires a shift from an individual to a collective perspective (Triandis, 1977, 1985).

Figure 11.6. Our attitude toward police officers would be more positive if we received deposits along with withdrawals.

Many people accept a collective or team attitude when it comes to work productivity and quality. Coaching for production or quality is part of the job. But coaching for personal safety is often perceived as meddling. People need to understand that safety-related behaviors require as much, if not more, interpersonal observation and feedback as any other job activity.

One way to convince people to accept and support safety coaching as a shared responsibility is to point out their plant's injury record for a certain period of time. While an injury did not happen to them, it did happen to someone, and everyone certainly cares about that. Given this underlying caring attitude, the challenge is to convince others that effective safety coaching by them will reduce injuries to their coworkers. This is enabled by a behavior-based measurement system, as discussed next.

"O" FOR OBSERVE

Safety coaches observe the behavior of others objectively and systematically, with an eye for supporting safe behavior and correcting at-risk behavior. Behavior that illustrates going beyond "the call of duty" for the safety of another person should be especially supported. This is the sort of behavior that contributes sig-

Figure 11.7. Safety coaches are up-front about their intentions, always asking permission before observing.

nificantly to safety improvement and can be increased through rewarding feedback. As illustrated in Figure 11.7, a safety observer does not hide or "spy," and always asks permission first. Only with permission should an observation process proceed.

Observing behavior for supportive and constructive feedback is easy if the coach: a) Knows exactly what behaviors are desired and undesired (an obvious requirement for athletic coaching); and b) takes the time to observe occurrences of these behaviors in the work setting. It is often advantageous—and usually essential—-to develop a checklist of safe and at-risk behaviors and to rank them in terms of risk. Ownership and commitment is increased when workers develop their own behavioral checklists.

DEVELOPING A CRITICAL BEHAVIORAL CHECKLIST. Observation checklists can be generic or job-specific. A generic checklist is used to observe behaviors that may occur during several jobs. A job-specific checklist is designed for one job. Deciding which items to include on a critical behavior checklist (CBC) is a very important part of the coaching process. A CBC enables coaches to look for critical behaviors. A critical behavior is a behavior that:

- Has led to a large number of injuries or near hits in the past
- Could potentially contribute to a large number of injuries or near hits because many people perform the behavior
- Has previously led to a serious injury or a fatality
- Could lead to a serious injury or fatality

If only a few behaviors will be observed in the beginning, which is often a good way to start a large-scale coaching process, a CBC should be designed for only the most critical behaviors. Several sources can be consulted to obtain behaviors for a CBC, including: a) injury records, b) near-hit reports, c) job hazard analyses, d) standard operating procedures, e) rules and procedural manuals, and f) the workers themselves. People already know a lot about their own safety performance. They know which safety rules they sometimes ignore, and they know when a near hit has occurred to themselves or to others because of at-risk behavior. In addition, it's often useful to obtain advice from the plant doctor, nurse, safety director, or anyone else who maintains injury statistics for the plant.

When starting out, don't develop an exhaustive checklist of critical behaviors. A list can get quite long in a hurry. A long list for one-on-one observations can appear overwhelming, and could inhibit the process. As with anything that is new and needs voluntary support, it's useful to start small and build. With practice, people find a CBC easy to use, and they accept additions to the list. They will also contribute in valuable ways to refine the CBC, from clarifying behavioral definitions to recommending behavioral additions and substitutions. The development and use of a CBC is really a continuous improvement process. Further development and refinement benefits coaching observations, and vice versa.

A work group on a mission to develop a CBC needs to meet periodically to select critical behaviors to observe. I have found the worksheet depicted in Figure 11.8 useful in beginning the development of a CBC. Through interactive discussions, work groups define safe and at-risk behaviors in their own work areas relevant to each category. The category on body positioning and protecting, for example, includes specif-

OPERATING PROCEDURES	SAFE OBSERVATION	AT-RISK OBSERVATION
Body Positioning/Protecting Positioning/protecting body parts (e.g., avoiding line of fire, using PPE, equipment guards, barricades, etc.)		
Visual Focusing Eyes and attention devoted to ongoing task(s)		
Communicating Verbal or nonverbal interaction that affects safety		
Pacing of Work Rate of ongoing work (e.g. spacing breaks appropriately, rushing)		
Moving Objects Body mechanics while lifting, pushing/pulling		
Complying with Lockout/Tagout Following procedures for lockout/tagout		
Complying with Permits Obtaining, then complying with permit(s) (e.g., confined space entry, hot work, excavation, open line, hot tap, etc.)		

Figure 11.8. A worksheet for developing a generic critical behavior checklist (CBC).

ic ways workers should protect themselves from environment or equipment hazards. This can range from using certain personal protective equipment to positioning their body parts in certain ways to avoid possible injury.

Some categories in Figure 11.8 may be irrelevant for certain work groups, like locking or tagging out equipment or complying with certain permit policies. A work group might add another general procedural category to cover particular work behaviors. Note that defining safe and at-risk behaviors results in safety training in the best sense of the word. Participants learn exactly what safe behaviors are needed for a particular work process.

A list of work behaviors covering all the generic categories in Figure 11.8 can be extensive and overwhelming. This gives numerous opportunities for coaching feedback, but remember, it takes time and practice to observe behaviors reliably—and to get used to being observed while working. I have found it useful to start the observation procedure with a brief CBC of four or five behaviors, and then build on the list with practice, group discussion, and more practice.

Each of the generic categories in Figure 11.8 could be used as a separate check-

Observer:____________________

Person 1
Date:__________________
Time:__________________
Department:______________
Building:________________
Floor:_______ Area:________

Person 2
Date:__________________
Time:__________________
Department:______________
Building:________________
Floor:_______ Area:________

OPERATING PROCEDURES	SAFE OPERATION	AT-RISK OBSERVATION
Body Positioning/Protecting		
Visual Focusing		
Communicating		
Pacing of Work		
Moving Objects		
Complying with Lockout/Tagout		
Complying with Permits		

1=observations for first person; 2= observations for second person

% Safe Behaviors: $\frac{\text{Total Safe Observations}}{\text{Total Safe and At-Risk Observations}}$ = ______%

Figure 11.9. A comprehensive critical behavior checklist (CBC) for one-on-one observation.

list at the initiation of a coaching process. The first category, for example, could lead to the development of a CBC for observation of personal protective equipment. Specific PPE behaviors for the work area could be listed in a lefthand column, with space on the right to check safe and at-risk observations. This kind of CBC could be used to record the observations of several individuals, by simply adding checks in the safe or at-risk columns for each observation of an individual's use or non-use of a particular PPE item.

SAMPLE CRITICAL BEHAVIOR CHECKLIST. Figure 11.9 depicts a comprehensive CBC for recording the results of a coaching observation. This kind of CBC recording sheet should be used after the participants (optimally, everyone in a particular work area) have derived precise behavioral definitions for each category, and have practiced rather extensively with shorter CBCs. Such practice enables careful refinement of behavioral definitions and builds confidence and trust in the process.

The CBC in Figure 11.9 allows for recording two or more one-on-one coaching observations. A "1" would be placed in either the "safe" or "at-risk" column for each behavioral observation of Person 1, and a "2" would be used to indicate specific "safe" and "at-risk" behaviors for Person 2. Note that only the name of the observer is included on the data sheet. When people realize that safety coaching is only to increase safe behavior and decrease at-risk behavior, not to identify unsafe workers, voluntary participation will increase, along with trust.

SCHEDULING OBSERVATION SESSIONS. There is no best way to arrange for coaching observations. The process needs to fit the setting and work process. This can only happen if workers themselves decide on the frequency and duration of the observations, and derive a method for scheduling the coaching sessions. I've seen the protocol for effective coaching observations vary widely across plants, and across departments within the same plant. And the success of those processes have not varied as a function of protocol.

The 350 employees at one Exxon Chemical plant, for example, designed a process calling for people to schedule their own coaching sessions with any two other employees. On days and at times selected by the person to be observed, two observers show up at the individual's worksite and use a CBC to conduct a systematic, 30-minute observation session. This plant started with only one scheduled observation per month, and observers were selected from a list of volunteers who had received special coaching training. One year later, employees scheduled two observations per month, and any plant employee could be called on to coach.

The Exxon procedure is markedly different than the "planned 60-second actively caring review" implemented at a Hoechst Celanese plant. For this one-on-one coaching process, all employees attempt to complete a one-minute observation of another employee's work practices in five general categories: body position, personal apparel, housekeeping, tools/equipment, and operating procedures. The initial plant goal was for each of the 800 employees to complete one 60-second behavioral observation every day. Results were entered into a computer file for a behavioral safety analysis of the work culture.

The CBC used for the one-minute coaching observations is shown in Figure 11.10. The front of each card includes the five behavioral categories, a column to check "safe" or "at risk" per category, and columns ("feedback targets") to write comments about the observations. These comments facilitate a feedback session following the observation session, if it is convenient. The back of this CBC includes examples ("memory joggers") related to each behavioral category on the front of the card. These examples summarize the category definitions developed by the CBC steering committee, and determine whether "safe" or "at risk" should be checked on the front of the card.

CRITICAL FEATURES OF THE OBSERVATION PROCESS. Duration, frequency, and scheduling procedures of CBC observations vary widely; still, there are a few common features. First and foremost, the observer must ask permission before beginning an observation process. And the name of the person observed must never be recorded. To build trust and increase participation, a "no" to a request to observe must be honored.

Observer:__________ Location:__________ Date:_______

AUDIT CATEGORY	SAFE	AT RISK	FEEDBACK TARGETS	
			SAFE	AT RISK
Position				
Safe Apparel				
Housekeeping				
Tools/Equipment				
Procedures				
Total				

FRONT OF ONE-MINUTE AUDIT CARD

OBSERVATION TARGETS	SAFE	AT RISK	OBSERVATION TARGETS	SAFE	AT RISK
Position * Line of Fire * Falling * Pinch Points * Lifting			**Tools/Equipment** * Condition * Use * Guards		
Safe Apparel * Hair * Clothes * Jewelry * PPE			**Procedures** * SOPs * JSAs * Permits * Lockout * Barricade * Equipment Release		
Housekeeping * Floor * Equipment * Storage of materials					

BACK OF ONE-MINUTE AUDIT CARD

Figure 11.10. The critical behavior checklist (CBC) used for one-minute observations.

Asking permission to observe serves notice to work safely, and thus biases the observation data, right? In other words, when workers give permission to be coached, their attention to safety will likely increase and they will try to follow all safety procedures. It's possible, though, for people to overlook safety precautions, even when trying their best. They could be unconsciously at risk. When people give permission to be coached, their willingness to accept and appreciate feedback, even when it is corrective, is maximized.

What if people sneak around and conduct behavioral observations with no warning? This is indeed an unbiased plant-wide audit of work practices. It might even be accepted if those observed were not identified. However, if one-on-one coaching is added to this procedure, an atmosphere of mistrust can develop.

Safety coaching should not be a way to enforce rules or play "gotcha." It needs to be seen as a process to help people develop safe work habits through supportive and constructive feedback. Giving corrective feedback after "catching" an individual off-guard performing an at-risk behavior will likely lead to defensiveness and lack of appreciation, even for a well-intentioned effort. It can also reduce interpersonal trust and alienate a person toward the entire safety coaching process.

Feedback is essential. Each observation process with a CBC provides for tallying and graphing results as "percent safe behavior" on a group feedback chart. The CBC shown in Figure 11.9 includes a formula at the bottom for calculating "percent safe behavior" per coaching session. In this case, all checks for safe observations are added and divided by the total number of checks (safe plus at-risk behaviors). The result is multiplied by 100 to yield "percent safe behavior."

Applying the formula in Figure 11.9 to checks written on the front of the CBC shown in Figure 11.10 results in a conservative estimate of "percent safe behavior." That's because a "safe" check mark on the CBC in this application meant that each separate behavior of a certain category was marked "safe" on the back. Thus, this calculation required all behaviors relating to a particular observation category to be safe for a "safe" designation. The CBC shown in Figure 11.9 does not use this all-or-nothing calculation, and generally results in higher percentages of safe behaviors.

Figure 11.11. Feedback from a critical behavior checklist can be given one-on-one and in groups.

There is no best way to do these calculations. What's important is for participants to understand the meaning of the feedback percentages. As shown in the lower half of Figure 11.11, these percentages can be readily displayed on a group feedback chart or graph. While feedback percentages are valuable, it is vital to realize that the process is more important than the numbers.

The true value of the coaching process is not in the behavioral data, which are no doubt biased by uncontrollable factors, but in the behavior-based interaction between employees. I've actually seen observers get so caught up in recording the numbers, such as frequency of safe and at-risk behaviors, that they let coworkers continue to perform an at-risk behavior while they observe and check columns on a CBC.

An individual's safety must come before the numbers in any observation process. When observers see an at-risk behavior that immediately threatens a person's health or safety, they should intervene at once. They can usually pick up the observation process afterwards. On the other hand, if the CBC was partially completed before they stepped in, it might be most convenient to communicate other observations, especially if there are some safe behaviors to report. This "deposit" will help compensate for the "withdrawal" that was probably implicated by the need to stop a risky behavior at once.

Observation procedures should always include a provision for one-on-one feedback. The employees who designed one protocol decided to make feedback optional in order to increase participation in the observation process, and at least obtain group feedback from all departments. The number of CBC cards collected per department in that instance was exhibited in a large display case at the plant entrance, along with daily percentages of observations resulting in one-on-one feedback.

"A" FOR ANALYZE

When interpreting observations, safety coaches draw on their understanding of the three-term contingency or ABC principle (for Activator, Behavior, Consequence) discussed earlier in Chapter 8. They realize observable reasons usually exist for why safe or risky behaviors occur. They know certain dangerous behaviors are triggered by activators such as work demands, risky example-setting by peers, and inconsistent messages from management. They also appreciate the fact that at-risk behaviors are often motivated by one or more natural consequences, including comfort, convenience, work breaks, and approval from peers or work supervisors.

This understanding is critical if safety coaching is to be a "fact finding" rather than "fault finding" process. And it leads to an objective and constructive analysis of the situations observed. This is how people discover the reasons behind at-risk behaviors, and design interventions to decrease them.

An ABC analysis can be done before giving feedback to the person observed, or during the one-on-one feedback process. Discussing the activators and consequences that possibly influenced certain work practices can lead to environment or system improvements for decreasing at-risk behavior.

- Was the behavior observed activated by a work demand or a desire to go on break or leave work early?
- Does the design of the equipment or environment, or the ergonomic design of the task, influence at-risk behavior?
- Is certain personal protective equipment uncomfortable or difficult to use?
- Are fellow workers or supervisors activating dangerous behavior by requesting or demanding an excessive work pace?
- Are certain people motivating at-risk behavior from others by giving rewarding consequences, like words of appreciation, for work done quickly at the expense of safety?

Answers to these and other questions are explored in the next phase of safety coaching—the heart of the process.

"C" FOR COMMUNICATE

A good coach is a good communicator. This means being an "active" listener and persuasive speaker. Since none of us are born with these skills, communication training sessions that incorporate role-playing exercises can be invaluable in developing the confidence and competence needed to send and receive behavioral feedback. Such training should emphasize the need to separate behavior (actions) from person factors (attitudes and feelings). This enables corrective feedback without "stepping on" feelings.

People need to understand that anyone can be at risk without even realizing it, as in "unconscious incompetence," and performance can only improve with behavior-specific feedback. Once this fact is established, corrective feedback that is appropriately given will be appreciated, regardless of who's giving the feedback. Work status is not a factor.

INITIATING FEEDBACK. I remember key aspects of effective verbal presentation with the ***SOFTEN*** acronym listed in Figure 11.12. First, it's important for the observer to initiate communication with a friendly smile and an open (flexible) perspective. "Territory" reflects the need to respect the fact that you are encroaching on another person's work area. You should ask the person where it would be appropriate and safe to talk. It's also important to maintain a proper physical distance during this interaction.

Standing too close to another person can cause interference and discomfort. Edward Hall (1959, 1966) coined the term "proxemics" to refer to how we manage space, and he researched the distances people keep from each other in various situ-

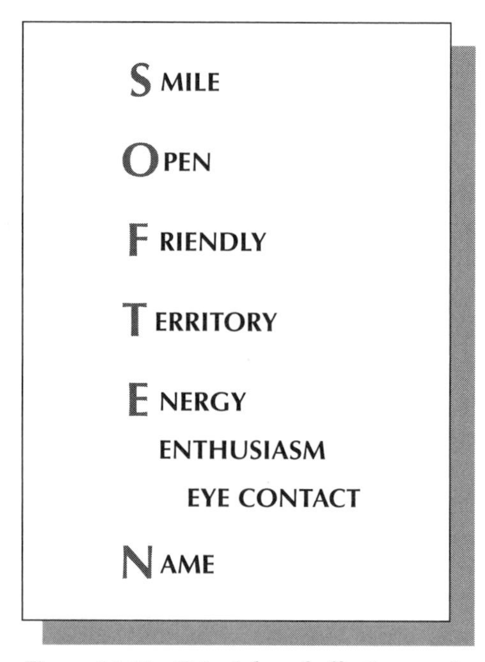

Figure 11.12. Principles of effective sending can be remembered with SOFTEN.

ations. There are prominent cultural differences in interpersonal distance norms. In the United States, communicating closer than 18 inches with another person—measured nose to nose—is considered an intimate distance (Hall, 1966), with 0 to 6 inches reserved for comforting, protesting, lovemaking, wrestling, and other full-contact behaviors. The far phase of the intimate zone (6 to 18 inches) is used by individuals who are on very close terms.

Safety coaches in the United States should most likely communicate at a "personal distance" (18 inches to 4 feet). According to Hall (1966), the near phase of the personal distance (18 to 30 inches) is reserved for those who are familiar with one another and on good terms. The far phase of the personal zone (2.5 to 4 feet) is typically used for social interactions between friends and acquaintances. This is likely to be the most common interaction "zone" for a workplace coaching session. Some coaching communication might occur at the near phase (4 to 7 feet) of Hall's "social distance," which is typical for unacquainted individuals interacting informally. These distance recommendations are not hard and fast rules of conduct, but rather personal territory norms we need to consider .

The "E" of our acronym represents three important directives to remember when coaching—energy, enthusiasm, and eye contact. Your energy and enthusiasm can activate concern and caring on the part of the person you're communicating with. We all know that excited, committed coaches can make "true believers" out of the troops—and that indifferent or distracted coaches can have the opposite effect. Actually, as depicted in Figure 11.13, our body language speaks louder than words.

Figure 11.13. Body language communicates more than words.

Proper eye contact is body language critically important to maintain throughout a coaching session. Everyone has experienced the uncomfortable feeling of talking or listening to someone who does not look at them. In contrast, there's the piercing stare we sometimes perceive with too much eye-to-eye contact. Thus, we learn through natural feedback, often unconsciously, the definition of "proper eye contact."

Finally, we need to remember that the dearest word to someone's ears is his or her own name (Carnegie, 1936). Refer to the other person by name, but make it clear the behavioral observations you have recorded will remain anonymous.

THE POWER OF FEEDBACK. A critical part of communicating is feedback. Its power is evident in the famous Hawthorne studies. Ask any safety manager, industrial consultant, or applied psychologist whether they have heard of the "Hawthorne Effect," and they probably will say, "yes." They might not be able to describe any details of the studies that occurred between 1927 and 1932 at the Western Electric plant in the Hawthorne community near Chicago that led to the classic "Hawthorne Effect." Most, however, will be able to paraphrase the infamous finding from these studies that the hourly output rates of the employees studied increased whenever an obvious environmental change occurred in the work setting.

The explanation of the Hawthorne results is also well-known, and recited as a potential confounding factor in numerous field studies of human behavior. Specifically, it is commonly believed that the Hawthorne studies showed that people will change their behavior in desired directions when they know their behavior is being observed. The primary Hawthorne sources (Mayo, 1933; Roethisberger & Dickson, 1939; Whitehead, 1938) leave us with this impression, and in fact this interpretation seems intuitive. This is really another common-sense fallacy. The fact is, this interpretation of the Hawthorne studies is not accurate—it is nothing but a widely disseminated myth.

H. McIlvaine Parsons conducted a careful reexamination of the Hawthorne data and interviewed eyewitness observers, as well as one of the five female relay assemblers who were the primary subjects of the studies. Parsons' findings were published in a seminal *Science* article entitled "What happened at Hawthorne?" (Parsons, 1974). What happened was that the five women observed systematically in the Relay Assembly Test Room received regular feedback about the number of relays each had assembled. "They were told daily about their output, and they found out during the working day how they were doing simply by getting up and walking a few steps to where a record of each output was being accumulated" (Parsons, 1980, p. 58).

From his scientific detective work, Parsons concluded that performance feedback was the principal extraneous, confounding variable that accounted for the Hawthorne Effect. The performance feedback was important to the workers (so they were apt to respond to it), because their salaries were influenced by an individual piecework schedule—the more relays each employee assembled, the more money each earned.

There's one other point I'd like to make about the Hawthorne studies. The five test subjects preferred working in the test room rather than in the regular department. But when asked why, they didn't mention anything about receiving feedback. Instead, their reasons included "smaller groups," "no bosses," "less supervision," "freedom," and "the way we are treated" (Roethlisberger & Dickson, 1939, pp. 66-67). This is worth noting, because it suggests you should not rely only on verbal reports to discover factors influencing work performance. Sometimes people are not aware of the basic contingencies controlling their behavior. Through systematic and objective observation these factors can be uncovered—and instructive feedback given.

FEEDBACK FOR SAFETY. For anyone who has studied the behavior-based approach to performance management, the only surprise in Parsons' research is that the critical role of performance feedback was not documented in the original reports of the Hawthorne studies. Numerous research studies have shown that posting results of behavioral observations related to safety, production, or quality has a positive impact on targeted work behaviors. If desired work behaviors are targeted, they increase in frequency; when undesired behaviors are targeted for observation and feedback, they decrease in probability (Geller *et al,* 1980; Kim & Hammer, 1982; Komaki *et al,* 1980; Krause *et al,* 1990; Sulzer-Azaroff, 1982).

INDIVIDUAL FEEDBACK. Whether the aim is to support or correct behavior, feedback should be specific and timely. It should specify a particular behavior, and occur soon after the target behavior is performed. Also, it should be private, given one-on-one to avoid any interference or embarrassment from others. Corrective feedback is most effective if the alternative safe behavior is specified and potential solutions for eliminating the at-risk behavior are discussed.

Anyone giving feedback must actively listen to reactions. This is how a safety coach shows sincere concern for the feelings and self-esteem of the person on the receiving end of feedback. The best listeners give empathic attention with facial cues and posture, paraphrase to check understanding, prompt for more details ("Tell me more"), accept stated feelings without interpretation, and avoid arrogance (such as, "When I worked in your department, I always worked safely").

Figure 11.14 reviews the critical characteristics of effective rewarding (or supportive) and correcting feedback. This figure can be used as a guide for group practice sessions. Because it's not easy to give safety feedback properly and because many people feel awkward or uncomfortable doing it, practice is important.

In training sessions, I ask groups of three to seven individuals to develop a brief skit that illustrates rewarding or correcting feedback. Each skit involves at least three participants: a safety coach, a worker receiving feedback, and a narrator who sets the scene. The exercise can be more fun if groups first demonstrate the wrong way to give feedback, and then show the correct procedure. Afterward, the audience should provide feedback. The presenters should hear about particular strengths of their demonstration and places where it could be improved. A good facilitator can draw out important lessons from this communication exercise, while keeping the atmosphere congenial and enjoyable. It has to be informative and rewarding for all involved, because

USE REWARDING FEEDBACK TO SUPPORT SAFE BEHAVIOR

- Give the feedback one-on-one and privately
- Give the feedback as soon as possible after the observation process
- Identify the safe behavior(s) observed
- Be sincere and genuine
- Express personal appreciation for setting the right example for others

Use Correcting Feedback to Decrease At-Risk Behavior

- Give the feedback one-on-one and privately
- Give the feedback as soon as possible after the observation process
- Begin with acknowledgment of safe behavior(s) observed
- Identify the at-risk behavior(s) observed
- Specify the safe alternative to the at-risk behavior(s)
- Indicate concern for the person's welfare
- Request commitment to avoid the at-risk behavior(s)
- Thank the individual for commitment to continuous improvement

Figure 11.14. Key characteristics of rewarding and correcting feedback.

everyone will need to participate in several demonstrations. It takes practice and peer support for participants to feel comfortable and effective when giving rewarding and correcting feedback. Fortunately, we learn much from watching others perform (Bandura, 1969), and so the audience learns by vicariously observing demonstrations by peers.

GROUP FEEDBACK. As we've discussed, observations recorded on the CBC can be summarized as a calculation of "percent safe behavior." These percentages can be calculated per day or per week or per month and displayed on a feedback graph (see Figure 11.11). When this graph is posted in a prominent place, perhaps next to the plant's statistical process control charts, employees can monitor their progress and be naturally rewarded for their efforts. This gives safety the same status as quality, and promotes group achievement. People can monitor the progress of their work team as its percentage of safe behaviors increases.

Percentages of safe observations can be calculated separately for various workplace activities (see Figures 11.9 and 11.10). Graphs can readily show the percentage of employees wearing safe apparel, using appropriate personal protective equipment, avoiding the line of fire, lifting or moving objects safely, keeping work areas neat and free of trip hazards, using tools and equipment safely, and complying with lockout/tagout procedures or work-permit requirements.

Monitoring behavioral categories separately lets you see what needs special attention. A variety of interventions may be called for, from implementing special training sessions to ergonomically rearranging a particular work area. Observation and feedback provide invaluable diagnostic information. The graphs hold people accountable for process numbers they can control on a daily basis—in contrast to outcome numbers like total recordable injury rate or workers' compensation costs. Improving these upstream numbers (percent safe behavior) will eventually reduce the outcome number (work injuries).

"H" FOR HELP

The word *help* summarizes what a safety coach is all about. The purpose is to help an individual prevent injury by supporting safe work practices and correcting at-risk practices. It's critical, of course, that a coach's help is accepted. The four letters in the word ***help*** outline strategies to increase the probability that a coach's advice, directions, or feedback will be appreciated.

***H*UMOR.** Safety is certainly a serious matter, but sometimes a little humor can add spice to our communications, increasing interest and acceptance. It can take the sting out of what some find to be an awkward situation. In fact, researchers have shown that laughter can reduce distress and even benefit our immune systems (Goodman, 1995; National Safety News, 1985).

***E*STEEM.** People who feel inadequate, unappreciated, or unimportant are not as likely to go beyond "the call of duty" to benefit the safety of themselves or others as are people who feel capable and valuable (see Chapter 13 for support of this argument).

The most effective coaches choose their words carefully, emphasizing the positive over the negative, to build or avoid lessening another person's self-esteem. Although Figure 11.15 is humorous, it is unfortunately an accurate portrayal of the atmosphere

Figure 11.15. Feedback communications more often depreciate than appreciate a person's self-esteem.

in many organizational cultures, including the university environment in which I've worked for almost 30 years.

LISTEN. One of the most powerful and convenient ways to build self-esteem is to listen attentively to another person. This sends the signal that the listener cares about the person and his or her situation. And it builds self-esteem: "I must be valuable to the organization because my opinion is appreciated." After a safety coach listens actively, his or her message is more likely to be heard and accepted. As Covey (1989) put it, "seek first to understand, then to be understood" (p.235).

PRAISE. Praising others for their specific accomplishments is another powerful way to build self-esteem. And if the praise targets a particular behavior, the probability of the behavior recurring increases. This reflects the basic principle of positive reinforcement, and motivates people to continue their safe work practices and look out for the safety of coworkers.

Behavior-focused praising is a powerful rewarding consequence which not only increases the behavior it follows, but also increases a person's self-esteem. This in turn increases the individual's willingness to actively care for the safety of others, as I discuss more completely in Part Four of this text.

Human nature directs more attention to mistakes than successes. As illustrated in Figure 11.16, errors stick out and disrupt the flow, so they attract reaction and attempts to correct them. In contrast, when things are going smoothly—and safely—there is usually no stimulus to signal success. A person's good performance is typically taken for granted. We need to resist the tendency to go with the flow, and sometimes express sincere appreciation for ongoing safe behavior.

Figure 11.16. We selectively see more negatives than positives.

What Can a Safety Coach Achieve?

The safety coaching process described here is founded on the basic premise of behavior-based safety: Injuries are a direct function of at-risk behaviors, and if these behaviors can be decreased and safe behaviors increased, injuries will be prevented. Indeed, the well-known Heinrich Law of Safety implicates "unsafe acts" as the root cause of most near hits and injuries (Heinrich *et al,* 1980).

Over the past 20 years, a variety of behavior-based research studies have verified this aspect of Heinrich's Law by systematically evaluating the impact of interventions designed to increase workers' safe behaviors and decrease their at-risk behaviors. Feedback from behavioral observations was a common ingredient in most of the successful interventions, whether it was delivered through tables, charts, interpersonal communication, congratulatory notes, or a reward following a particular behavior (see, for example, the comprehensive review by Dan Petersen, 1989; or individual research articles by Chhokar & Wallin, 1984; Geller *et al,* 1980; Komaki *et al,* 1980; Sulzer-Azaroff & De Santamaria, 1980).

The behavior-based feedback and coaching process described here is analogous to the behavior-based safety process detailed by Krause *et al* (1990) and McSween (1995), and taught in training videotapes and workbooks developed by Tel-A-Train, Inc. (1995). In addition to the research referred to above, there are hundreds of real-world case studies that provide evidence of the injury prevention impact of behavior-based coaching. I have personally been teaching variations of this approach to industry for more than 20 years and have never seen the process fail to work when implemented properly. But I have witnessed numerous cases of companies receiving less than desired benefits due to incomplete or inadequate implementation.

There is no quick-fix substitute for this process, no effective step-by-step cookbook. Achievements from safety coaching are a direct function of the effort put into it. The guidelines presented in this chapter need to be customized. Who knows best what step-by-step coaching procedures will succeed in a given work area? The people employed there know best, and they need to be empowered to develop their own safety coaching process.

I recently returned from visiting an Exxon Chemical facility in Texas that demonstrated exemplary success with an actively caring coaching process. By the end of two years, they had almost 100 percent participation, and have reaped extraordinary benefits. From an outcome perspective, they started with a baseline of 13 OSHA recordable injuries in 1992 (TRIR = 4.11), and progressed to 5 OSHA recordables in 1993 (TRIR = 1.60). They sustained only one OSHA recordable in 1994 (TRIR = 0.35). At the time of this writing (mid-1995), they have yet to have a single OSHA recordable this year. Obviously, many factors contributed to this performance, but there is little doubt their safety coaching process played a critical role. At the end of 1994, for example, 98 percent of the workforce had participated as observers to complete a total of 3,350 documented safety coaching sessions. They identified 51,048 behaviors, of which 46,659 were safe and 4,389 were at risk.

In 1992, a safety culture survey was administered before the safety coaching process was initiated. It was repeated again in 1994. Results revealed statistically significant improvement in:

1. Perceptions and attitudes toward industrial safety
2. Intentions to actively care for other workers' safety
3. Feelings of belongingness and group cohesion throughout the work culture

It's important to realize that these dramatic improvements in safety perceptions, attitudes, and intentions occurred while Exxon and the petrochemical industry experienced significant downsizing.

The self-survey in Figure 11.17 reflects attributes of ideal safety coaches. Several of the concepts—particularly self-esteem, self-confidence, optimism, and teamwork—are discussed in more detail in Part Four of this book. By rating how often you accomplish the ideal coaching characteristic implied by each item in the questionnaire, you will review key points of this chapter.

If you're honest and frank, you'll gain important insight from this exercise. Define your strengths and weaknesses, then apply what you've learned here to improve your competence as an actively caring safety coach.

	NEVER	RARELY	SOMETIMES	OFTEN	MOST TIMES
1. I take a balanced approach to long-term goals and short-term results.	1	2	3	4	5
2. I give credit for a job well done.	1	2	3	4	5
3. I refer to specific observable actions when discussing a worker's performance.	1	2	3	4	5
4. I explain the rationale for policies and procedures.	1	2	3	4	5
5. I provide both direction and encouragement when requesting behavior change.	1	2	3	4	5
6. I avoid talking down to other workers I supervise.	1	2	3	4	5
7. I practice active listening.	1	2	3	4	5
8. I display a sense of humor.	1	2	3	4	5
9. I only make promises I can keep.	1	2	3	4	5
10. I work with others to set performance standards.	1	2	3	4	5
11. I treat others fairly.	1	2	3	4	5
12. I express interest in the career growth of workers I supervise.	1	2	3	4	5
13. I ask others for ideas and opinions.	1	2	3	4	5
14. I promote feelings of ownership among team members.	1	2	3	4	5
15. I demonstrate personal integrity when dealing with others.	1	2	3	4	5

Figure 11.17. Self-appraisal of coaching skills ***(continued on the next page)***

Continued from previous page

	NEVER	RARELY	SOMETIMES	OFTEN	MOST TIMES
16. I practice principles of appropriate rewarding feedback.	1	2	3	4	5
17. I practice principles of appropriate correcting feedback.	1	2	3	4	5
18. I take ownership and responsibility for personal decisions.	1	2	3	4	5
19. I treat others with dignity and respect.	1	2	3	4	5
20. I encourage and accept performance feedback from others.	1	2	3	4	5
21. I find ways to celebrate others' accomplishments.	1	2	3	4	5
22. I accept others' failures as opportunities to learn.	1	2	3	4	5
23. When appropriate I challenge higher level management.	1	2	3	4	5
24. I create an atmosphere of interpersonal trust.	1	2	3	4	5
25. I evaluate others' performance as objectively as possible.	1	2	3	4	5
26. I help team members solve problems constructively.	1	2	3	4	5
27. I act to support the value that "people are our most important asset."	1	2	3	4	5
28. I encourage others to participate actively in conversations, discussions, and meetings.	1	2	3	4	5
29. I show sensitivity to the feelings of others.	1	2	3	4	5
30. I promote synergy among team members.	1	2	3	4	5
31. I promote a win/win approach to problem solving.	1	2	3	4	5
32. I promote others' self-esteem.	1	2	3	4	5
33. I promote others' sense of personal control.	1	2	3	4	5
34. I promote others' perceptions of self-confidence.	1	2	3	4	5
35. I am optimistic.	1	2	3	4	5
36. I encourage teamwork.	1	2	3	4	5

In Conclusion

Safety coaching is a key intervention process for developing and maintaining a Total Safety Culture. In fact, the more employees who effectively apply the principles of safety coaching discussed here, the closer an organization will come to achieving a Total Safety Culture. The same is true for preventing injury in the community and among our immediate family members at home. Indeed, we need to practice the principles of safety coaching in every situation where an injury could occur following at-risk behavior.

Systematic safety coaching throughout a work culture is certainly feasible in most settings. Large-scale success requires time and resources to develop materials, train necessary personnel, establish support mechanisms, monitor progress, and continually improve the process and support mechanisms whenever possible. For example, the following questions need to be answered at the start of developing an initial action plan:

- Who will develop the critical behavior checklist (CBC)?
- How extensive will the first CBC be?
- What information will be used to define critical behaviors?
- How will safety coaches be trained and receive practice and feedback?
- How many coaches will be trained initially, and how can additional people volunteer to participate as safety coaches?
- How will the coaching sessions be scheduled, how often will people be coached, and how long will the coaching sessions last?
- Where will the group feedback graphs be posted, and who will be responsible for preparing the displays of safe behavior percentages?
- Who will be on the steering committee to oversee the safety coaching process, answer these and other questions about process implementation, maintain records, monitor progress, and refine procedural components whenever necessary?

This doesn't cover all the issues, yet the list might appear overwhelming at first. Organizational cultures vary widely according to personnel, history, policy, the work process, environmental factors, and current contingencies. So implementation procedures need to be customized. There must be significant input from the people protected by the coaching process and from whom long-term participation is needed. It's likely to take significant time, effort, and resources to achieve a plant wide safety coaching process. With this end in mind, I recommend starting small to build confidence and optimism on small-win accomplishments. Then with patience and diligence, set long-term goals for continuous improvement. Remember to celebrate achievements that reflect successive approximations of your vision—an organization of people who consistently coach each other effectively to increase safe work practices and decrease at-risk behaviors.

Safety coaching is a critically important intervention approach, but keep in mind the many other ways you can contribute to the health and safety of a work culture. In other words, safety coaching is one type of intervention for the ***I*** stage of the ***DO IT*** process. Any variety of activator and consequence strategies explained in Chapters 9 and 10, respectively, can be used as a behavior-based intervention. These activities require people to go beyond their normal routine to help another person.

The next three chapters address the challenge of increasing people's willingness to intervene on behalf of another person's safety or health. As you know, I call this "actively caring for safety." First, you'll learn what psychological research has revealed regarding conditions and person characteristics that influence the willingness to actively care. Then I'll discuss ways of integrating behavior-based and person-based psychology to increase active caring throughout an organization, community, neighborhood, and family. When we teach people the appropriate tools for improving behavior, as presented in Part Three of this text, and show them how to increase their willingness to use the tools as actively caring intervention agents, we are en route to achieving a Total Safety Culture.

REFERENCES

Bandura, A. (1969). *Principles of behavior modification*. New York: Holt, Rinehart and Winston.

Carnegie, D. (1936). *How to win friends and influence people*. New York: Simon and Schuster.

Chhokar, J. S., & Wallin, J. A. (1984). A field study of the effects of feedback frequency on performance. *Journal of Applied Psychology*, 69, 524-530.

Cone, J. D., & Hayes, S. C. (1980). *Environmental problems: Behavioral solutions*. Monterey, CA: Brooks/Cole.

Covey, S. R. (1989). *The seven habits of highly effective people*. New York: Simon and Schuster.

Eysenck, H. J. (1976). *The structure of human personality* (Third Edition). London: Methuen.

Eysenck, H. J., & Eysenck, M. W. (1985). *Personality and individual differences: A natural science approach*. New York: Plenum Press.

Geller, E. S. (1991). If only more would actively care. *Journal of Applied Behavior Analysis*, 24. 607-612.

———. (1992). Applications of behavior analysis to prevent injuries from vehicle crashes. Cambridge Center for Behavioral Studies Monograph Series: *Progress in Behavioral Studies*, Monograph #2. Cambridge, MA: Cambridge Center for Behavioral Studies.

———. (1994). Ten principles for achieving a total safety culture. *Professional Safety*, 39(9), 18-24.

———. (1995). Safety coaching: Key to achieving a total safety culture. *Professional Safety*, 40 (7), 16-22.

———. (1996). Managing the human element of occupational health and safety. In R. W. Lack (Ed.) *Essentials of safety and health management*. Boca Raton, FL: CRC Press/Lewis Publishers.

Geller, E. S., Berry, T. D., Ludwig, T. D., Evans, R. E., Gilmore, M. R., & Clarke, S. W. (1990). A conceptual framework for developing and evaluating behavior change interventions for injury control. *Health Education Research: Theory & Practice*, 5(2), 125-137.

Geller, E. S., Eason, S. L., Phillips, J. A., & Pierson, M. D. (1980). Intervention to improve sanitation during food preparation. *Journal of Organizational Behavior Management*, 2(3), 229-240.

Geller, E. S., Winett, R. A., & Everett, P. B. (1982). *Preserving the environment: New strategies for behavior change*. New York: Pergamon Press.

Glenwick, D. S., & Jason, L. A. (1980) (Eds.). *Behavioral community psychology: Progress and prospects*. New York: Praeger.

——— & ———. (1993) (Eds.). *Promoting health and mental health in children, youth, and families*. New York: Springer Publishing Company.

Goodman, J. B. (1995). Laughing matters: Taking your job seriously and yourself lightly. *Journal of the American Medical Association,* 267(13), 1858.

Greene, B. F., Winett, R. A., Van Houten, R., Geller, E. S., & Iwata, B. A. (1987) (Eds.). *Behavior analysis in the community: Readings from the Journal of Applied Behavior Analysis.* Lawrence, KS: University of Kansas Press.

Hall, E. T. (1959). *The silent language.* Garden City, NY: Doubleday.

———. (1966). *The hidden dimension.* Garden City, NY: Doubleday.

Heinrich, H. W., Petersen, D., & Roos, N. (1980). *Industrial accident prevention: A safety management approach* (Fifth Edition). New York: McGraw Hill.

Katz, E., & Lazarfeld, P.E. (1955). *Personal influence: The part played by people in the flow of mass communication.* Glencoe, Il: Free Press.

Kim, J., & Hamner, C. (1982). Effect of performance feedback and goal setting on productivity and satisfaction in an organizational setting. *Journal of Applied Psychology,* 61, 48-57.

Komaki, J., Heinzmann, A. T., & Lawson, L. (1980). Effect training and feedback: Component analysis of a behavioral safety program. *Journal of Applied Psychology,* 65 (3), 261-270.

Krause, T. R., Hidley, J. H., & Hodson, S. J. (1990). *The behavior-based safety process: Managing involvement for an injury-free culture.* New York: Van Nostrand Reinhold.

Mayo, E. (1933). *The human problems of an industrialized civilization.* Boston, MA: Harvard University Graduate School of Business Administration.

McSween, T. E. (1995). *The values-based safety process: Improving your safety culture with a behavioral approach.* New York: Van Nostrand Reinhold.

National Safety News. (1985). Laughter could really be the best medicine, p. 15.

Parsons, H. M. (1980). Lessons for productivity from the Hawthorne studies. *Proceedings of Human Factors Symposium sponsored by the Metropolitan Chapter of the Human Factors Society,* pp. 57-67. New York: Columbia University.

Parsons, H. M. (1974). What happened at Hawthorne? *Science,* 183, 922-932.

Petersen, D. (1989). *Safe behavior reinforcement.* New York: Aloray, Inc.

Phares, E. J. (1973). *Locus of control: A personality determinant of behavior.* Morristown, NJ: General Learning Press.

Rothlisberger, F. J., & Dickson, W. J. (1939). *Management and the worker.* Cambridge, MA: Harvard University Press.

Skinner, B. F. (1974). *About behaviorism.* New York: Alfred A. Knopf.

Strickland, B. R. (1989). Internal-external control of expectances: From contingency to creativity. *American Psychologist,* 44, 1-12.

Sulzer-Azaroff, B. (1982). Behavioral approaches to occupational health and safety. In L. W. Frederiksen (Ed.) *Handbook of organizational behavior management.* New York: John Wiley & Sons.

Sulzer-Azaroff, B., & De Santamaria, M. C. (1980). Industrial safety hazard reduction through performance feedback. *Journal of Applied Behavior Analysis,* 13, 287-295.

Tel-A-Train, Inc. (1995). *Actively caring for safety.* Dallas, TX: Wescott Communications.

Triandis, H. C. (1977). *Interpersonal behavior.* Monterey, CA: Brooks/Cole.

——— (1985). The self and social behavior in differing cultural contexts. *Journal of Personality and Social Psychology,* 96, 506-520.

Whitehead, T. N. (1938). *The industrial worker.* Cambridge, MA: Harvard University Press.

Part Four

Actively Caring for Safety

Chapter 12
Understanding Actively Caring

Actively caring is planned and purposeful behavior, directed at environment, person, or behavior factors. It is reactive or proactive, and direct or indirect. Direct, proactive, and behavior-focused caring is most challenging, but in many situations it is most important for large-scale injury prevention. This chapter discusses conditions and situations that inhibit actively caring behavior. We need to understand why people resist opportunities to actively care for safety. Then we can develop interventions to increase this desired behavior, which is critical for achieving a Total Safety Culture.

We cannot live only for ourselves. A thousand fibers connect us with our fellow men; and among those fibers, as sympathetic threads, our actions run as causes; and they come back to us as effects.

This quotation from Herman Melville appeared in a popular paperback entitled *Random Acts of Kindness* (p. 31). Here the editors of Conari Press (1993) introduced the idea of randomly showing kindness or generosity toward others for no ulterior motive except to benefit humanity. This notion seems quite analogous to the actively caring concept I've discussed earlier in various contexts. Indeed, a recurring theme in this book is that a Total Safety Culture can only be achieved if people intervene regularly to protect and promote the safety and health of others. But actively caring is not usually random. It is planned and purposeful. Plus, as implied in Melville's quote, actively caring behaviors (actions) are supported by positive consequences (effects). Sometimes the consequences are immediate, as when someone expresses their appreciation for an act of caring. Or they are delayed but powerful, as in working with care to develop a safer work setting and prevent injuries.

Part Four of this text addresses the need to increase actively caring behavior throughout a culture, and to get the maximum safety and health benefits from this type of behavioral intervention. Psychologists have identified conditions and person states that influence people's willingness to actively care for the safety or health of others. I'll present these and link them to practical things we can do to increase the occurrence of active caring.

While the concept of "random acts of kindness" is thoughtful, benevolent, and clearly related to actively caring behavior, I propose a more systematic goal-directed approach with this concept. I suggest we define actively caring behaviors that give us the "biggest bang for our buck" in particular situations, and then manage situations and response-consequence contingencies to increase the frequency of such behaviors.

In Part Three I presented techniques that actively caring intervention agents could use to increase safe behaviors and reduce at-risk behaviors. I propose that we practice systematic and purposeful acts of kindness to keep other people safe and healthy. We clearly need more of this in our society. Before examining ways to increase actively caring behaviors, though, it's necessary to define the concept more precisely and objectively.

What is Actively Caring?

Figure 12.1 presents a simple flow chart summarizing the basic approach to culture change presented up to this point. We start a culture-change mission with a vision or ultimate purpose—for example, to achieve a Total Safety Culture. With group consensus supporting the vision, we develop procedures or action plans to accomplish our mission. These are reflected in process-oriented goals which hopefully activate goal-related behaviors.

It's revealing that many consultants and pop psychologists stop here. As I've indicated earlier, the popular writings of Covey (1989,1990), Peale (1952), Kohn (1993), and Deming (1986,1993) suggest that behavior is activated and maintained by self-affirmations, internal motivation and personal principles or values. For example, I heard Joel Barker (1993), the futurist who convinced us to change the dictionary meaning of "paradigm," proclaim that "vision alone is only dreaming and behavior alone is only marking time." Mr. Barker explained, however, that turning vision into goals that specify behaviors will lead to positive organization change.

Appropriate goal-setting, as I described in Chapter 9, self-affirmations, and a positive attitude can activate behaviors to achieve goals and visions. But we must not forget one of B. F. Skinner's most important legacies—"selection by consequences" (Skinner, 1981). As depicted in Figure 12.1, consequences are needed to support the right behaviors and correct wrong ones. Without support for the "right stuff," good

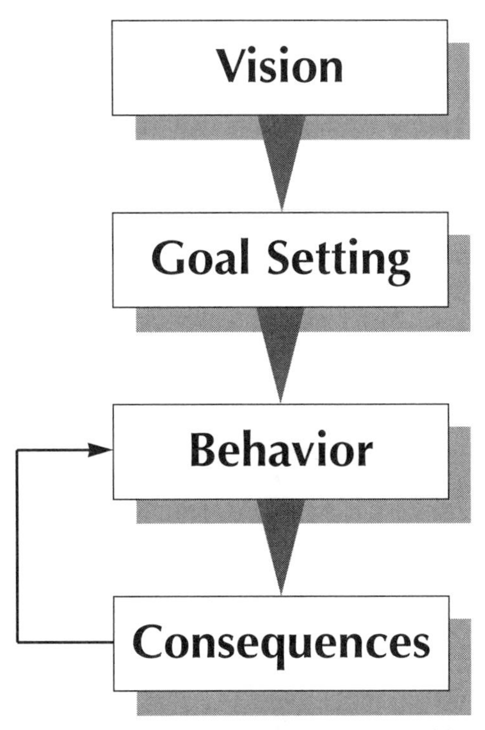

Figure 12.1. A Total Safety Culture requires vision and behavior management.

intentions and initial efforts can fade away. Sometimes natural consequences are available to motivate desired behaviors, but often—especially in safety—consequence-contingencies need to be managed to motivate the behavior needed to achieve our goals. As discussed in Chapter 10, we might be able to eliminate positive consequences that motivate undesired behavior. More often, however, it's necessary to add some positive consequences for desired behavior.

In Figure 12.2 a new box is added to the basic flow diagram in Figure 12.1. My point is simple but extremely important: Vision, goals and consequence-contingencies are not sufficient for culture change. People need to actively care about goals, action plans, and consequences. They need to believe in and own the vision. They need to feel obligated to work toward attaining goals that support the vision. And they need to give rewarding or corrective feedback to increase behaviors consistent with vision-relevant goals. This is the key to continuous improvement and achieving a Total Safety Culture.

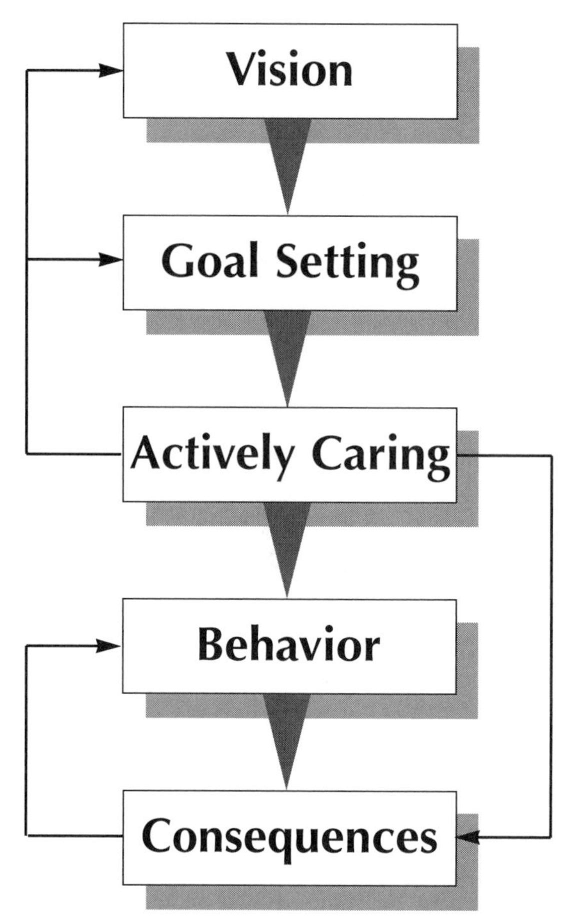

Figure 12.2. Continuous improvement requires actively caring.

THREE WAYS TO ACTIVELY CARE

The "Safety Triad" (Geller, 1989) introduced in Chapter 3 is useful to categorize actively caring behaviors. These behaviors can address environment factors, person factors, or behaviors. When people alter environmental conditions, or reorganize or redistribute resources in an attempt to benefit others, they are actively caring from an environment perspective. Actively caring safety behaviors in this category include: attending to housekeeping details, posting a warning sign near an environmental hazard, designing a guard for a machine, locking out the energy source to production equipment, and cleaning up a spill.

Person-based active caring occurs when we attempt to make other people feel better. We address their emotions, attitudes or mood states. Actively listening to others, inquiring with concern about another person's difficulties, complimenting an individual's personal appearance, and sending a get-well card are examples. This type of active caring is likely to boost a person's self-esteem, optimism, or belongingness—which in turn increases their propensity to actively care. I'll discuss this in detail in Chapter 13. Also included here are reactive behaviors performed in crisis situations.

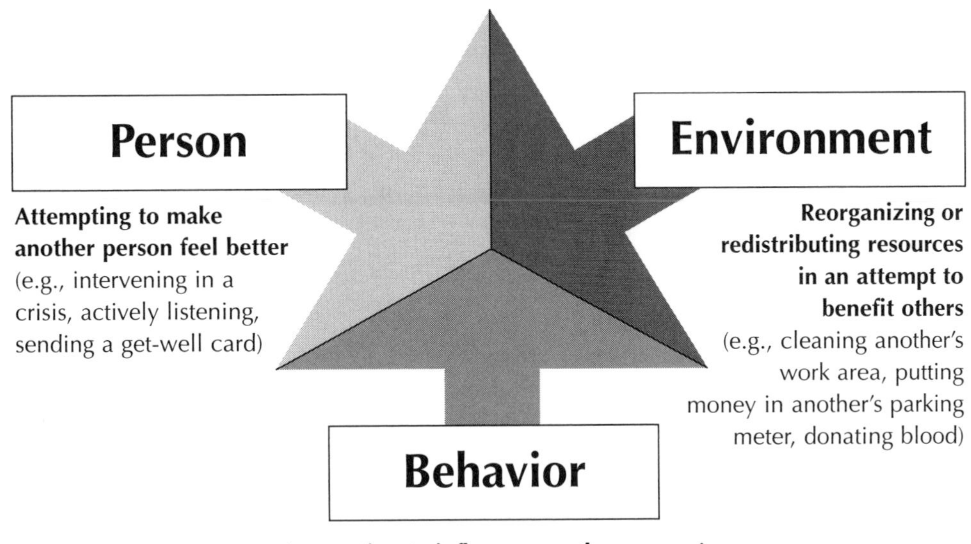

Figure 12.3. Actively caring can target three factors.

For example, if you pull someone out of an equipment pinch point or administer cardiopulmonary resuscitation, you're actively caring from a person-based perspective.

Behavior-focused active caring is most useful, and most challenging, from a proactive perspective. This happens when you attempt to direct another person's safe behavior. When we teach others about safe work practices or provide rewarding or corrective feedback regarding observed behavior, we are actively caring from a behavior focus. Obviously, the one-on-one coaching process described in Chapter 11 represents behavior-focused actively caring.

However, when we give feedback on the results of a critical behavior checklist we need to consider the feelings of the recipient. We should make more deposits than withdrawals and actively listen to reactions and suggestions. This is actively caring from a person perspective. Thus, a good safety coach practices both behavior-focused and person-focused active caring.

Figure 12.3 categorizes actively caring behaviors. Obviously, this concept applies to behaviors outside the safety field. In the fall of 1991, H. Jackson Brown (1991) gave his son, who was leaving home to begin his freshman year at college, a list of 511 principles to live by. Later that year, these principles were published in a best seller, entitled *Life's Little Instruction Book.* Two years later, Brown (1993) included 517 more tips in a sequel.

Figure 12.4 lists several of the life tips Mr. Brown gave his son which I consider actively caring behaviors. Can you categorize them according to the schema in Figure 12.3? In other words, is each item environment-focused, person-focused, or behavior-

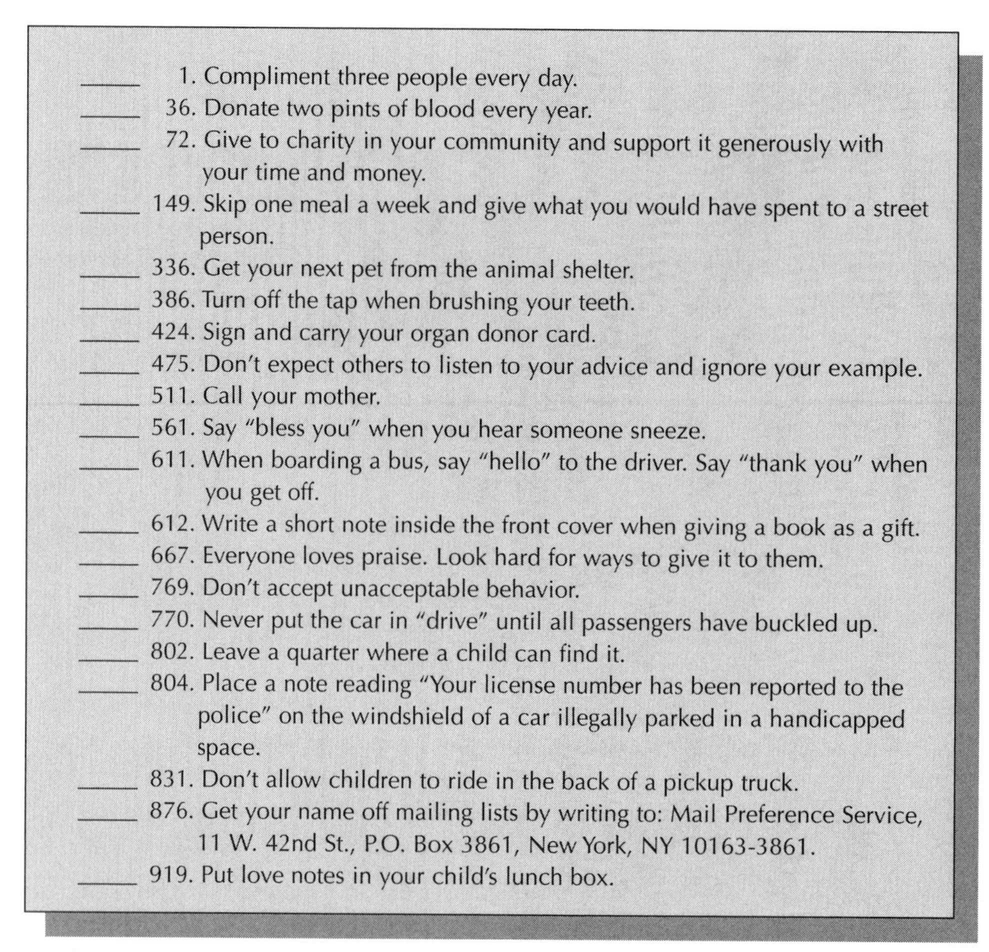

_____ 1. Compliment three people every day.
_____ 36. Donate two pints of blood every year.
_____ 72. Give to charity in your community and support it generously with your time and money.
_____ 149. Skip one meal a week and give what you would have spent to a street person.
_____ 336. Get your next pet from the animal shelter.
_____ 386. Turn off the tap when brushing your teeth.
_____ 424. Sign and carry your organ donor card.
_____ 475. Don't expect others to listen to your advice and ignore your example.
_____ 511. Call your mother.
_____ 561. Say "bless you" when you hear someone sneeze.
_____ 611. When boarding a bus, say "hello" to the driver. Say "thank you" when you get off.
_____ 612. Write a short note inside the front cover when giving a book as a gift.
_____ 667. Everyone loves praise. Look hard for ways to give it to them.
_____ 769. Don't accept unacceptable behavior.
_____ 770. Never put the car in "drive" until all passengers have buckled up.
_____ 802. Leave a quarter where a child can find it.
_____ 804. Place a note reading "Your license number has been reported to the police" on the windshield of a car illegally parked in a handicapped space.
_____ 831. Don't allow children to ride in the back of a pickup truck.
_____ 876. Get your name off mailing lists by writing to: Mail Preference Service, 11 W. 42nd St., P.O. Box 3861, New York, NY 10163-3861.
_____ 919. Put love notes in your child's lunch box.

Figure 12.4. Selected items from Brown (1991, 1993) which typify actively caring.

focused? This might not seem like a straightforward exercise. It might not be clear, for example, whether an actively caring behavior focuses on a person's feeling states or behaviors, or perhaps both. In some cases, it would be necessary to assess the intentions of the actively caring agent. The categorizations I recommend for this list are given in Figure 12.5.

WHY CATEGORIZE ACTIVELY CARING BEHAVIORS?

So why go to the trouble of categorizing actively caring behaviors? Good question! I think it's useful to consider what these behaviors are trying to accomplish, and realize the relative difficulty in performing each of them. Environment-focused active caring might be easiest for some people because it usually does not involve interpersonal interaction. When people contribute to a charity, donate blood, or complete a organ donor card, they do not interact personally with the recipient of the contribution. These behaviors are certainly commendable and may represent significant commitment and effort, but the absence of personal encounters between giver and receiver warrants separate study from other types of actively caring behavior.

Certain conditions and person characterisitics might facilitate or inhibit one type of actively caring behavior and not the other. For example, communication skills are needed to actively care on the personal or behavioral level. And different skills usually come into play. Behavior-focused active caring is more direct and usually more intrusive than person-focused active caring. It's more risky and potentially confrontational to attempt to direct or motivate another person's behavior than it is to demonstrate concern, respect or empathy for someone.

Helping someone in a crisis situation certainly takes effort and requires special skills, but there is rarely a possibility of rejection. On the other hand, attempting to correct someone's behavior could lead to negative, even hostile, reactions. Indeed, effective behavior-based active caring, as in safety coaching, often requires both person-based skills to gain the person's trust and behavior-based skills to maintain desired behavior or correct undesired behavior.

Classifying actively caring behaviors also provides insight into their benefits and liabilities. Both Brown (1991) and the editors of Conari Press (1993) recommend we feed expired parking meters to keep people from paying excessive fines. Let's consider the behavioral impact of this environment-focused "random act of kindness" (Conari Press, 1993). What will the vehicle owner think when finding an unexpired parking meter? Could this lead to a belief that parking meters are unreliable—and to further mismanagement of time? Is there a price to pay in people becoming less responsible about sharing public parking spaces?

Considering the long-term and large-scale impact of some actively caring strategies might suggest other approaches. In the parking meter situation, for example, the potential impact would be improved by adding some behavior-focused active caring. Along with feeding the expired parking meter, why not place a note under the vehicle's windshield wiper explaining the act? The note might also include some time management hints. This additional step might not only improve behavior, but set an example. The recipient of the note is probably more inclined to actively care for someone else.

Each type of actively caring behavior can be direct or indirect, with direct behavior requiring effective communication strategies. For instance, leaving a note to explain an actively caring act does not involve interpersonal confrontation.

ITEM	FOCUS	INTERACTION
1	P	D
36	E	I
72	E	I
149	E	D
336	E	I
386	E	I
424	E	I
475	B	I
511	P	D
561	P	D
611	P	D
612	P	I
667	P	D
769	B	D
770	B	D
802	E	I
804	B	I
831	B	D
876	E	I
919	P	I

Figure 12.5. Categorization of the items in Figure 12.4 with regard to the focus: environment (E), person (P) or behavior (B); and whether interaction was direct (D) or indirect (I).

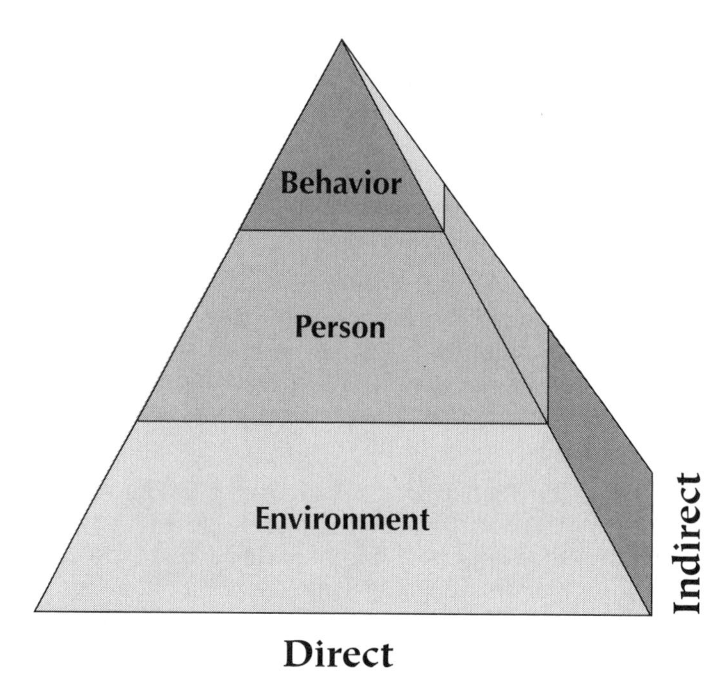

Figure 12.6. Actively caring is usually most challenging and useful when direct and behavior-focused.

Similarly, you can report an individual's safe or at-risk behavior to a supervisor, and eliminate the need for special communication skills. In the same vein, person-focused actively caring does not always involve interpersonal dialogue. You can send someone a get-well card or leave a friendly uplifting statement on an answering machine or e-mail. It's also possible that environment-focused acts will include personal confrontation, say if you deliver a contribution to a needy individual. This additional category for actively caring behavior is illustrated in Figure 12.6. You can assess your understanding by assigning a D (for direct) or I (for indirect) to each item in Figure 12.4. Then compare your answers with those in Figure 12.5.

AN ILLUSTRATIVE ANECDOTE

Several years ago I was driving on a toll road in Norfolk, Virginia, en route to the Fort Eustis Army Base, where a transportation safety conference was being held. Three of my students were with me, each scheduled to give a 15-minute talk at the conference. This was to be their first professional presentation, and they appeared quite distressed. Each was paging frantically through their notes making last-minute adjustments.

"Were you this nervous, Doc, when you gave your first professional address?" one student asked.

"No, I don't think so," I replied in jest, "I was obviously better prepared."

"Can we just read our paper?" asked another.

"Absolutely not." I retorted, "Anyone could read your paper. It's much more professional and instructive to just talk about your paper informally with the audience."

Naturally, this conversation just caused more anxiety and distress for my students. Something had to be done to distract them, to break the tension. As we approached the first of several toll booths along the highway, I thought of an actively caring solution. After paying a quarter for my vehicle, I handed the attendant another quarter and said, "This is for the vehicle behind us, the driver is using a safety belt and deserves the recognition." My students put down their papers and watched the attendant explain to the driver that we paid her toll because she was buckled up. Because we slowed down to observe this, the driver caught up with us, pulled next to us in the right lane, and acknowledged our actively caring behavior with a "shoulder belt salute"—a smile and tug on her shoulder strap.

At the next toll booth, the driver of the vehicle behind us was not buckled up, but that didn't stop me. I gave the attendant an extra quarter and said, "This is for the vehicle behind us, please ask the driver to buckle up." Again we slowed down to watch, and to our delight the driver buckled up on the spot. When the vehicle passed us, the driver gave us a smile and a "thumbs up" sign.

I kept doing this at every toll booth until exiting the highway, by which time my students had almost forgotten about their papers. They seemed relaxed and at ease when entering the conference room, and each gave an excellent presentation. Later we discussed how the toll booth intervention actively took their minds off their papers and their anxiety.

Brown (1991) recommended that his son occasionally pay the toll for the vehicle behind him. This is redistribution of resources. It's also actively caring with an environment focus. By adding a safety-belt message, I was able to accomplish more than the "random act of kindness" suggested by Brown (1991) and the editors of Conari Press (1993). I was able to support and reward those who were already buckled up, and influence other drivers to buckle up. In other words, realizing the special value of behavior-based actively caring enabled me to get more benefit from an environment-focused strategy with very little extra effort. This behavior-based effort was particularly convenient and effortless because it was indirect. You can see how the system for categorizing actively caring behavior allows us to analyze real and potential acts, and then consider ways to increase their impact.

A HIERARCHY OF NEEDS

The hierarchy of needs proposed by the humanist Abraham Maslow (1943, 1954) is probably the most popular theory of motivation. It's taught in a variety of college courses, including introductory classes in psychology, sociology, economics, marketing, human factors, and systems management. It's considered a stage theory: categories of needs are arranged hierarchically, and we don't attempt to satisfy needs at one stage or level until the needs at the lower stages are satisfied.

First, we are motivated to fulfill our physiological needs which include basic survival requirements for food, water, shelter and sleep. After these needs are under control, we are motivated by safety and security needs—the desire to feel secure and protected from future dangers. When we prepare for future physiological needs, we are proactively working to satisfy our need for safety and security.

The next motivational stage includes our social acceptance needs—the need to have friends and to feel like we belong. When these needs are gratified, our concern focuses on self-esteem, the desire to develop self-respect, gain the approval of others, and achieve success.

When I ask audiences to tell me the highest level of Maslow's Hierarchy of Needs, several people usually shout "self-actualization." When I ask for the meaning of "self-actualization," however, I receive limited or no reaction. This is probably because the concept of being self-actualized is rather vague and ambiguous. In general terms, we reach a level of self-actualization when we believe we have become the best we can be, taking the fullest advantage of our potential as human beings. We are working to reach this level when we strive to be as productive and creative as possible. Once accomplished, we possess a feeling of brotherhood and affection for all human beings, and a desire to help humanity as members of a single family—the

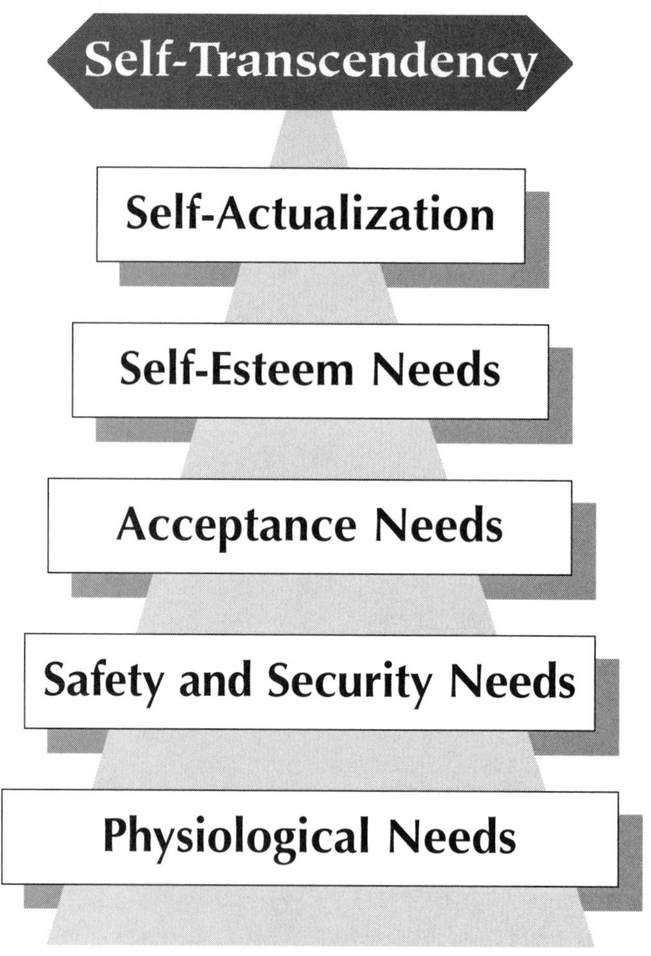

Figure 12.7. The highest need in Maslow's hierarchy reflects actively caring.

human race (Schultz, 1977). Perhaps it's fair to say that these individuals are ready to actively care.

Maslow's Hierarchy of Needs is illustrated in Figure 12.7, but self-actualization is not at the top. Maslow (1971) revised his renowned hierarchy shortly before his death in 1970 to put self-transcendence above self-actualization. Transcending the self means going beyond self-interest and is quite analogous to the actively caring concept. According to Viktor Frankl (1962), for example, self-transcendence includes giving ourselves to a cause or another person, and is the ultimate state of existence for the healthy person. Thus, after satisfying needs for self-preservation, safety and security, acceptance, self-esteem, and self-actualization, people can be motivated to reach the ultimate state of self-transcendence by reaching out to help others—to actively care.

It seems intuitive that various self-needs require satisfaction before self-transcendent or actively caring behavior is likely to occur. Actually, there is little research support for ranking needs in a hierarchy. In fact, it's possible to think of a number of examples where individuals actively care for others before satisfying all of their own needs. Mahatma Gandhi is a prime example of a leader who put the concerns of others before his own. He suffered imprisonment, extensive fasts, and eventually assassination in his 50-year struggle to help his poor and downtrodden compatriots. Figure 12.8 includes a story about one of Gandhi's actively caring behaviors. Notice how quickly Gandhi reacted in order to leave his second shoe next to the one he accidentally lost from the train. Actively caring was obviously habitual for Gandhi, developed over a lifetime of active public service. Gandhi focused on the most fundamental of human responsibilities—our responsibility to treat others as ourselves (Nair, 1995).

I'm sure you can think of individuals in your life, including perhaps yourself, who reached the level of self-transcendence before satisfying needs in the lower stages. I'll demonstrate in Chapter 13, however, that while satisfying lower need levels might not be *necessary* for actively caring behavior, people are generally more willing to actively care after satisfying the lower need levels in Maslow's hierarchy.

From a behavior-based perspective, you can see that these different need levels simply define the kinds of consequences that motivate our behavior. When we are at the first stage of the hierarchy, for example, we're working to achieve consequences—or avoid losing consequences—necessary to sustain life. We need money to buy food and pay the rent or mortgage. Then, consequences that imply safety and security are reinforced: Money is needed to buy insurance or feed a savings account, for example.

As Ghandi stepped aboard a train one day, one of his shoes slipped off and landed on the track. To the amazement of his companions, Ghandi calmly took off his other shoe and threw it back along the track to land close to the first. Asked by a fellow passenger why he did so, Ghandi smiled. "The poor man who finds the shoe lying on the track will now have a pair he can use."

Figure 12.8. Actively caring was a habit for Mohandas Karamchand Ghandi. (From Fadiman, 1985)

At the social acceptance level we perform to receive peer support or to avoid negative peer pressure.

Consequences (rewards) that recognize our efforts build our self-esteem and eventually enable us to be self-actualized. At the highest stages of Maslow's Hierarchy of Needs—self-actualization and self-transcendence—we are presumably rewarded by the realization that we have helped another person. At these levels we truly believe it's better to give than to receive. How can we help people get to this motivation level? Let's see how psychologists have attempted to answer this important question.

The Psychology of Actively Caring

Walking home on March 13, 1964, Catherine (Kitty) Genovese reached her apartment in Queens, New York, at 3:30 a.m. Suddenly, a man approached with a knife, stabbed her repeatedly, and then raped her.

When Kitty screamed "Oh my God, he stabbed me! Please help me!" into the early morning stillness, lights went on and windows opened in nearby buildings. Seeing the lights, the attacker fled; but when he saw no one come to the victim's aid, he returned to stab her eight more times and rape her again. The murder and rape lasted more than 30 minutes, and was witnessed by 38 neighbors. One couple pulled up chairs to their window and turned off the lights so they could get a better view. Only after the murderer and rapist departed for good did anyone phone the police. When the neighbors were questioned about their lack of intervention, they couldn't explain it.

The reporter who first publicized the Kitty Genovese story, and later made it the subject of a book (Rosenthal, 1964), assumed this bystander apathy was caused by big city life. He presumed that people's indifference to their neighbors' troubles was a conditioned reflex in crowded cities like New York. After this incident, hundreds of experiments were conducted by social psychologists in an attempt to determine causes of this "bystander apathy" (Latané & Darley, 1968). This research has actually discredited the reporter's common-sense conclusion. Several factors other than big city life contribute to bystander apathy. Actually, common sense suggests that if more people are present during a crisis, there's a greater chance that a victim will receive help.

At the time I was preparing the first draft of this chapter, a tragedy occurred in Detroit, Michigan, that paralleled the Kitty Genovese story and, unfortunately, many others just like it. On Saturday morning, August 20, 1995, Deletha Word (age 33) leaped off the Detroit River Bridge to escape Martell Welch (age 19) who had smashed her car with a tire iron after a fender bender. Dozens of people just watched as Ms. Word was attacked. There were reports that some spectators actually cheered, presumably encouraging Ms. Word to jump. Two men did dive into the river in an attempt to save her, but the victim reportedly resisted their efforts. City Council President Maryann Mahaffey interpreted this resistance as indicating "She was apparently so frightened that she couldn't trust anyone" (Curley, 1995, p. 3A).

An editorial appearing in *USA Today* (1995) reflects concern for the bystander apathy in this incident. This editorial is given verbatim in Figure 12.9. It refers to psy-

LACK OF HEROES

We all complain about crime. Tragically, though, when some of us have a chance to do something about it, we fail miserably.

Weekend revelers in Detroit had a chance to stop a crime and save a life. Instead, they apparently gawked. And Deletha Word, 33, working mother, is dead.

She was pulled from her car by a teenager who tore off most of her clothes, hit and chased her until she jumped off a bridge to her death in the Detroit River.

Unfortunately, there were no heroes in that part of Detroit on that tragic night. No one answered Word's pleas for help.

Missing were the kinds of bystanders who tackled Francisco Duran after he shot at the White House last fall. Or Don Lanini, 40 who in June jumped before a Manhattan subway train to rescue a woman.

Two men did dive into the river in a futile attempt to save Word. But apparently no one acted sooner.

Defenders of the Detroit crowd say maybe the bystanders weren't sure what was happening. Maybe they thought someone else would assume responsibility. Or maybe they were just afraid to interfere with the young toughs, one weighing about 200 pounds and brandishing a crowbar.

In fact, psychologists say individuals are more likely than crowds to risk helping in an emergency. Individuals alone tend to act, then think; they can't wait for someone else. Crowds tend to inhibit their members.

But that's no excuse to tolerate violence and inhumanity. Indifference encourages evildoers. Someone could have rallied the crowd to rush the assailant. Or yelled and stopped more motorists to help.

Some of the Detroit spectators undoubtedly are tormented by guilt. They deserve it. Others doubtless distanced themselves from the tragedy and feel nothing.

Ask yourself: What would you do?

Figure 12.9. Is a sole observer more likely than a crowd to intervene? (Excerpted from *USA Today*, 1995, p. 10A)

chological research to interpret the tragedy. Should the editorial have said more about the relevance of psychological research? Was the reference to psychological research accurate and complete? Will this editorial reduce future bystander apathy? Use your common sense to answer these questions, then read on for research-based answers.

LESSONS FROM RESEARCH

Professors Bibb Latané, John Darley and colleagues studied bystander apathy by staging emergency events observed by varying numbers of individuals. Then they sys-

Figure 12.10. Subjects in the Latané and Darley experiment could not see each other and thought they were conversing with one, two, or five other individuals.

tematically recorded the speed at which one or more persons came to the victim's rescue. In the most controlled experiments, the observers sat in separate cubicles (as depicted in Figure 12.10) and could not be influenced by the body language of other subjects.

In the first study of this type, the subjects introduced themselves and discussed problems associated with living in an urban environment. In each condition, the first individual introduced himself and then casually mentioned he had epilepsy and that the pressures of city life made him prone to seizures. During the course of the discussion over the intercom, he became increasingly loud and incoherent, choking, gasping, and crying out before lapsing into silence. The experimenters measured how quickly the subjects left their cubicles to help him.

When a subject believed that he or she was the sole witness, 85 percent left their cubicles within three minutes to intervene. But only 62 percent of the subjects who believed one other witness was present left their cubicle to intervene, and only 31 percent of those who thought four other witnesses were available attempted to intervene. Within three to six minutes after the seizure began, 100 percent of the lone subjects, 81 percent of the subjects with one presumed witness, and 62 percent of the subjects with four other bystanders left their cubicles to intervene.

The reduced tendency of observers of an emergency to help a victim when they believe other potential helpers are available has been termed the *bystander effect* and has been replicated in several situations (Latané & Nida, 1981). Researchers have systematically explored reasons for the bystander effect and have identified conditions influencing this phenomenon. The results most relevant to safety management are reviewed here. Some suggest ways to prevent bystander apathy—a critical barrier to achieving a Total Safety Culture. Keep in mind this research only studied reactions in crisis situations, what we would categorize as reactive, person-focused actively caring. It seems intuitive, though, that the findings are relevant for both environment-focused and behavior-focused actively caring in proactive situations.

Many years ago, my students and I studied the bystander effect in a situation requiring environment-focused actively caring behavior (Jenkins *et al,* 1978). We planted litter (a small paper bag and sandwich wrappings from a fast-food restaurant) next to a 50-gallon trash barrel located along a busy sidewalk of our university campus. Then we watched people walk by to see if anyone would pick up the litter. Several people used the trash barrel, but only one person (a female) of 598 people who walked past the trash barrel, alone or in groups, stopped to pick up and deposit the litter. The fact that this actively caring person was alone lends some minuscule support to the bystander effect. But the more remarkable finding was that almost everyone walked around or over the litter without stopping to perform a relatively convenient act of caring. Those who noticed the litter, and several did look down as they walked, probably assumed that someone else will take care of the problem. They presumed it was someone else's responsibility. Let's consider this and other factors affecting our inclination to actively care.

DIFFUSION OF RESPONSIBILITY. As in our litter example, a key contributor to the bystander effect is a presumption that someone else should assume the responsibility. It's likely, for example, many observers of the Kitty Genovese rape and murder assumed that another witness would call the police, or attempt to scare away the assailant. Perhaps some observers waited for a witness more capable than they to rescue Kitty.

Does this factor contribute to lack of intervention for occupational safety? Do people overlook environmental hazards or at-risk behaviors in the workplace because they presume someone else will make the correction? Perhaps some people assume, "If the employees who work in the work area don't care enough to remove the hazard or correct the risk, why should I?"

Social psychology research suggests that teaching people about the bystander effect can make them less likely to fall prey to it themselves (Beaman *et al,* 1978). Also, eliminating a "we-they" attitude or a territorial perspective ("I'm responsible for this area; you're responsible for that area") will increase willingness to look out for others (Hornstein, 1976).

A HELPING NORM. Many, if not most, U.S. citizens are raised to be independent rather than interdependent. But intervening for the benefit of others, whether reactively in a crisis situation or proactively to prevent a crisis, requires sincere belief and commitment toward interdependence. Social psychologists refer to a "social responsibility norm" as the belief that people should help those who need help. Subjects who scored high on a measure of this norm, as a result of upbringing during childhood or special training sessions, were more likely to intervene in a bystander intervention study, regardless of the number of other witnesses (Bierhoff *et al,* 1991).

Some cultures are more interdependent, or collectivistic, than others, promoting social responsibility and group welfare. Chinese and Japanese children, for example, learn collectivism early on. American and British children are raised to be more individualistic. Figure 12.11 contrasts the slogans or common phrases repeated in our culture with those found in the Japanese culture. The difference between an individualistic and collectivistic perspective (Triandis *et al,* 1990) is clearly shown here, and suggests that an interdependence or helping norm is stronger in Japan than the U.S.A survey by Chinese psychologist Hing-Keung Ma (1985) supported this prediction by showing greater concern and responsiveness for other people's problems among residents in Hong Kong versus London.

KNOWING WHAT TO DO. When people know what to do in a crisis, they do not fear making a fool out of themselves and do not wait for another, more appropriate person to intervene. The bystander effect was eliminated when observers had certain competencies, such as training in first-aid treatment, which enabled them to take charge of the situation (Shotland & Heinold, 1985). In other words, when observers believed

INDIVIDUALISTIC	COLLECTIVISTIC
The squeaky wheel gets the grease.	Still waters run deep.
You have to blow your own horn.	The nail that stands above the board invites a hammering down.
He travels fastest who travels alone.	A wise man does not speak, for it is the shallow water that makes the noise.
Nice guys finish last.	One does not make the wind but is blown by it.
Where there is a will there is a way.	What is possible depends upon the circustances.
I am the captain of my soul, captain of my fate.	The greatness of a person may be measured by one's humility, not by one's assertiveness.

Figure 12.11. Expressions reflect socialization and cultural norms.

they had the appropriate tools to help, bystander apathy was decreased or eliminated.

This conclusion is also relevant for proactive or preventive action, as in safety intervention. When people receive tools to improve safety, and believe the tools will be accepted and effective to prevent injuries, bystander apathy for safety will decrease. This implies, of course, the need to (a) promote a social responsibility or interdependence norm throughout the culture, and (b) teach and support specific intervention strategies or tools to prevent workplace injuries.

IT'S IMPORTANT TO BELONG. Researchers demonstrated reduced bystander apathy when observers knew one another and had developed a sense of belonging or mutual respect from prior interactions (Rutkowski *et al*, 1983). Most, if not all, of the witnesses to Kitty Genovese's murder did not know her personally, and it's likely the neighbors did not feel a sense of comradeship or community with one another. Situations and interactions that reduce a "we-they" or territorial perspective and increase feelings of togetherness or community will increase the likelihood of people looking out for each other.

MOOD STATES. Several social psychology studies have found that people are more likely to offer help when they are in a good mood (Carlson *et al*, 1988). And the mood states that facilitated helping behavior were created very easily—by arranging for potential helpers to find a dime in a phone booth, giving them a cookie, showing them a comedy film, or providing pleasant aromas. Are these findings relevant for occupational safety?

Daily events can elevate or depress our moods. Some events are controllable, some are not. Clearly, the nature of our interactions with others can have a dramatic impact on the mood of everyone involved. As depicted in Figure 12.12, even a telephone conversation can lift a person's spirits and increase his or her propensity to actively care. Perhaps remembering the research on mood and its effect will motivate us to adjust our interactive behaviors with coworkers. We should also interact in a way that could influence a person's beliefs or expectations in certain directions, as explained next.

Figure 12.12. Telephone conversations can lift moods and increase one's propensity to actively care.

BELIEFS AND EXPECTATIONS. Social psychologists have shown that certain personal characteristics or beliefs influence one's inclination to help a person in an emergency. Specifically, individuals who believe the world is fair and predictable, a world in which good behavior is rewarded and bad behavior is punished, are more likely to help others in a crisis (Bierhoff *et al,* 1991). Also, people with a higher sense of social responsibility and the general expectation that people control their own destiny showed greater willingness to actively care (Schwartz & Clausen, 1970; Staub, 1974).

The beliefs and expectancies that influence helping behaviors are not developed overnight and obviously cannot be changed overnight. But a work culture, including its policies, appraisal and recognition procedures, educational opportunities and approaches to discipline, can certainly increase or decrease perceptions or beliefs in a just world, social responsibility, and personal control, and in turn influence people's willingness to actively care for the safety of others.

Deciding to Actively Care

As a result of their seminal research, Latané and Darley (1970) proposed that an observer makes five sequential decisions before helping a victim. The five decisions (depicted in Figure 12.13) are influenced by the situation or environmental context in which the emergency occurs, the nature of the emergency, the presence of other bystanders and their reactions, and relevant social norms and rules. While the model was developed to evaluate intervention in emergency situations—where there is need for direct, reactive, person-focused actively caring—it is quite relevant for the other types of actively caring. Actually, the model has been used effectively in a variety of intervention situations, ranging from preventing a person from driving drunk (behavior-focused actively caring) to making an environment-focused decision to donate a kidney to a relative (Borgida *et al,* 1992; Rabow *et al,* 1990).

STEP 1: IS SOMETHING WRONG?

The first step in deciding whether to intervene is simply noticing that something is wrong. Some situations or events naturally attract more attention than others. This

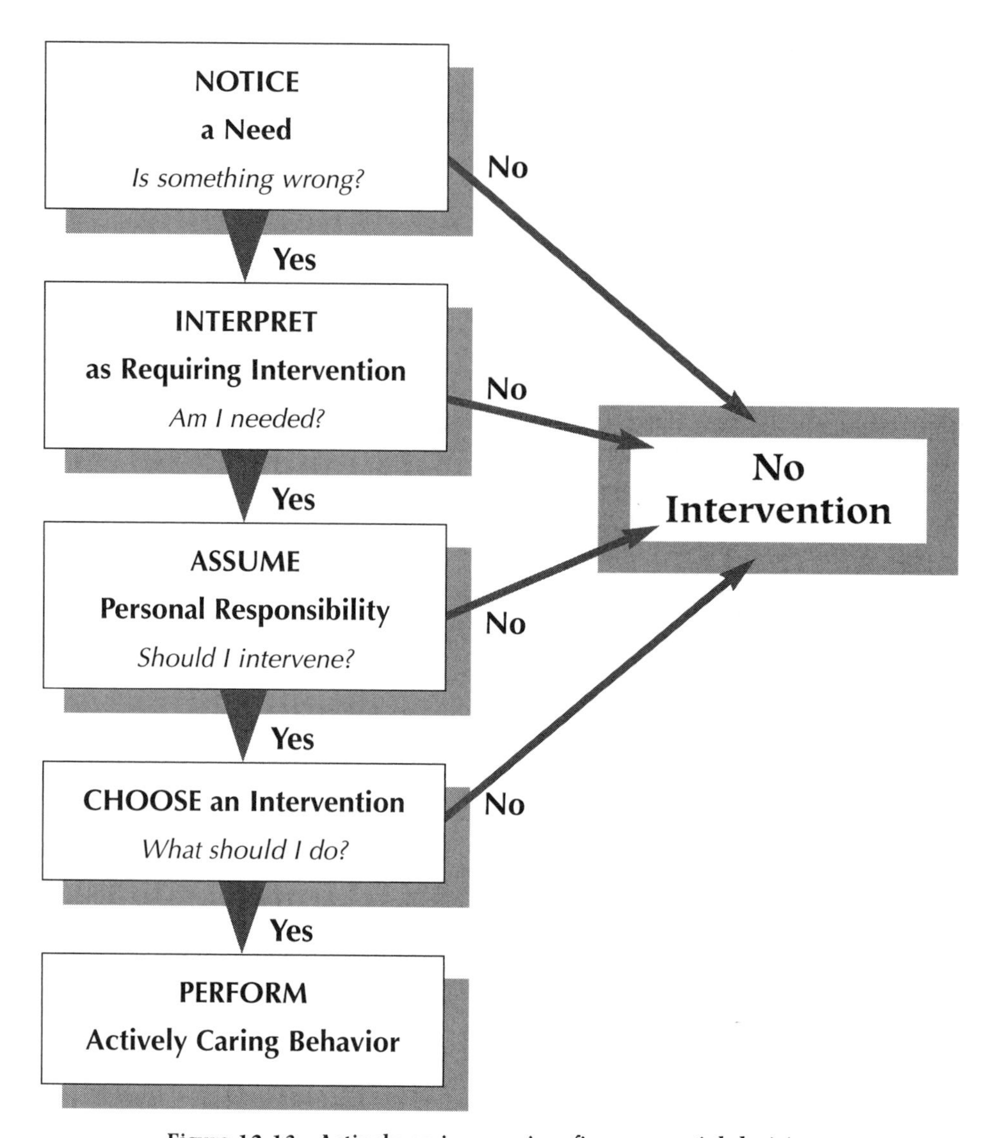

Figure 12.13. Actively caring requires five sequential decisions.

point relates to the discussion in Chapter 9 about relative attention and habituation to various activators. Most emergencies are novel and upset the normal flow of events. However, as shown by Piliavin *et al,* (1976), the onset of an emergency such as a person slipping on a spill or falling down a flight of stairs will attract more attention and helping behavior than the aftermath of an incident, as when a victim is regaining consciousness or rubbing an ankle after a fall. Of course, we should expect much less attention to a nonemergency situation.

Context also plays a role here. A significant amount of research has shown, for example, that people are more helpful in rural than urban settings (Steblay, 1987), and this difference may be partly due to context (Schroeder *et al,* 1995). The stimulus overload of the city may lead to people not noticing a need to intervene. Indeed, in active and noisy environments, like various work settings, many people narrow their focus to what is personally relevant. They learn to tune out irrelevant stimuli. Bickman *et al* (1973) used this stimulus overload theory to explain their finding that university students living in high-density, high-rise dormitories were less likely to return a lost letter than were students residing in less densely populated buildings.

Matthews and Canon (1975) tested the stimulus overload theory directly in a real-world field study. On several trials a research accomplice, wearing a wrist-to-shoulder cast, dropped several boxes of books a few feet in front of a potential helper. Researchers observed systematically whether the potential helper intervened. Environmental stimulation was manipulated by running a power lawn mower nearby on half of the trials. In the noisy condition, only 15 percent of the potential helpers showed helping behavior; however, without the excessive noise, 80 percent of the subjects stopped to help pick up the dropped boxes.

What's going on here? It's possible the loud noise may have had a negative effect on the mood of the potential helpers. In fact, mood state may be a critical factor in stimulus overload studies. Environmental stressors like noise, pollution, and crowding usually have a negative impact on mood states (Bell *et al,* 1990), with depressed moods leading to self-centeredness and lower awareness of other persons' needs.

If stimulus overload can affect people's attention to an emergency, it can certainly reduce attention to common everyday situations that are not very obtrusive, but nevertheless require actively caring behavior. Consider, for example, the various needs for proactive behavior that can prevent an injury. Environmental hazards are easy to overlook, especially in a busy and noisy workplace requiring focused attention on a demanding task. Even less noticeable and attention-getting are the ongoing safe and at-risk behaviors of people around us. Yet these behaviors need proactive support or correction, as in the safety coaching approach described in Chapter 11.

Now, even if the need for proactive intervention is noticed, actively caring behavior will not necessarily occur. The observer must interpret the situation as requiring intervention. Which leads us to the next question that must be answered before deciding to intervene.

STEP 2: AM I NEEDED?

As shown in Figure 12.14, people can come up with a variety of excuses for not helping. Distress cues, such as cries for help, and the actions of other observers can clarify an event as an emergency. When people are confused, they look to other people for information and guidance. In other words, through observational learning (Chapter 7), people figure out how to interpret an ambiguous event and how to react to it. Thus, the behavior of others is

Figure 12.14. People give a variety of excuses for not helping.

especially important when stimulus cues are not present to clarify a situation as requiring intervention (Clark & Word, 1972). This was illustrated in one of the early seminal experiments by Latané and Darley (1968).

Professors Latané and Darley invited male students to discuss problems they experienced at a large urban university. While the students were completing a questionnaire, pungent smoke began puffing through a vent into the testing room. Smoke quickly filled the room. The danger of the situation was rather ambiguous, however, because real smoke was not used. The experimenters expected the subjects to rely on others when deciding what to do.

The social context of the situation varied. Some subjects were alone in the room. Others filled out the questionnaire with two other subjects who were strangers. Some subjects were with two accomplices of the researchers who shrugged their shoulders and acted as if nothing were wrong. Social context had a dramatic impact on whether the subjects left the room, presumably to save their lives. Of the students who were alone, 75 percent left the room to report the smoke, but only 10 percent of the subjects with two passive strangers left the room. In fact, many of the subjects in this circumstance later reported that they believed nothing was wrong. Some concluded that the smoke was "truth gas." Thus, the passive behavior of others led most subjects to interpret the situation as safe and requiring no intervention.

Is this relevant to many work situations? How often are environmental hazards or at-risk behavior overlooked or ignored because the social context—other people—gives the signal that nothing is wrong? What about the situation with three naive subjects? Does your common sense tell you that at least one of these subjects left the room to inquire about the smoke? With three people uninformed about the risk, the probability that someone will take action should be high, right? Wrong! Only 38 percent of the time did any one person leave the room to inquire about the smoke. Each subject tried to "stay cool." Thus, when looking around for social cues each subject saw two other individuals remaining calm, cool, and collected. The group developed a shared illusion of safety. The investigators labeled this phenomenon *pluralistic ignorance.*

Follow-up research to pluralistic ignorance caused by mutual passive reaction to potential dangers has demonstrated the critical value of people's reactive words in the situation (Wilson, 1976). We know these words to be verbal activators. Ervin Staub (1974), for example, varied systematically what his accomplice said after pairs of bystanders, a subject and the accomplice, heard a crash in an adjoining room and a female's cry for help. When the accomplice said, "That sounds like a tape-recording. Maybe they are trying to test us," only 25 percent of the subjects left the room to help. On the other hand, when the accomplice reacted with, "That sounds bad. Maybe we should do something," 100 percent of the subjects intervened.

Thus, in situations where the need for intervention or corrective action is not obvious, people will seek information from others to understand what is going on and to receive direction. This is the typical state of affairs when it comes to safety in the workplace. In fact, the need for proactively caring behavior is rarely as obvious as smoke entering a room or the sound of a crash. If activators like these occurred in the workplace, many people would likely react in a hurry. Such events would be noticeable and likely interpreted as needing attention.

But would you assume personal responsibility and respond? Surely the bystanders in the Kitty Genovese and Deletha Word cases described earlier noted the events and interpreted them as requiring assistance. Steps 1 and 2 of Latané and Darley's decision model were likely satisfied. The breakdown probably occurred at Step 3, perceiving personal responsibility.

STEP 3: SHOULD I INTERVENE?

In this stage you ask yourself, "Is it my responsibility to intervene?" The answer would be obvious if you were the only witness to a situation you perceive as an emergency. But you might not answer "yes" to this question when you know that other people are also observing the emergency, or the safety hazard. In this case, you have reason to believe that someone else will intervene, perhaps a person more capable than you. This perception relieves you of personal responsibility. But what happens when everyone believes the other guy will take care of it? This is likely what happened in the Kitty Genovese and Deletha Word tragedies, as well as many similar incidents.

A breakdown at this stage of the decision model doesn't mean the observers don't care about the welfare of the victim. Actually, it's probably incorrect to call lack of intervention "bystander apathy" (Schroeder *et al*, 1995). The bystanders might care very much about the victim, but defer responsibility to others because they believe other observers are more likely or better qualified to intervene. Similarly, people might care a great deal about the safety and health of their coworkers, but they might not feel capable of acting on their caring. People might resist taking personal responsibility to actively care because they don't believe they have the most effective tools to make a difference. This can be remedied by teaching employees the principles and procedures presented in Chapters 9, 10, and 11.

In addition to having a "can do" attitude, people need to believe it is their personal responsibility to intervene. In many work situations, it's easy to assume that safety is someone else's responsibility—the safety director or a team safety captain. After all, these individuals have "safety" in their job titles, and they meet regularly to discuss safety issues. And they get to go off-site now and then to attend a safety conference—where they learn the techniques that make them the most capable to intervene. So it's their responsibility.

Psychologists have shown that people will take responsibility, even among strangers, if their responsibility is clearly specified (Baumeister *et al*, 1988). In an interesting field study, for example, researchers staged a theft on a public beach and then observed whether assigning responsibility to some individuals increased their frequency of intervention (Moriarty, 1975). Researchers posing as vacationers randomly asked individual sunbathers to watch their possessions, including a radio, while they went for a walk. In the control condition, the researchers only asked sunbathers for a match and then left for a walk. A short time later, a second researcher approached the unoccupied towel, snatched the radio, and ran down the beach.

How often did the individual sunbather intervene? Surprisingly, 94 percent of the sunbathers assigned the "watchdog" responsibility intervened, often with dramatic and physical displays of aggressive protection. In contrast, only 20 percent of the bystanders in the control condition reacted in an attempt to retrieve the radio.

Perhaps your common sense predicted the correct answer this time, and that's why you have asked strangers in public places to watch your possessions for a short period of time.

The challenge in achieving a Total Safety Culture is to convince everyone they have a responsibility to intervene for safety. Indeed, a social norm or expectancy must be established that everyone shares equally in the responsibility to keep everyone safe and healthy. Furthermore, safety leaders or captains need to accept the special responsibility of teaching others any techniques they learn at conferences or group meetings that could increase a person's perceived competence to intervene effectively. All this is easier said than done, of course. Unfortunately, if we don't meet this challenge many people are apt to decide that actively caring safety intervention is not for them. They could feel this way even after viewing an obvious at-risk behavior or condition that would benefit from their immediate action.

STEPS 4 AND 5: WHAT SHOULD I DO?

These last two steps of Latané and Darley's decision model point out the importance of education and training. Education gives people the rationale and principles behind a particular intervention approach. It gives people information to design or refine intervention strategies, leading to a sense of ownership for the particular tools they help to develop. Through training, people learn how to translate principles and rules into specific behaviors or intervention strategies. Safety coach training, for example, should include role-playing exercises so people practice certain communication techniques and receive specific feedback regarding their strengths and weaknesses. The bottom line here is that people who have learned how to intervene effectively through relevant education and training are likely to be successful agents of intervention.

Research by Shotland and Heinold (1985) showed that bystanders without first-aid training were just as likely to intervene for a victim with obvious arterial bleeding as were bystanders who previously received first-aid training. But the choice and execution of an intervention varied significantly, depending on prior training. Those with training intervened with much greater competence, with some untrained helpers doing more harm than good. Similarly, Clark and Word (1974) demonstrated that people without proper information regarding electricity would sometimes impulsively touch a victim who was holding a "live" electrical wire, jeopardizing their own lives.

SUMMARY OF THE DECISION FRAMEWORK

In this section I have reviewed the decision process model Latané and Darley proposed as a sequence of choices people make before actively caring on behalf of another person. Although developed to understand the bystander effect in emergency situations, this decision framework is certainly relevant for proactive situations and for each type of actively caring behavior defined in this chapter (direct versus indirect, and environment-focused, person-focused or behavior-focused). The model can help us understand why an individual might not actively care for another person's health and safety, and it can be used to guide the development of strategies to increase the frequency of actively caring behaviors. For example, conditions (activators) will increase the probability of this behavior if they increase the likelihood a person will notice and perceive a need for intervention, and assume personal responsibility for helping. Plus, education and training sessions that increase skills and self-confidence to intervene *effectively* will increase the amount of constructive actively caring behavior occurring throughout a culture.

More strategies for increasing these behaviors are entertained in Chapter 14. Before turning to a discussion of ways to increase active caring, however, we need to consider another approach to interpreting bystander intervention, or the lack of it. Since behavior is motivated by consequences, a person's decision to actively care can be analyzed according to the perceived positive versus negative consequences one expects to receive. If people are motivated to maximize positive consequences and minimize negative consequences, actively caring behavior will only occur if perceived rewards outweigh perceived costs. This framework suggests strategies for increasing actively caring not prompted by Latané and Darley's sequential decision model.

A Consequence Analysis of Actively Caring

When I related the cases of Kitty Genovese and Deletha Word to my family and asked for opinions, I received a unanimous reaction that I could not readily explain with the decision model discussed above. My wife and two daughters proclaimed that most observers didn't help these women because they feared for their own safety. The perpetrator was armed with a knife in the 1964 case and a tire iron in the 1995 tragedy. Each was obviously dangerous. The onlookers could certainly see there was an emergency requiring specific assistance from anyone who would take responsibility. According to an interpretation based on our understanding of the power of consequences, people resisted taking responsibility because they perceived that it could mean more trouble—or potential harm—than it was worth. It was safer to assume that someone else more capable would intervene. According to this consequence model, people hesitated to intervene because they perceived more potential costs than benefits, not because they were apathetic or failed to interpret a need to take personal responsibility.

Dr. Jane Piliavin and colleagues (1969, 1981) have developed a cost-reward model to interpret people's propensity to help others in various emergency situations. Two basic categories of potential negative consequences for helping include: (1) personal costs, including effort, inconvenience, potential injury, and embarrassment; and (2) costs to the victim if no intervention occurs. This latter category includes two subcategories: the personal costs of not helping, including criticism, guilt, or shame; and empathic costs from internalizing the victim's distress and physical needs. The authors combine these negative consequences for direct intervention and for not intervening in order to predict whether actively caring behavior will occur under certain circumstances.

The matrix in Figure 12.15 combines two levels of cost (low versus high) to the potential intervention agent and the victim in order to predict when actively caring behavior will occur. It's most likely (lower left cell of Figure 12.15) when costs for helping are low, for example, convenient and not dangerous; and costs to the victim for

Cost to Bystander for Intervening

Cost to Victim for Not Intervening	Low	High
Low	Depends on person factors	Intervention improbable
High	Intervention probable	Diffusion of responsibility

Figure 12.15. Costs to bystanders for intervening, and costs to a victim if there is no intervention, determine the probability of intervention. (Adapted from Piliavin *et al*, 1981)

not helping are high, as when the victim is seriously injured. On the other hand, intervention is least likely when the perceived personal costs for intervening are high, for example, effortful and risky; and the apparent costs to the victim for no intervention are low, as when an experienced worker is performing at-risk behavior with no negative consequences.

The Kitty Genovese and Deletha Word cases fit the lower right cell of Figure 12.15: high perceived cost for both helper and victim. Although the costs for not helping these individuals were extremely high, resulting ultimately in their deaths, the costs for helping were also high, in fact potentially fatal. This means significant conflict for the person deciding what to do. They can resolve the conflict by helping indirectly, say by telephoning police or an ambulance; or by reinterpreting the situation (Schroeder *et al*, 1995). This can be done by presuming someone else will intervene—diffusion of responsibility—or perhaps by rationalizing that the person does not deserve help.

A bystander might rationalize, for example, that Ms. Genovese should not have been walking the streets in that neighborhood at 3:30 in the morning and Ms. Word brought on the attack by crashing into the assailant's vehicle. Rationalization reduces the perceived costs for not intervening, and enables the bystander to ignore the situation without excessive shame or guilt. According to this cost-reward interpretation, when bystanders perceive high costs both for intervening and for not intervening in a crisis, they recognize the need for action, hesitate because of perceived personal costs, and then search for an excuse to do nothing (Schroeder *et al*, 1995; Schwartz & Howard, 1981).

The upper left quadrant of Figure 12.15 represents situations most analogous to actively caring for injury prevention. Although a simple low-cost intervention might be called for to correct an environmental hazard or an at-risk behavior, there is no immediate emergency and thus no need for immediate action. There is low perceived cost if no action is taken: "We've been working under these conditions for months and no one has been hurt." Piliavin *et al* (1981) presume that intervention in situations represented by this cost quadrant is most difficult to predict. Many factors can influence perceived consequences that are positive and negative, and small changes in these factors can tilt the cost-reward balance in favor of stepping in or standing back.

Through testimonials and constructive discussions, employees can be convinced that the potential cost of not intervening is higher than they initially thought. This can occur, for example, by considering the large degree of plant-wide exposure to a certain uncorrected hazard. Also, it might be worthwhile to remind people of the large-scale detrimental learning that could occur from the continuous performance of risky behavior. Furthermore, education and role-playing exercises can reduce the perceived personal costs of actively caring. It's also true that person factors, such as mood states discussed earlier, determine whether intervention occurs.

Figure 12.16 illustrates the cost-reward approach of a rational potential helper, as described by Professors Lawrence Wrightsman and Kay Deaux (1981, p. 261). Note that the potential helper in the story is considering the rewards as well as the costs for intervening and for not intervening. Although the matrix in Figure 12.15 focused entirely on negative consequences, it's important to consider that positive consequences can also play a prominent role in determining one's decision to get involved. In occupational safety, for example, proactive caring behavior can not only prevent a serious injury, it also sets the right example for others to follow. It can also increase

Old man, I see you slumped in the gutter, looking half dead. I realize that you need help. Will I be the one?

What will it cost me if I help? Maybe you're drunk and will slobber all over me. Maybe you'll get surly. What if I make things worse and am held liable? Call that 10 cost points.

Besides, if I just pass by, I'll be on time for my job interview and will probably get the position. Rewarding indeed—at least 5 reward points, which I'll lose if I help.

Then again, maybe there's something in it for me if I help. I feel for you, old man, and I can imagine your joy if I help. I like to think of myself as a helpful person. Maybe the job interviewer will be snowed when I explain why I was late. I know my friends will think well of me if I help. That adds up to 6 reward points.

It's going to hurt me some if I don't help, too. Maybe somebody I know will see me callously walk by and tell everyone. I'll feel guilty, I know; I'll wonder if I killed you, old man. Those costs are worth 3 points.

Time for the final tally. It will cost me 10 points to help, and I'll also lose 5 points I could have if I walk by. The total cost is 15 points. Helping is worth 6 reward points to me, and I'll avoid the 3-point cost of failing to help. The total reward value is 9 points.

Goodbye, old man. Perhaps when I'm in a better mood.

Figure 12.16. The mind of the rational helper: cost-reward analysis. (From Wrightsman & Deaux, 1981, p. 261)

certain positive person states in both the doer and beneficiary of the act, which in turn increases the probability of actively caring behavior by both in the future. These positive outcomes from giving and receiving are detailed in the next chapter.

In Conclusion

Actively caring behavior is planned and purposeful. It can be direct or indirect. And its focus is environment, person, or behavior. Active caring that addresses the environment is usually easiest to perform because it doesn't involve interpersonal confrontation. Behavior-focused actively caring is often most proactive, but is most difficult to carry out effectively because it attempts to influence another person's behavior in a nonemergency situation. Practically all of the research related to this concept has studied crisis situations in which a victim needs immediate assistance. This is essentially person-focused and reactive caring. Psychologists have determined factors that influence the probability of actively caring behavior in emergencies, and the results are relevant for both environment-focused and behavior-focused active caring. Understanding the conditions that lead to an increase or decrease in reactive caring behavior can help us find ways to facilitate proactive caring for safety.

The finding that people often refuse to act in a crisis, especially when they can share the responsibility of intervening with others, is quite analogous to most work settings. Hence, it's important to understand the factors that can influence this resistance, referred to as "bystander apathy." For example, people with a sense of social responsibility and comradeship for others at work, and who believe they have personal control in a just world, are more apt to intervene for the safety of others. It's possible to increase these personal characteristics among people through policy, procedures, and personal interaction. Increasing these states, and thus the willingness to actively care for safety, is key to achieving a Total Safety Culture and is addressed in the next two chapters.

A decision model (developed by Bibb Latané and John Darley) helps understand why we don't see more actively caring behavior. Before we step in, either reactively or proactively, we presumably make five sequential decisions: 1) Is something wrong? 2) Is my help needed? 3) Is it my responsibility to intervene? 4) What kind of intervention strategy should I use? 5) Exactly when and how should I intervene?

This decision logic suggests certain methods for increasing the likelihood that people will get involved. For example, the model shows the importance of teaching employees how to recognize and correct environmental hazards and at-risk behaviors. It's also critical to promote the ultimate aim or corporate mission to make safety a value. For this to happen, everyone must assume responsibility for safety and never wait for someone else to act.

A consequence, or cost-benefit, model offers more guidance for increasing actively caring behavior. It enables us to analyze motivational factors that shape decisions to actively care. Conditions and situations that increase perceptions of costs to victims (for not intervening) and reduce perceptions of personal costs to the intervention agent (for intervening) increase the probability of action being taken. In addition, it's important to help people realize the potential positive consequences or rewards available to both the giver and receiver of an actively caring intervention. When these perceived internal and external rewards outweigh the rewards for doing nothing, people will probably actively care.

Most safety situations involve relatively low costs and rewards to both the recipient and deliverer of the intervention. Although the relative costs to an individual for intervening may be low, the recipient of a proactive safety intervention is only a potential victim, so the perceived cost for not intervening is also low. Education and training can reduce these perceived costs to the intervention agent and increase the perceived costs to potential victims. The result: more frequent actively caring behavior for safety. However, education and training are not sufficient to achieve the amount of active caring needed for a Total Safety Culture. The next two chapters deal specifically with how to develop and implement strategies to increase actively caring behavior for occupational and community safety.

REFERENCES

Barker, J. A. (1993). *The power of vision* (videotape). New York: Chart House International Learning Corp.

Baumeister, R. F., Chesner, S. P., Sanders, P. S., & Tice, D. M. (1988). Who's in charge here? Group leaders do lend help in emergencies. *Personality and Social Psychology Bulletin*, 14, 17-22.

Beaman, A. I., Barnes, P. J., Klentz, B., & McQuirk, B. (1978). Increasing helping rates through informational dissemination: Teaching pays. *Personality and Social Psychology*, 37, 1835-1846.

Bell, P. A., Fisher, J. D., Baum, A., & Greene, T. (1990). *Environmental psychology* (3rd edition). Fort Worth, TX: Holt, Rinehart, & Winston.

Bickman, L., Teger, A., Gabriele, T., McLaughlin, C., Berger, M., & Sunaday, E. (1973). Dormitory density and helping behavior. *Environment and Behavior*, 5, 465-490.

Bierhoff, H. W., Klein, R., & Kramp, P. (1991). Evidence for the altruistic personality from data on accident research. *Journal of Personality*, 59, 263-280.

Borgida, E., Conner, C., & Manteufel, L. (1992). Understanding living kidney donation: A behavioral decision-making perspective. In S. Spacapan & S. Oskamp (Eds.), *Helping and being helped*. Newbury Park, CA: Sage.

Brown, Jr., H. J. (1991). *Life's little instruction book*. Nashville, TN: Rutledge Hill Press.

———. (1993). *Life's little instruction book*, Volume II. Nashville, TN: Rutledge Hill Press.

Carlson, M., Charlin, V., & Miller, N. (1988). Positive mood and helping behavior: A test of six hypotheses. *Journal of Personality and Social Psychology*, 55, 211-229.

Clark, R. D., III, & Word, L. E. (1972). Why don't bystanders help? Because of ambiguity? *Journal of Personality and Social Psychology*, 24, 392-400.

——— & ——— (1974). Where is the apathetic bystander? Situational characteristics of the emergency. *Journal of Personality and Social Psychology*, 29, 279-287.

Covey, S. R. (1990). *Principle-centered leadership*. New York: Simon and Schuster.

——— . (1989). *The seven habits of highly effective people*. New York: Simon and Schuster.

Curley, T. (1995). Anger, disbelief swirling after Detroit attack. *USA Today*, Wednesday, August 23, p. 3A.

Deming, W. E. (1986). *Out of the crisis*. Cambridge, MA: Massachusetts Institute of Technology, Center for Advanced Engineering Study.

———. (1993). *The new economics for industry, government, education*. Cambridge, MA: Massachusetts Institute of Technology, Center for Advanced Engineering Study.

Editors of Conari Press (1993). *Random acts of kindness*. Emeryville, CA: Conari Press.

Fadiman, C., ed. (1985) *The little brown book of anecdotes*. Boston, MA: Little, Brown and Company.

Frankl, V. (1962). *Man's search for meaning: An introduction to logotherapy*. Boston: Beacon Press.

Geller, E. S. (1989). Managing occupational safety in the auto industry. *Journal of Organizational Behavior Management,* 10(1), 181-185.

Hornstein, H. A. (1976). *Cruelty and kindness: A new look at aggression and altruism.* Englewood Cliffs, NJ: Prentice-Hall.

Jenkins, E., Cuddiky, K., Hearn, K., & Geller, E. S. (1978). *When will people pick up and pitch in?* Paper presented at the Virginia Academy of Science meeting, Blacksburg, VA.

Kohn, A. (1993). *Punished by rewards: The trouble with gold stars, incentive plans, A's, praise, and other bribes.* Boston: Houghton Mifflin.

Latané, B., & Darley, J. M. (1968). Group inhibition of bystander intervention. *Journal of Personality and Social Psychology,* 10, 215-221.

——— & ———. (1970). *The unresponsible bystander: Why doesn't he help?* New York: Appleton-Century-Crofts.

Lantané, B., & Nida, S. A. (1981). Ten years of research on group size and helping. *Psychological Bulletin,* 89, 308-324.

Ma, H. (1985). Cross-cultural study of altruism. *Psychological Reports,* 57, 337-338.

Maslow, A. H. (1943). A theory of human motivation. *Psychological Review,* 50, 370-396.

———. (1954). *Motivation and personality.* New York: Harper.

———. (1971). *The farther reaches of human nature.* New York: Viking.

Mathews, K. E., & Canon, L. K. (1975). Environmental noise level as a determinant of helping behavior. *Journal of Personality and Social Psychology,* 32, 571-577.

Moriarty, T. (1975). Crime, commitment, and the responsive bystander: Two field experiments. *Journal of Personality and Social Psychology,* 31, 370-376.

Nair, K. (1995). A clue from Gandhi. *Sky,* May, 26-31.

Peale, N. V. (1952). *The power of positive thinking.* New York: Prentice-Hall.

Piliavin, I. M., Rodin, J., & Piliavin, J. A. (1969). Good samaritanism: An underground phenomenon? *Journal of Personality and Social Psychology,* 13, 289-299.

Piliavin, J. A., Dovidio, J. F., Gaertner, S. L., & Clark, R. D., III. (1981). *Emergency intervention.* New York: Academic Press.

Piliavin, J. A., Piliavin, I. M., & Broll, L. (1976). Time of arousal at an emergency and likelihood of helping. *Personality and Social Psychology Bulletin,* 2, 273-276.

Rabow, J., Newcomb, M. D., & Hernandez, A. C. R. (1990). Altruism in drunk driving situations: Personal and situational factors in helping. *Social Psychology Quarterly,* 53, 199-213.

Rosenthal, A. M. (1964). *Thirty-eight witnesses.* New York: McGraw-Hill.

Rutkowski, G. K., Gruder, C. L., & Romer, D. (1983). Group cohesiveness, social norms, and bystander intervention. *Journal of Personality and Social Psychology,* 44, 545-552.

Schroeder, D. A., Penner, L. A., Dovidio, J. F., & Piliavin, J. A. (1995). *The psychology of helping and altruism.* New York: McGraw-Hill, Inc.

Schwartz, S. H., & Clausen, G. T. (1970). Responsibility, norms, and helping in an emergency. *Journal of Personality and Social Psychology,* 16, 299-310.

Schwartz, S. H., & Howard, J. A. (1981). A normative decision-making model of altruism. In J. P. Rushton & R. M. Sorrentino (Eds.), *Altruism and helping behavior: Social, personality, and developmental perspectives.* Hillsdale, NJ: Erlbaum.

Schultz, D. (1977). *Growth psychology: Models of the healthy personality.* New York: D. Van Nostrand.

Shotland, R. L., & Heinold, W. D. (1985). Bystander response to arterial bleeding: Helping skills, the decision-making process, and differentiating the helping response. *Journal of Personality and Social Psychology,* 49, 347-356.

Skinner, B. F. (1981). Selection by consequences. *Science,* 213, 502-504.

Staub, E. (1974). Helping a distressed person: Social, personality, and stimulus determinants. In L. Berkowitz (Ed.), *Advances in experimental social psychology,* Vol. 7, New York: Academic Press.

Steblay, N. M. (1987). Helping behavior in rural and urban environments: A meta-analysis. *Psychological Bulletin,* 102, 346-356.

Triandis, H. C., McCusker, C., & Hui, C. H. (1990). Multimethod probes of individualism and collectivism. *Journal of Personality and Social Psychology,* 59, 1006-1020.

USA Today (1995). Lack of heroes. Wednesday, August 23, p. 10A.

Wilson, J. P. (1976). Motivation, modeling, and altruism: A person X situation analysis. *Journal of Personality and Social Psychology,* 34, 1078-1086.

Wrightsman, L. S., & Deaux, K. (1981). *Social psychology in the 1980's* (Third Edition). Monterey, CA: Brooks/Cole Publishing Company.

Chapter 13
The Person-Based Approach to Actively Caring

Our willingness to actively care for others is affected by certain feelings and states of mind. If we have a strong sense of self-esteem, self-efficacy, personal control, optimism, and belongingness, there's a greater chance we'll go beyond the call of duty to help others. Each of these person states are explained in this chapter, and the research supporting direct relationships between these states and actively caring behavior is reviewed. Understanding these connections enables us to design conditions and interventions to increase actively caring behavior throughout an organization or community.

Our deeds determine us, as much as we determine our deeds.

This quotation from George Eliot indicates that our behaviors influence something about us and implies that good deeds or actively caring behaviors are good for us. They change something about us, and this in turn affects subsequent behavior. Does this mean our actively caring behaviors influence us to actively care even more? It's a nice thought and seems intuitive, but what does it really mean?

This chapter explores a host of questions arising from the concept reflected in Eliot's words: What is it about us that changes as a result of our good deeds, and will this change lead to more good deeds? Can making people more willing to actively care be influenced in ways other than managing activators and consequences to directly change behavior? In other words, can we change something about people that will make them more willing to actively care for the safety and health of others? If answers to these questions can be turned into practical procedures, we'll know how to increase actively caring behaviors throughout a culture.

PAIR BREATHES LIFE INTO DRIVER

Tywanii Hairston was on the way to pay her water bill Tuesday when she pulled up behind a Roanoke city truck at a red light on Campbell Avenue.

The light turned green, but the truck didn't budge. After a moment the driver of another car honked the horn, and still the truck didn't move. Hairston thought it had stalled.

As she slowly drove around the truck, she saw a man slumped over the wheel. A former nursing assistant, Hairston parked her car and jumped out to see what was wrong.

Meanwhile, John McKee, a driver for Alert Towing on his way to a job, also had seen the man and stopped. He radioed his dispatcher to call 911, and with the help of another passerby, pulled the man from the truck and laid him on the street.

Together, Hairston and McKee went into action. Neither had ever administered cardiopulmonary resuscitation, or CPR, before, although both knew the procedure. Hairston started giving mouth-to-mouth, and McKee pumped the man's chest until paramedics arrived. Their swift action quite possible saved Don Arthur's life. On Thursday, Arthur, 60, was in stable condition at Community Hospital, one of only a few who make it back after venturing so close to death.

Figure 13.1. Actively caring behavior saved a life. (From *The Roanoke* [Virginia] *Times*, Friday, Sept. 8, 1995)

At the time I was preparing the first draft of this chapter, a heartwarming story appeared in our local newspaper, *The Roanoke Times and World Report*. The newspaper report of the incident is reprinted in Figure 13.1 and it is clearly opposite to the Kitty Genovese and Deletha Word tragedies reviewed in Chapter 12. In this case, two individuals, Tywanii Hairston and John McKee, went out of their way to save the life of a truck driver named Don Arthur, whose truck was blocking traffic because he had blacked out. Several individuals had already driven around his truck without intervening—for a variety of possible reasons discussed in Chapter 12. But two individuals did interrupt their routine to actively care, and the result was a life saved.

Ms. Hairston and Mr. McKee went into action, rather than succumb to bystander apathy. Why? Were there special characteristics of the two heroes? Both individuals did have some life-saving training in the past, so as discussed in Chapter 12, they might have felt more responsible than others because they knew what to do. This chapter examines additional person-based reasons (or internal characteristics of people) that could have contributed to this success story. Did the two intervening agents have personality traits conducive to helping? Or did recent experiences influence their personality state in some beneficial way? Obviously, we'll never know the answer to these questions, but exploring the possibilities can help us better grasp the person factors contributing to actively caring behavior. And if these factors can be altered systematically, then it's possible to increase people's willingness to actively care for the safety and health of others.

Actively Caring from the Inside

Perhaps you recall earlier discussions in this text about "outside" versus "inside" aspects of people. In Chapter 3, for example, I distinguished between behaviors (outside) versus intentions, attitudes and values (inside), and emphasized that we should start with behaviors. A prime principle of behavior-based psychology is that it's easier, especially for large-scale culture change, to "act a person into safe thinking" than it is to address attitudes and values directly in an attempt to "think a person into safe acting." Another key principle of behavior-based psychology is that the consequences of our behavior influence how we feel about the behavior (see Figure 3.4 in Chapter 3). Generally, positive consequences lead to good feelings or attitudes; negative consequences lead to bad feelings or attitudes.

Long-term behavior change requires people to change "inside" as well as outside. The promise of a positive consequence or the threat of a negative one can maintain the desired behavior while the response-consequence contingencies are in place. But what happens when they are withdrawn? What happens when people are in situations, like at home, when no one is holding them accountable for their behavior? If people do not believe in the safe way of doing something and do not accept safety as a value or a personal mission, don't count on them to choose the safe way when they have the choice. In addition, if people are not self-motivated to keep themselves safe, don't expect them to actively care for the safety of others.

Figure 13.2 illustrates how person factors interact with the basic activator-behavior-consequence model of behavior-based psychology (adapted from Kreitner, 1982).

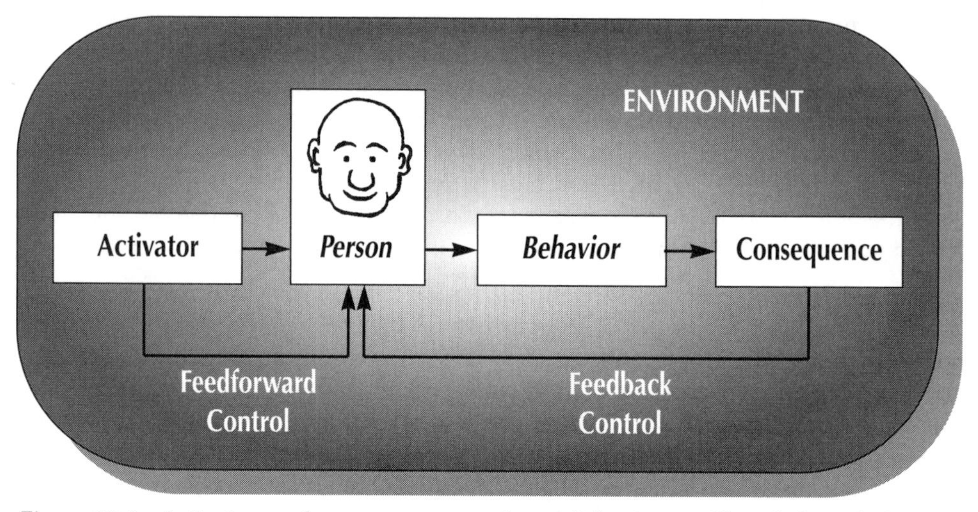

Figure 13.2. Activators and consequences to change behavior are filtered through the person. (Adapted from Kreitner, 1982)

As detailed earlier, activators direct behavior (Chapter 9) and consequences motivate behavior (Chapter 10). However, as shown in Figure 13.2, these events are first filtered through the person.

As discussed in Chapter 5, numerous internal and situational factors influence how we perceive activators and consequences. For example, if we see activators and consequences as nongenuine ploys to control us, our attitude about the situation will be negative. If we believe the external contingencies are genuine attempts to help us do the right thing, our attitude will be more positive. Thus, person, or internal dynamics determine how we receive activator and consequence information. This can influence whether environmental events enhance or diminish what we do.

Let's keep in mind that people operate within a system of environmental factors that have complex and often unmeasurable effects on perceptions, intentions, beliefs, attitudes, values, and behaviors. This is represented by the environment side of The Safety Triad (Geller, 1989) and the "environment" designation in Figure 13.2. Such complex interactions between personal, behavioral, and environmental dimensions of everyday existence often make it extremely difficult—sometimes impossible—to predict or influence what people will do. However, certain changes in external and internal conditions can influence people's behaviors consistently and substantially.

In Chapters 9 and 10, I showed how direct manipulations of activators and consequences can influence large-scale behavior change. Now let's see if changes in internal person factors can benefit behavior change. In particular, how do "inside" factors affect actively caring for safety?

PERSON TRAITS VERSUS STATES

Some person factors are presumed to be *traits,* while others are *states.* Theoretically, traits are relatively permanent characteristics of people, and do not vary appreciably over time or across situations. The popular Myers-Briggs Type Indicator (Myers & McCaulley, 1985), for example, was designed to measure where individuals fall along four dichotomous personality dimensions: extroversion vs. introversion, sensing vs. intuition, thinking vs. feeling, and judgment vs. perception. The various combinations of these dimensions allow for 16 different personality types, each with its special personality characteristics. These traits are presumed permanent and unchangeable, as determined largely by physiological or biological factors.

In contrast, states are characteristics that can change moment-to-moment depending on circumstances and personal interactions. When our goals are thwarted, for example, we can be in a state of frustration. And when experience leads us to believe we have little control over events around us, we can be in a state of apathy or helplessness. These states can influence certain behaviors.

Frustration often provokes aggressive behavior; perceptions of helplessness can inhibit constructive behavior or facilitate inactivity. In contrast, certain life experiences can affect positive person states, such as optimism, personal control, self-confidence and belongingness. This in turn increases constructive behavior, including actively caring.

SEARCHING FOR THE ACTIVELY CARING PERSONALITY

Beginning as early as 1928, psychologists have attempted to identify stable personality traits of helpful people. Although some psychologists might disagree (Huston & Korte, 1976; Rushton, 1984; Staub, 1974), the search has not been particularly successful. For example, Hugh Hartshorne and Mark May (1928) developed 33 different tests to measure positive social behaviors, including helping, honesty, and resistance to temptation; then gave these tests to hundreds of children. Although some children reported a greater willingness to actively care than others, inconsistencies were more common than consistencies. For instance, children who stated they would work as a Red Cross volunteer or send get-well cards to hospitalized peers were not necessarily more willing to share money with classmates, and vice versa. The authors concluded that kind behavior is determined more by situational factors than personality differences. Follow-up research has generally supported this conclusion (Bar-Tal, 1976; Schwartz, 1977).

In one follow-up study, Gergen *et al* (1972) administered a battery of personality tests to 72 college students and subsequently recorded the students' responses to five different requests for help from the psychology department. Results were neither simple nor straightforward. A personality trait that correlated significantly with one actively caring behavior did not relate to another type of helping behavior. Plus, relationships between traits and behaviors were not consistent for males and females. Again, the researchers had to conclude that personality traits interact in complex ways with both situational factors and the nature of helping behavior.

There is some evidence that certain people are consistently more generous, kind, and helping than others (Rushton, 1984), and that these people score higher on certain personality scales. Specifically, individuals with a high propensity to actively care tend to demonstrate empathy or emotional concern or compassion for the welfare of others (Batson, 1991; Oliner & Oliner, 1988). They score relatively high on measures of moral development and social responsibility (Staub, 1974), and tend to be self-confident (Aronoff & Wilson, 1985), idealistic rather than pragmatic (Gilligan, 1982; Waterman, 1988), and possess a sense of self-control, self-directedness, flexibility, self-acceptance, and independence (Oliner & Oliner, 1988; Rosenhan, 1970).

It's possible, though, that these characteristics are states rather than traits. They might vary within people according to situations. If so, these states can be changed with planned intervention. We can influence them through the use of response-consequence contingencies, and by changing the culture of interpersonal relationships. People can even be taught to be more empathic toward others and interventions can be set up to increase and support empathic or other-directed behavior. It makes sense to treat most person factors related to actively caring as changeable "states" rather than permanent "traits." Now we can consider ways to change these states to facilitate active caring.

ACTIVELY CARING STATES

Using animals, psychologists have influenced marked changes in performance by altering certain physiological states of their subjects through food, sleep, or activity deprivation. Similarly, behavioral scientists have demonstrated significant behavior change in both normal and developmentally disabled children by altering aspects of the social context (Gewirtz & Baer, 1958a,b) or the temporal proximity of lunch and response-consequence contingencies (Vollmer & Iwata, 1991). Behavioral scientists typically refer to these manipulations of physiological conditions or psychological states as "establishing operations" (Michael, 1982). They set the stage or establish circumstances to facilitate the impact of an intervention program.

Likewise, certain past or present situations or environmental conditions can influence or establish physiological or psychological states within individuals. This in turn can affect their behavior. From the behavior-based perspective, a basic mechanism for doing this is to use the power of positive consequences.

I contend that actively caring characteristics internal to people are states, not traits. Plus, certain conditions—including activators and consequences—can influence these psychological states (Geller, 1991, 1995). These states are illustrated in Figure 13.3, a model my associates and I have used many times to stimulate discussions among employees. We talk of specific situations, operations, or incidents that influence their willingness to actively care for the achievement of a Total Safety Culture. Let's examine these influential states in more detail.

SELF-ESTEEM ("I AM VALUABLE"). How do you feel about yourself? Generally good or generally bad? Your level of self-esteem is determined by the extent to which you generally feel good about yourself. If we don't feel good about ourselves, it's unlikely we'll care about making a difference in the lives of others. The better we feel about ourselves, the more willing we are to actively care for the safety and health of other people.

One's self-concept, or feeling of worth, is the central theme of most humanistic therapies (Rogers, 1957, 1977). According to Carl Rogers and his followers, we have a real and ideal self-concept. That is, we have notions or dreams of what we would like to be (our ideal self) and what we think we are (our real self). The greater the gap between our real and ideal self-concepts, the lower our self-esteem. Thus, a prime goal of many humanistic therapies is to help a person reduce this gap. This can be done by raising a person's perceptions of their real self-concept: "Count your strengths and blessings and you'll see that you're much better than that." The alternative is to lower one's aspirations or ideal self-concept: "You expect too much; no one is perfect; take life one step at a time and you'll eventually get there."

It's important to maintain a healthy level of self-esteem, and to help others raise their self-esteem. Research shows that people with high self-esteem report fewer negative emotions and less depression than people with low self-esteem (Straumann & Higgins, 1988). Those with higher self-esteem also handle life's stresses better (Brown & McGill, 1989). Recall the discussion of stress versus distress in Chapter 6. Higher self-esteem turns stress into something positive, rather than negative distress.

Researchers have also found that individuals who score higher on measures of self-esteem are less susceptible to outside influences (Wylie, 1974), more confident of achieving personal goals (Wells & Marwell, 1976), and make more favorable impressions on others in social situations (Baron & Byrne, 1994). And supporting the actively caring model depicted in Figure 13.3, people with higher self-esteem help others more frequently than those scoring lower on a self-esteem scale (Batson *et al*, 1986).

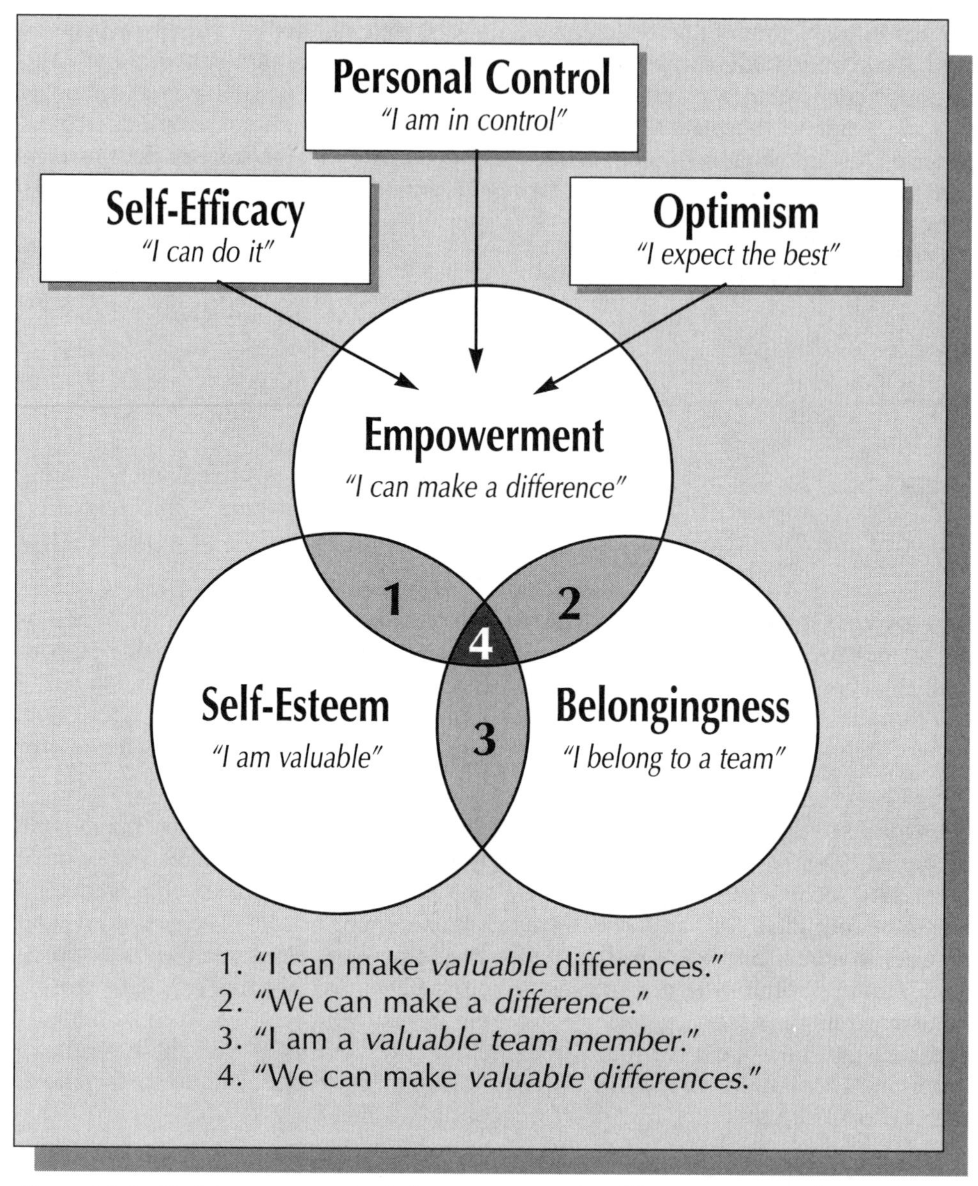

Figure 13.3. Certain person states influence a person's willingness to actively care for the safety and health of others.

Here's something to keep in mind, though. It's also been found that people with high self-esteem are less willing to ask for help than people with low self-esteem (Nadler & Fisher, 1986; Weiss & Knight, 1980). Later in this chapter, I discuss in more detail research that shows a direct relationship between self-esteem and actively caring behavior.

EMPOWERMENT ("I CAN MAKE A DIFFERENCE"). In management literature, empowerment typically refers to delegating authority or responsibility, or sharing decision making (Conger & Kanungo, 1988). In contrast, the person-based perspective of empowerment focuses on how the person who receives more power or influence reacts. From a psychological perspective, empowerment is a matter of feeling—how does it feel to be more responsible? Can you handle it? This view of empowerment requires the personal belief that "I can make a difference." Perceptions of personal control (Rotter, 1966), self-efficacy (Bandura, 1977), and optimism (Scheier & Carver, 1985; Seligman, 1991) strengthen this belief. An empowered state is presumed to increase motivation to "make a difference," perhaps by going beyond the call of duty. As I discuss later in this chapter, there is empirical support for this intuitive hypothesis (Bandura, 1986; Barling & Beattie, 1983; Ozer & Bandura, 1990; Phares, 1976).

Once a little boy went to school. It was quite a big school, but when the boy found he could go right to his room from the playground outside he was happy, and the school didn't seem quite so big anymore. One morning when the little boy had been in school for awhile, the teacher said, "Today we are going to make a picture."

"Good," thought the little boy. He liked to make pictures. He could draw *lions* and *tigers* and *trains* and *boats*. But the teacher said, "Wait. It's not time to begin." And she waited until everyone looked ready. "Now," said the teacher, "We are going to make flowers." "Good," thought the little boy, and he began to make beautiful flowers with his orange and pink and blue crayons. But the teacher said, "Wait." She drew a picture on the blackboard. **It was red with a green stem.** "There, now you may begin." The little boy looked at the teacher's flower. He liked his better, but he did not say this. He just turned his paper over and made a flower like the teacher's. **It was red with a green stem.**

On another day the teacher said, "Today we are going to make something with clay." "Good," thought the little boy. He could make all kinds of things with clay—*snakes* and *snowmen* and *elephants* and *mice*—and he began to pinch and pull his ball of clay. But again the teacher said, "Wait, I will show you how." And she showed everyone how to make one deep dish. The little boy just rolled his clay into a round ball and made a dish like the teacher's. And pretty soon the little boy learned to wait and to watch, and to make things just like the teacher's. And pretty soon he didn't make things of his own anymore.

And then it happened that the little boy and his family moved to another city and the boy had to go to another school. On the very first day he went to school the teacher said, "Today we are going to make a picture." "Good," thought the boy and he waited for the teacher to tell him what to do. But the teacher didn't say anything. She just walked around the room. When she came to the boy she said, "Don't you want to make a picture?" "Yes," said the boy. "What are we going to make?" "Well, I don't know until you make it," said the teacher. "How should I make it?" said the boy. "Why, any way you like!" "And any color?" "Any color," said the teacher. "If everyone made the same thing in the same color, how would I know who made what and which was which?" "I don't know," said the boy, and he began to draw a flower. **It was red with a green stem.**

Figure 13.4. A simple story illustrates the all-too-common sapping of empowerment and its unfortunate consequences. (Adapted from Canfield and Hansen, 1993)

Figure 13.4 includes an instructive and provocative story about the loss of empowerment in a simple but typical school situation. Many readers will relate empathetically with the young boy, having been in similar situations themselves. They know what it's like to have their empowerment sapped (Byham, 1988). The first teacher takes too much personal control over the situation. As a result, the student loses a sense of self-efficacy ("I can do it myself"), personal control ("I am in control"), and even optimism ("I expect the best"). All of this diminishes one's sense of being able to contribute. Instead, the individual learns to wait for top-down instructions and is motivated to do only what is required. This is sad and all too common in home, school, and individual settings. What's even more sad is that many people, like the first teacher in this story, sap empowerment from others and don't even realize it.

Let's look more closely at these three factors affecting our sense of worth and ability—and our propensity to actively care:

Self-efficacy is the idea that "I can do it." This is a key factor in social learning theory, determining whether a therapeutic intervention will succeed over the long term (Bandura, 1990, 1994). We're talking about your self-confidence. Dozens of studies have found that subjects who score relatively high on a measure of self-efficacy perform better at a wide range of tasks. They show more commitment to a goal and work harder to pursue it. They demonstrate greater ability and motivation to solve complex problems at work. They have better health and safety habits; and they are more apt to handle stressors positively, rather than with negative distress (Bandura, 1982; Betz & Hackett, 1986; Hackett *et al*, 1992).

Self-efficacy contributes to self-esteem, and vice versa; but these constructs are different. Simply put, self-esteem refers to a general sense of self-worth; self-efficacy refers to feeling successful or effective at a particular task. Self-efficacy is more focused, and can vary markedly from one task to another. One's level of self-esteem remains rather constant across situations.

When I'm losing to an opponent on the tennis court, my self-efficacy usually drops considerably. However, my self-esteem might not change at all. I might protect my self-esteem by rationalizing that my opponent is younger and more experienced, or that I'm more physically tired and mentally preoccupied than usual. My damaged self-efficacy will undoubtedly lead to reduced optimism about winning the match. If I continue to lose at tennis and run out of excuses, my self-esteem could suffer if I think it's important for me to play tennis well. Now there would be a prominent gap between my real self, a loser at tennis, and ideal self, a winner on the court.

Personal control is the feeling that "I am in control." Professor J. B. Rotter (1966) used the term *locus of control* to refer to a general outlook regarding the location of forces controlling a person's life—internal or external. Those with an internal locus of control believe they usually have direct personal control over significant life events as a result of their knowledge, skill, and abilities. They believe they are captain of their life's ship. In contrast, persons with an external locus of control believe factors like chance, luck or fate play important roles in their lives. In a sense, externals believe they are victims, or sometimes beneficiaries, of circumstances beyond their direct personal control (Rotter, 1966; Rushton, 1984). As depicted in Figure 13.5, however, there are times when everyone likes to feel that their successes resulted from their own efforts.

" COULD I TAKE A MOMENT TO CHANGE MY SHIRT?"

Figure 13.5. There are times when we all want credit for our performance.

Personal control has been one of the most researched individual difference dimensions in psychology. Since Dr. Rotter developed the first measure of this construct in 1966, more than 2,000 studies have investigated the relationship between perceptions of personal control and other variables (Hunt, 1993). Internals are more achievement-oriented and health conscious than externals. They are less prone to distress, and more likely to seek medical treatment when they need it (Nowicki & Strickland, 1973; Strickland, 1989). In addition, having an internal locus of control helps reduce chronic pain, facilitates psychological and physical adjustment to illness and surgery, and hastens recovery from some diseases (Taylor, 1991). Internals perform better at jobs that allow them to set their own pace, whereas externals work better when a machine controls the pace (Eskew & Riche, 1982; Phares, 1991).

Optimism is reflected in the statement, "I expect the best." It's the learned expectation that life events, including personal actions, will turn out well (Scheier & Carver, 1985; Seligman, 1991). Optimism relates positively to achievement. Professor Martin Seligman (1991) reported, for example, that world-class swimmers who scored high on a measure of optimism recovered from defeat and swam even faster compared to those swimmers scoring low. Following defeat, the pessimistic swimmers swam slower.

Compared to pessimists, optimists maintain a sense of humor, perceive problems or challenges in a positive light, and plan for a successful future. *They focus on what they can do rather than on how they feel* (Carver *et al,* 1989; Sherer *et al,* 1982; Peterson & Barrett, 1987). As a result, optimists handle stressors constructively and experience positive stress rather than negative distress (Scheier *et al,* 1986). Optimists essentially expect to be successful at whatever they do, and so they work harder than pessimists to reach their goals. As a result, optimists are beneficiaries of the self-fulfilling prophecy (Tavris & Wade, 1995).

The self-fulfilling prophecy (Merton, 1948) starts with a personal expectation about one's future performance and ends with that expectation coming true because the individual performs in such a way to make it happen. An experiment by Feather (1966) demonstrated how quickly the self-fulfilling prophecy can take effect. For 15 trials, he asked female college students to unscramble letters to make a word. Prior to

Figure 13.6. When our boss asks to see us, we expect the worst.

each trial, the subjects predicted their chances of solving the puzzle or anagram. For the first five trials, half of the women received easy anagrams, while the other subjects received five anagrams with no solution. As you might expect, the group that started with easy anagrams increased their estimates of success on subsequent trials; those who received the five insoluable puzzles became pessimistic about their future success. The optimistic subjects performed markedly better on the last ten anagrams, which were soluble and the same for both groups. The higher a person's expectation for success, the more anagrams solved. When people are optimistic and expect the best, they work hard to make their prediction come true. As a result, they often achieve the best.

What do you expect when your boss or supervisor asks to see you? Do you expect the best? Our past experiences with top-down control and the use of negative consequences to influence our behavior often results in pessimistic rather than optimistic expectations. Moreover, our approach to this situation, illustrated in Figure 13.6, can support our negative expectations.

If you expect to be punished or reprimanded every time your boss or supervisor calls you into the office, then your body language and demeanor will subtly reflect that expectation. You'll "telegraph" these signals to your boss, who might think, "Scott sure looks guilty, I wonder what he's done that needs to be punished?" However, if you approach the interaction with an optimistic attitude, reflected in your body language and verbal behavior, the results could be more positive. You could, for example, write a different internal script: "No one is perfect, and I might have missed something. Everyone can improve with specific behavioral feedback. If I help to make the interaction constructive, the outcome can only be positive."

It's important to understand that fulfilling a pessimistic prophecy can depreciate our perceptions of personal control, self-efficacy, and even self-esteem. Realizing this should motivate us to do whatever we can to make interpersonal interactions positive and constructive. This will not only increase optimism in a work culture, but also promote a sense of group cohesiveness or belongingness—another person state that facilitates actively caring behavior.

BELONGINGNESS. In his best seller, *The different drum: Community making and peace,* Dr. M. Scott Peck (1979) challenges us to experience a sense of true community with others. We need to develop feelings of belongingness with one another regardless of our political preferences, cultural backgrounds and religious doctrine. We need to transcend our differences, overcome our defenses and prejudices, and develop a deep respect for diversity. He proclaims that we must develop a sense of community or interconnectedness with one another if we are to accomplish our best and ensure our survival as human beings.

It seems intuitive that building a sense of community or belongingness among our coworkers will improve organizational safety. Safety improvement requires interpersonal observation and feedback, and for this to happen, people need to adopt a collective win/win perspective instead of the individualistic win/lose orientation common in many work settings. A sense of belongingness or interconnectedness leads to interpersonal trust and caring—essential features of a Total Safety Culture.

In my numerous group discussions with employees on the belongingness concept, someone inevitably raises the point that a sense of belongingness or community at their plant has decreased over recent years. "We used to be more like family around here" is a common theme. For many companies, growth spurts, continuous turnover—particularly among managers—or "lean and mean" cutbacks have left many employees feeling less connected and trusting. It seems, in some cases, people's need level on Maslow's hierarchy (see Figure 12.7) has regressed from satisfying belongingness needs to concentrating on maintaining job security, in order to keep food on the table.

Figure 13.7 lists a number of special attributes prevalent in most families, where interpersonal trust and belongingness are often optimal. We are willing to actively care in special ways for the members of our immediate family. The result is optimal trust, belongingness, and actively caring behavior for the safety and health of our family members. To the extent we follow the guidelines in Figure 13.7 among members of our "corporate family," we will achieve a Total Safety Culture. Following the principles in Figure 13.7 will develop trust and belongingness among people, and lead to the quantity and quality of actively caring behavior expected among family members—at home and at work.

The psychological construct most analogous to the actively caring concept of belongingness is group cohesion—the sum of positive and negative forces attracting group members to each other (Wheeless *et al,* 1982). Satisfaction is considered a key determinant of group cohesiveness. The more cohesive a group, the more satisfied are

- We use more rewards than penalties with ***family*** members.
- We don't pick on the mistakes of ***family*** members.
- We don't rank one ***family*** member against another.
- We brag on the accomplishments of ***family*** members.
- We respect the property and personal space of ***family*** members.
- We pick up after ***family*** members.
- We correct the at-risk behavior of ***family*** members.
- We accept the corrective feedback of ***family*** members.
- We are our brother/sisters keepers of ***family*** members.
- We actively care because they're ***family***.

Figure 13.7. Incorporating an actively caring *family* perspective in an organization will help to cultivate a Total Safety Culture.

members with belonging to the group. Also, the greater the member satisfaction with the group, the greater the group cohesiveness. Wheeless identified two beneficial levels of satisfaction in interpersonal relationships: independence and involvement. Independence refers to an internal locus of control in group decision-making, and group involvement reflects the level of interpersonal concern, respect, and warmth present in the group.

Ridgeway (1983) defined five benefits of group cohesiveness, including increased:

- Quantity and quality of communication
- Individual participation
- Group loyalty and satisfaction
- Ability to enforce group norms and focus energy toward goal attainment
- Elaboration of group culture, typified by special behavioral routines that increase the group's sense of togetherness

From this conceptualization, it follows that members of a cohesive group should demonstrate actively caring behavior for each other. The actively caring model also predicts that group cohesiveness will increase this behavior for targets—persons, behaviors, and environments—outside the group.

Measuring Actively Caring States

Surveys that measure workplace safety cultures are quite popular these days (Geller, 1992; Geller & Roberts, 1993; Simon & Simon, 1992). Some proponents recommend their use to discriminate between "safe" and "unsafe" employees (Krause, 1992, 1995). To justify these surveys, consultants often teach that individuals have stable personality traits determining both their motivation level for particular tasks and their propensity to have an injury. This perspective seems to be on the rise today, as safety consultants peddle their "quick fix" measurement devices. But I believe the idea of a stable personality bias can interfere with the more practical, cost-effective behavior-based approach to managing human resources, which is what this text is all about. If you recall, in Chapter 1 the comprehensive research comparisons by Professor Guastello (1993) revealed this personnel selection approach to be most popular—but quite ineffective at reducing industrial injuries.

At my industrial safety workshops, I explain that valid individual difference scales do not exist to reliably predict an individual's propensity to get hurt on the job (Geller, 1994). Even if they were available, you still must account for the influence of environmental conditions, management systems, response contingencies, and peer interactions—in addition to individual factors. I have found, though, that assessing individual differences, including different lifestyles, personality factors, perceptions, and cognitive strategies, can be useful in an employee education and training program to teach the concept of individual diversity, and to increase employees' awareness of their own idiosyncrasies that relate potentially to injury proneness.

Safety Culture Survey. The Safety Culture Survey which Steve Roberts, Mike Gilmore, and I developed for culture assessment and corporate training programs

(Geller & Roberts, 1993; Geller *et al,* 1992, 1996) includes sub-scales to measure safety-related perceptions and risk propensity, including cognitive failures (Broadbent *et al,* 1982), sensation seeking (Zuckerman, 1979), psychological reactance (Tucker & Byers, 1987), and extroversion (Eysenck & Eysenck, 1985)

The most useful sub-scale of our Safety Culture Survey, from both a training and culture-change perspective, is the actively caring scale, which includes adaptations from standard measures of self-esteem (Rosenberg, 1965), self-efficacy (Scheier *et al,* 1982), personal control (Nowicki & Duke, 1974), optimism (Scheier & Carver, 1985), and group cohesion (Wheeless *et al,* 1982). The survey also includes direct measures of willingness to actively care from an environment focus ("I am willing to pick up after another employee in order to maintain good housekeeping"), a person focus ("If an employee needs assistance with a task, I am willing to help even if it causes me inconvenience"), and a behavior-change focus ("I am willing to observe the work practices of another employee in order to provide direct feedback to him/her"). Respondents' reactions to each of the 154 items of the survey are given on a 5-point Likert-type scale ranging from "Highly Disagree" to "Highly Agree."

SUPPORT FOR THE ACTIVELY CARING MODEL. Analyzing Safety Culture Survey results from three large industrial complexes shows remarkable support for the actively caring model (Geller *et al,* 1996; Roberts & Geller, 1995). The personal control factor was consistently most influential in predicting willingness to actively care. Belongingness scores predicted significant differences in actively caring propensity at two of three plants. Self-esteem and optimism always correlated highly with each other, and with willingness to actively care, but only one or the other predicted independent variance in actively caring propensity. For these tests, our survey did not include a measure of self-efficacy. The multiple regression coefficients and sample sizes were .54 (n=262), .57 (n=307), and .71 (n=207) at the three plants, respectively (see Geller *et al,* 1996, for details).

These regression results were not of much interest to the plant managers, supervisors, and trainers at the three facilities. But there is a practical value to classifying actively caring attributes according to various work groups, including managers, operators, secretaries, contractors, and laboratory personnel. At another plant, for example, relatively high levels of willingness to actively care convinced the plant manager to support an actively caring training and intervention process. In another case, extreme differences in the inclination to help across work areas prompted the development of special intervention programs for certain work groups.

The twenty questions included in Figure 13.8 were selected from the Actively Caring Person Scale of our Safety Culture Survey. Each of the five actively caring states discussed in this chapter are assessed. There are only four questions per person state, so this should not be considered a reliable nor valid measure of these factors. In other words, don't read too much into this survey. Just respond to each query according to the instructions and then check the answer key in Figure 13.9 to increase your understanding of the five actively caring person states. Comparing the items that measure self-esteem with those that assess self-efficacy, for example, help you understand the distinct difference between these two constructs. Remember, these person factors are presumed to be states that fluctuate from day to day and from situation to situation. The score you get today might be quite different than the one you would obtain on another day under a different set of circumstances.

This is a questionnaire about your beliefs and feelings. Read each statement, then circle the number that best describes your current feelings. There are no "right" or "wrong" answers. This questionnaire only asks about your personal opinions.

	HIGHLY DISAGREE	DISAGREE	NOT SURE	AGREE	HIGHLY AGREE
1. I feel I have a number of good qualities.	1	2	3	4	5
2. Most people I know can do a better job than I can.	1	2	3	4	5
3. On the whole, I am satisfied with myself.	1	2	3	4	5
4. I feel I don't have much to be proud of.	1	2	3	4	5
5. When I make plans, I am certain I can make them work.	1	2	3	4	5
6. I give up on things before completing them.	1	2	3	4	5
7. I avoid challenges.	1	2	3	4	5
8. Failure just makes me try harder.	1	2	3	4	5
9. People who never get injured are just plain lucky.	1	2	3	4	5
10. People's injuries result from their own carelessness.	1	2	3	4	5
11. I am directly responsible for my own safety.	1	2	3	4	5
12. Wishing can make good things happen.	1	2	3	4	5
13. I hardly ever expect things to go my way.	1	2	3	4	5
14. If anything can go wrong for me, it probably will.	1	2	3	4	5
15. I always look on the bright side of things.	1	2	3	4	5
16. I firmly believe that every cloud has a silver lining.	1	2	3	4	5
17. My work group is very close.	1	2	3	4	5
18. I distrust the other workers in my department.	1	2	3	4	5
19. I feel like I really belong to my work group.	1	2	3	4	5
20. I don't understand my coworkers.	1	2	3	4	5

Figure 13.8. Sample items from the Actively Caring Person Scale.

Self-Esteem *(items 1-4)* = feelings of self worth and value
Actual scale = 16 items
(a) Add numbers for items 1 & 3 — Total 1 = _____
(b) Add numbers for items 2 & 4 and subtract from 14 — Total 2 = _____

Self-Efficacy *(items 5-8)* = general levels of belief in one's competence
Actual scale = 23 items
(a) Add numbers for items 5 & 8 — Total 1 = _____
(b) Add numbers for items 6 & 7 and subtract from 14 — Total 2 = _____

Personal Control *(items 9-12)* = the extent to which a person believes he or she is responsible for his/her life situation
Actual scale = 25 items
(a) Add numbers for items 10 & 11 — Total 1 = _____
(b) Add numbers for items 9 & 12 and subtract from 14 — Total 2 = _____

Optimism *(items 13-16)* = the extent to which a person expects the best will happen for him/her
Actual scale = 8 items
(a) Add numbers for items 15 & 16 — Total 1 = _____
(b) Add numbers for items 13 & 14 and subtract from 14 — Total 2 = _____

Belongingness *(items 17-20)* = the perception of group cohesiveness or feelings of togetherness
Actual scale = 20 items
(a) Add numbers for items 17 & 19 — Total 1 = _____
(b) Add numbers for items 16 & 20 and subtract from 14 — Total 2 = _____

"ACTIVELY CARING" SCORE = Sum of Self-Esteem, Self-Efficacy, Optimism, Personal Control, and Belongingness Totals — **Total Score =** _____

Figure 13.9. Scoring your answers to the twenty person-state items will increase your understanding of the Actively Caring model.

The Actively Caring Model

THEORETICAL SUPPORT FOR THE MODEL

The actively caring model depicted in Figure 13.3 certainly makes intuitive sense. Doesn't your willingness to help others often change, depending on the person states given in the model? When we feel better about ourselves, we're less preoccupied with personal problems, and more likely to do something nice for someone else. And, of course, we feel better about ourselves when reaching our aspirations through self-efficacy and personal control. In turn, this satisfaction can lead to optimistic expectations and heightened self-esteem. Plus, when it comes to helping others, we're more apt to help those we like and feel close to. Theorizing a direct relationship between the probability of actively caring behavior and the degree of belongingness between the helper and the victim should not surprise anyone.

The common-sense appeal of the actively caring model is also supported by comparing it with Maslow's popular and intuitive motivation theory. Abraham Maslow's hierarchy of needs was discussed in Chapter 12 (see Figure 12.7), with particular reference to Maslow's (1971) later addition of "self-trancendency" to the top of his hierarchy. I bet you see the similarity between self-trancendency and actively caring.

Figure 13.10 depicts a hierarchy of the concepts in the actively caring model, along with the need levels of Malsow's hierarchy. The similarities are noteworthy. Indeed, half of the actively caring concepts—belongingness, self-esteem, and actively caring—are exactly the same as three of Maslow's need levels. The other three person states—self-efficacy, personal control, and optimism—can be readily linked to the remaining three need levels. It makes sense to relate self-efficacy or self-effectiveness

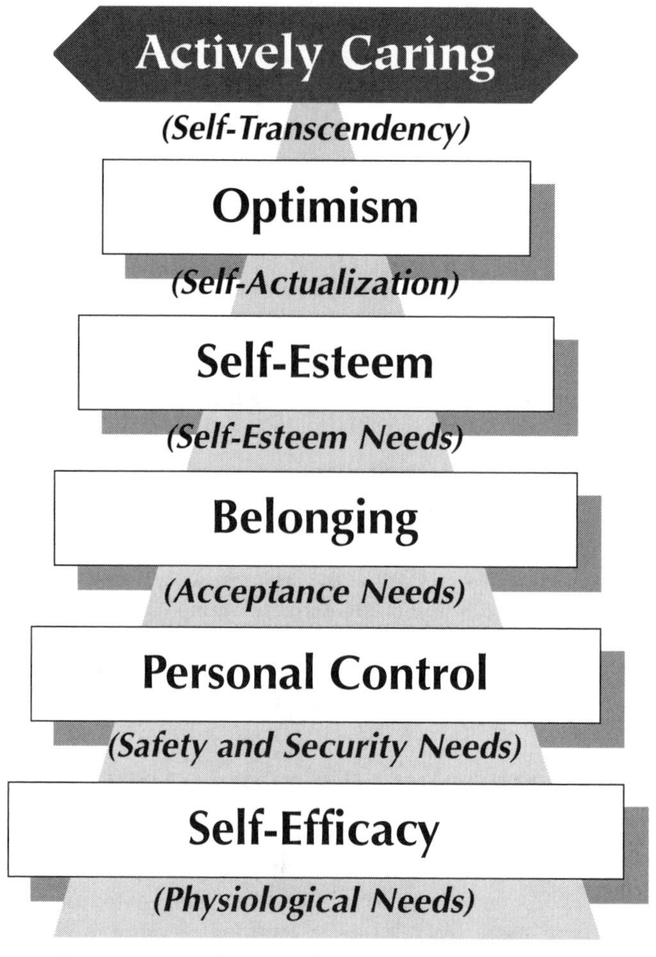

Figure 13.10. The actively caring person factors are reflected in Maslow's hierarchy of needs.

to an individual's drive to satisfy basic physiological needs. And increasing one's sense of personal control is basic to feeling safe and secure. Indeed, increasing employees' personal control of safety is fundamental to achieving a Total Safety Culture.

Linking optimism with self-actualization might be a bit of a stretch. But doesn't it seem that optimism reflects self-actualization, or vice versa? When people believe they are the best they can be, when they are self-actualized, they are happiest and most optimistic about the future. Their self-fulfilling prophesy comes true, and they continue to be self-actualized.

All of this sounds good, but I can't allow common sense to determine the value of a theory; after all, that would be inconsistent with a primary theme of this book: Principles and procedures must be based on valid results from research, not on common sense.

There's a practical benefit to connecting a theory or model to common sense—it scores points for what you're trying to prove. But the accuracy or applicability of a theory or concept can't be based on whether it sounds good, or seems acceptable. As discussed earlier, Malsow's hierarchy of needs has intuitive charm, but limited empirical research has been conducted to support the concept of a motivational hierarchy. On the other hand, there has been substantial research on helping behavior that can be related to the actively caring model. Let's see if research supports the theory. Perhaps some of it will suggest practical applications for injury prevention.

RESEARCH SUPPORT FOR THE MODEL

I have found a number of empirical studies, mostly in social psychology literature, supporting individual components of the actively caring model. Although these studies did not address more than one factor at a time, the combined evidence gives substantial empirical support for the model. The bystander intervention paradigm (as described in Chapter 12) has been the most common and rigorous laboratory method used to study person factors related to actively caring behaviors. This research measures or manipulates self-esteem, empowerment, and belongingness among a group of subjects. Then these individuals are placed in a situation where they have an opportunity to help another person who is presumably encountering some kind of a personal crisis, like falling off a ladder, dropping belongings, or feigning a heart attack or illness. The delay in coming to the rescue is studied as a measure of a subject's social situation or personality state. As pointed out in Chapter 12, the actively caring behaviors studied in these experiments were reactive and person-focused. They were never proactive and behavior-focused.

SELF-ESTEEM. According to Coopersmith (1967), self-esteem can be relatively general and enduring—and also situational and transitory. Often considered a general trait, self-esteem can be affected markedly by situations, response-consequence contingencies, and personal interactions. When circumstances return to "normal," self-esteem usually returns to its chronic level. But permanent changes in circumstances or perceived personal competencies can have a lasting impact. Maturity, for example, can shape or reshape personal aspirations, and so change the gap between a person's perceived real and ideal self.

Michelini *et al* (1975) and Wilson (1976) measured subjects' self-esteem with a sentence completion test (described in Aronoff, 1967). Then they observed whether these subjects helped out in a bystander intervention test. High self-esteem subjects were significantly more likely than those with low self-esteem to help another person pick up dropped books (Michelini *et al,* 1975), and to leave an experimental room to assist a person in another room who screamed he had broken his foot following a mock "explosion" (Wilson, 1976). Similarly, subjects with higher self-esteem scores were more likely to help a stranger (the experimenter's accomplice) by taking his place in an experiment where they would presumably receive electric shocks (Batson *et al,* 1986).

In a naturalistic field study of actively caring behavior, Bierhoff *et al* (1991) compared individual differences among people who helped or only observed at vehicle crashes. People who stopped at the crash scene were identified by ambulance workers, and were later given a questionnaire measuring certain personality constructs. Those who helped scored significantly higher on self-esteem, personal control, and social responsibility.

PERSONAL CONTROL. Some studies have measured subjects' locus of control and then observed the probability of actively caring behavior in a bystander intervention test. Another study manipulated subjects' perceptions of personal control prior to observing their actively caring behaviors. The field study discussed above by Bierhoff *et al* (1991) found more active caring at vehicle-crash scenes by bystanders with an internal locus of control. Also, those high-self-esteem subjects who showed more active caring than low-self-esteem subjects in Wilson's (1976) bystander intervention study (discussed above) were also characterized as *internals,* in contrast to the lower self-esteem *externals,* who were less apt to actively care. In addition, Midlarsky (1971) found more internals than externals willing to help a confederate perform a motor coordination task that involved the reception of electric shocks.

Sherrod and Downs (1974) asked subjects to perform a task in the presence of loud, distracting noise. They manipulated subjects' perception of personal control by telling half the subjects they could terminate the noise, if necessary, by notifying them through an intercom. The subjects who could have terminated the noise but did not were significantly more likely to comply with a later request by an accomplice to help solve math problems requiring extra time and resulting in no extrinsic benefits.

OPTIMISM. As mentioned in Chapter 12, researchers have manipulated optimistic states or moods by giving test subjects unexpected rewards or positive feedback, and then observing the frequency of actively caring behaviors. Isen and Levin (1972), for example, observed that 84 percent of those individuals who found a dime in the coin-return slot of a public phone (placed there by researchers) helped an accomplice pick up papers he dropped in the subject's vicinity. In contrast, only 4 percent of those who did not find a dime helped the man pick up his papers. Similarly, students given a cookie while studying at a university library were more likely than those not given a cookie to agree to help another person by participating in a psychology experiment (Isen & Levin, 1972).

Isen *et al* (1976) delivered free samples of stationery to people's homes and then called them later to request an act of caring. Specifically, the caller said he had dialed the wrong number but since he'd used his last dime, he needed the subject to call a garage to tow his car. Subjects who had received the stationery were more likely to make the call than subjects who had not received this gift.

Carlson *et al* (1988) reviewed these and other studies that showed direct relationships between mood—or optimism—and actively caring behavior. They reported that these pleasant experiences increased active caring, purportedly by inducing a positive mood or optimistic outlook: finding a dime, receiving a packet of stationery, listening to soothing music, being on a winning football team, imagining a vacation in Hawaii, and being labeled a charitable person. The authors suggested that the state or mood caused by the pleasant experiences may have increased the perceived rewarding value of helping others.

Berkowitz and Connor (1966) found a direct relationship between perceived success and actively caring behavior. Their subjects were instructed to complete certain puzzles in less than two minutes. The task was manipulated to allow half of the subjects to succeed and half to fail. Afterward, successful subjects made more boxes for the researcher's accomplice than did the unsuccessful subjects. In a series of analogous laboratory studies, Isen (1970) manipulated performance feedback on a perceptual-motor task. Subjects who were told that they had performed extremely well were more likely to donate money to charity, pick up a dropped book, and hold a door open for a confederate than those who were told that they had performed very poorly.

These later studies that manipulated the outcome of a task illustrated a potential overlap between optimism, self-efficacy, and personal control. It's reasonable to assume performance feedback increases one's perception of self-efficacy and personal control, as well as one's optimism. Indeed, Scheier and Carver's (1985) measure of optimism correlated significantly with locus of control. Optimism, self-efficacy, and personal control determine feelings of empowerment, according to the actively caring model. Thus, these performance-feedback studies support the general hypothesis that we can increase the chances for active caring by boosting individual perceptions of empowerment.

BELONGINGNESS. Staub (1978) reviewed studies showing that people are more likely to help victims who belong to their own group, with *group* defined as race, nationality, or an arbitrary distinction defined by preference for a particular artist's paintings. Similarly, Batson *et al* (1986) found subjects more likely to help a confederate if they rated her as similar to them. In a bystander intervention experiment, pairs of friends intervened faster to help a female experimenter who had fallen from a chair than did pairs of strangers. Thus the bystander intervention effect, which holds that victims are less likely to be helped as the number of observers increases, may not occur when friends are involved. Group cohesiveness or belongingness counteracts the diffusion of responsibility that presumably accounts for bystander inaction.

By experimentally manipulating group cohesion in groups of two and four, Rutkowski *et al* (1983) tested whether group cohesion can reverse the usual bystander intervention effect. Cohesiveness was created by having the groups discuss topics and feelings they had in common related to college life. Then the researchers studied the number of subjects who left the experimental room to assist a "victim" who had presumably fallen off a ladder, and measured how long it took to respond. Findings indicated that group cohesiveness increased actively caring behaviors, perhaps because of reduced diffusion of responsibility. Both frequency and speed of helping was greater for the cohesive groups. Subjects in the high-cohesion/four-person group were most likely to respond quickly.

Blake (1978) studied real-world relationships between group cohesion and the ultimate act of caring—altruistic suicide. He gathered his data from official records of Medal of Honor awards given during World War II and Vietnam. The independent variable was the cohesiveness of combat units, estimated by group training and size; the dependent variable was percentage of "grenade acts"— voluntarily using one's body to shield others from exploding devices. Results revealed that the smaller, more elite, specially trained combat units—the Marine Corps and Army airborne units—accounted for a substantially larger percentage of "grenade acts" than larger, less specialized units—Army non-airborne units. These findings also supported the hypothesis that group cohesion increases actively caring behavior.

DIRECT TEST OF THE MODEL

As I indicated, none of these empirical studies were designed to test the actively caring model. But our research with the Safety Culture Survey has shown a direct relationship between employees' scores on the five person states and their self-reported actively caring behavior (Geller & Roberts, 1993; Geller *et al,* 1996). People who scored high on measures of self-esteem, self-efficacy, personal control, optimism, and belongingness reported that they had performed more acts of caring in the past, and they reported a significantly greater willingness to actively care in the future. A major weakness of this research was its complete reliance on verbal report. Actual behavior was not observed. Two field tests overcame this weakness.

Roberts and Geller (1994) studied relationships between on-the-job actively caring behaviors of 65 employees and prior measures of their self-esteem, optimism, and group cohesion. Self-efficacy and personal control were not assessed. Employees were told to give coworkers "Actively Caring Thank-You Cards" redeemable for a beverage in the cafeteria whenever they saw a coworker going beyond the call of duty for safety. Employees were trained to look for proactive actively caring behavior that removed a hazard, supported safe behavior, or corrected at-risk behavior. Employees who gave or received a thank-you card scored significantly higher on measures of self-esteem and group cohesion than those who did not give nor receive a card.

Five of my students (Buermeyer *et al,* 1994) tested the entire actively caring model by asking 156 of their peers (75 males and 86 females) who had just donated blood at a campus location to complete a 60-item survey that measured each of the five person factors in the model. The high return rate of 92 percent was consistent with an actively caring profile. Most remarkable, though, was that the blood donors scored significantly higher on *each* of the five subscales than did a group of 292 randomly selected students from the same university population. The blood donors also scored significantly higher than the others on self-report measures of willingness to actively care.

The prominently higher survey scores from blood donors could have resulted from the immediate effects of donating blood. Actually, as reflected in the opening quotation of this chapter, actively caring might very well increase a person's sense of self-esteem, self-efficacy, personal control, optimism, and belongingness. Whether the differences between blood donors and the control group were due to preexisting states of those who gave blood or to the impact of giving blood, the entire actively caring model was supported by this research.

Actively caring person states are probably present before acts of caring, serving as establishing conditions that activate the caring behavior, and these states are likely affected in positive directions after performing an act of caring. Obviously, more research is needed to study this model as a predictor of when people will actively care and as a predictor of changes in person states following actively caring behavior. In the next chapter, I use the model as a framework for exploring ways to increase actively caring behavior throughout an organization or culture.

In Conclusion

In this chapter we continued to develop an understanding of actively caring behavior as it relates to injury prevention. A person-based approach was emphasized; we considered subjective factors inside people as potential determinants of active caring. The notion of a general actively caring personality was entertained at first, but discarded because of relatively limited empirical support. Even if some people do have permanent traits which make them more or less prone to actively care, it's unlikely we could use this information to improve safety or prevent injuries. However, we could benefit safety if certain person states influence people's willingness to actively care, and if these states can be manipulated by controllable outside factors.

Five person states were proposed as influencing people's willingness to actively care—self-esteem, self-efficacy, personal control, optimism, and belongingness. Each of these person variables has a prosperous research history in psychology and some of this research relates directly to the actively caring model. Research that tested relationships between person states and actual behavior was reviewed. Although the results strongly supported the model, none of the actively caring behavior studied was proactive or behavior-based.

A few direct tests support the model, but suggest the need for further research. A particularly important question is whether actively caring states are both antecedents and consequences of a caring act. It seems intuitive that performing an act of kindness that is effective, accepted, and appreciated could increase a helper's self-esteem, self-efficacy, personal control, optimism, and belongingness. This, in turn, should increase the probability of more actively caring behavior. In other words, one act of caring, properly appreciated, should lead to another . . . and another. A self-supporting actively caring cycle is likely to occur. You can see how important it is to get actively caring behavior started and accepted among a large group of individuals. This challenge is addressed in the next chapter.

REFERENCES

Aronoff, J. (1967). *Psychological needs and cultural systems.* NJ: Van Nostrand.

Aronoff, J., & Wilson, J.P. (1985). *Personality in the social process.* Hillsdale, NJ: Erlbaum.

Bandura, A. (1977). Self-efficacy: Toward a unifying theory of behavioral change. *Psychological Review,* 84, 191-215.

———. 1982). Self-efficacy mechanism in human agency. *American Psychologist, 37,* 122-147.

———. 1986). *Social foundations of thought and action.* Englewood Cliffs, NJ: Prentice-Hall.

———. (1990). Self-regulation of motivation through goal systems. In R. A. Dienstbier (Ed.), *Nebraska symposium on motivation,* Vol. 38. Lincoln, NE: University of Nebraska Press.

———. (1994). Self-efficacy. In *Encyclopedia of Human Behavior,* Vol. 4. Orlando, FL: Academic Press.

Bar-Tal, D. (1976). *Prosocial behavior: Theory and research.* Washington: Hemisphere Publishing Corp.

Barling, J., & Beattie, R. (1983). Self-efficacy beliefs and sales performance. *Journal of Organizational Behavior Management, 5,* 41-51.

Baron, R. A., & Byrne, D. (1994). *Social psychology: Understanding human interaction* (Seventh Edition). Boston: Allyn and Bacon.

Batson, C. D. (1991). *The altruism question: Toward a social-psychological answer.* Hillsdale, NY: Erlbaum.

Batson, C. D., Bolen, M. H., Cross, J. A., & Neuringer-Benefiel, H. E. (1986). Where is altruism in the altruistic personality? *Journal of Personality and Social Psychology, 1,* 212-220.

Berkowitz, L., & Connor, W. H. (1966). Success, failure, and social responsibility. *Journal of Personality and Social Psychology, 4,* 664-669.

Betz, N. E., & Hackett, G. (1986). Applications of self-efficacy theory to understanding career choice behavior. *Journal of Social and Clinical Psychology, 4,* 279-289.

Bierhoff, H. W., Klein, R., & Kramp, P. (1991). Evidence for altruistic personality from data on accident research. *Journal of Personality, 59*(2), 263-279.

Blake, J. A. (1978). Death by hand grenade: Altruistic suicide in combat. *Suicide and Life-threatening Behavior, 8,* 46-59.

Broadbent, D., Cooper, P. F., Fitzgerald, P., & Parker, K. (1982). The cognitive failures questionnaire (CFQ) and its correlates. *British Journal of Clinical Psychology, 21,* 1-16.

Brown, J. D., & McGill, K. L. (1989). The cost of good fortune: When positive life events produce negative health consequences. *Journal of Personality and Social Psychology, 57,* 1103-1110.

Buermeyer, C.M., Rasmussen, D., Roberts, D.S., Martin, C., & Gershenoff, A.G. (1994, May). *Red Cross blood donors vs. a sample of students: An assessment of differences between groups on "actively caring" person factors.* Paper presented at the Virginia Academy of Science, Harrisonburg, VA.

Byham, W. C. (1988). *Zapp! The lightning of empowerment.* New York: Fawcett Columbine.

Canfield, J., & Hansen, M. V. (1993). *Chicken soup for the soul.* Deerfield Beach, FL: Health Communications, Inc.

Carlson, M., Charlin, V., & Miller, N. (1988). Positive mood and helping behavior: A test of six hypotheses. *Journal of Personality and Social Psychology, 55,* 211-229.

Carver, C. S., Scheier, M. F., & Weintraub, J. K. (1989). Assessing coping strategies: A theoretically based approach. *Journal of Personality and Social Psychology, 56,* 267-283.

Conger, J. A., & Kanungo, R. N. (1988). The empowerment process: Integrating theory and practice. *Academy of Management Review, 13,* 471-482.

Coopersmith, S. (1967). *The antecedents of self-esteem.* San Francisco, CA: W. H. Freeman.

Eskew, R. T., & Riche, C. V. (1982). Pacing and locus of control in quality control inspection. *Human Factors, 24,* 411-415.

Eysenck, H. J., & Eysenck, N. W. (1985). *Personality and individual differences: A natural science approach.* New York, NY: Plenum.

Feather, N. T. (1966). Effects of prior success and failure on expectations of success and subsequent performance. *Journal of Personality and Social Psychology, 3,* 287-298.

Geller, E. S. (1989). Managing occupational safety in the auto industry. *Journal of Organizational Behavior Management, 101,* 181-185.

———. (1991). If only more would actively care. *Journal of Applied Behavior Analysis, 24,* 601-612.

———. (1992). *A critical review of human dimension sessions at the 1992 National Safety Council Congress and Exposition.* Technical Report for Exxon Chemical Company. Blacksburg, VA: Virginia Polytechnic Institute and State University.

———. (1994). Survey reliability vs. validity: Digging into important technical issues. *Industrial Safety and Hygiene News,* 12, 12-13.

———. (1995). Actively caring for the environment: An interaction of behaviorism and humanism. *Environment and Behavior,* 27, 184-195.

Geller, E. S., & Roberts, D. S. (1993, January). *Beyond behavior modification for continuous improvement in occupational safety.* Paper presented at the FABA/OBM Network Conference, St. Petersburg, FL.

Geller, E. S., Roberts, D. S., & Gilmore, M. R. (1992). *Achieving a total safety culture through employee involvement.* Newport, VA: Make-A-Difference, Inc.

———, ———, & ———. (1996). Predicting propensity to actively care for occupational safety. *Journal of Safety Research, 27,* 1–8.

Gergen, K. J., Gergen, M. M., & Meter, K. (1972). Individual orientation to prosocial behavior. *Journal of Social Issues,* 28(3), 105-130.

Gewirtz, J. L., & Baer, D. M. (1958a). Deprivation and satiation of social reinforcers as drive conditions. *Journal of Abnormal and Social Psychology,* 57, 165-172.

——— & ———. (1958b). The effect of brief social deprivation on behaviors for a social reinforcer. *Journal of Abnormal and Social Psychology,* 56, 49-56.

Gilligan, C. (1982). *In a different voice: Psychological theory and women's development.* Cambridge, MA: Harvard University Press.

Guastello, S.S. (1993). Do we really know how well our occupational accident prevention programs work? *Safety Science,* 16, 445-463.

Hackett, G., Betz, N. E., Casas, J. M., & Rocha-Singh, I. A. (1992). Gender, ethnicity, and social cognitive factors predicting the academic achievement of students in engineering. *Journal of Counseling Psychology,* 39, 527-538.

Hartshorne, H., & May, M. A. (1928). *Studies in the nature of character* (Vol. 1). *Studies in deceit.* New York: MacMillan.

Hunt, M. M. (1993). *The story of psychology.* New York: Doubleday.

Huston, T. L., & Korte, C. (1976). The responsive bystander: Why he helps. In T. Lickona (Ed.), *Moral development and behavior: Theory, research, and social issues.* New York: Holt, Rhinehart, & Winston.

Isen, A. M. (1970). Success, failure, attention, and reaction to others: The warm glow of success. *Journal of Personality and Social Psychology,* 15, 294-301.

Isen, A. M., & Levin, P. F. (1972). Effect of feeling good on helping: Cookies and kindness. *Journal of Personality and Social Psychology,* 21, 384-388.

Isen, A. M., Clark, M., & Schwartz, M. (1976). Duration of the effect of good mood on helping: "Footprints in the sands of time." *Journal of Personality and Social Psychology,* 34, 385-393.

Krause, T. R. (1992, October). *Behavior science methods for accident prevention.* Presentation at the National Safety Council Congress and Exposition, Orlando, FL.

———. (1995). *Employee-driven systems for safe behavior: Integrating behavioral and statistical methodologies.* New York: Van Nostrand Reinhold.

Kreitner, R. (1982). The feedforward and feedback control of job performance through organizational behavior management (OBM). *Journal of Organizational Behavior Management,* 4(2), 3-21.

Maslow, A. H. (1971). *The farther reaches of human nature.* New York: Viking.

Merton, R. (1948). The self-fulfilling prophecy. *Antioch Review,* 8, 193-210.

Michael, J. (1982). Distinguishing between discriminative and motivational functions of stimuli. *Journal of the Experimental Analysis of Behavior,* 37, 149-155.

Michelini, R. L., Wilson, J. P., & Messe, L. A. (1975). The influence of psychological needs on helping behavior. *The Journal of Psychology,* 91, 253-258.

Midlarsky, E. (1971). Aiding under stress: The effects of competence, dependency, visibility, and fatalism. *Journal of Personality,* 39, 132-149.

Myers, I. B., & McCaulley, M. H. (1985). *Manual: A guide to the development and use of the Myers-Briggs Type Indicator.* Palo Alto, CA: Consulting Psychologists Press.

Nadler, A., & Fisher, J. D. (1986). The role of threat to self-esteem and perceived control in recipient reaction to help: Theory development and empirical validation. In L. Berkowitz (Ed.), *Advances in experimental social psychology* (Vol. 19). San Diego, CA: Academic Press.

Nowicki, S., & Duke, M. P. (1974). A locus of control scale for college as well as noncollege adults. *Journal of Personality Assessment, 38*, 136-137.

Nowicki, S., & Strickland, B. R. (1973). A locus of control scale for children. *Journal of Consulting Psychology, 40*, 148-154.

Oliner, S. P., & Oliner, P. M. (1988). *The altruistic personality: Rescuers of Jews in Nazi Europe.* New York: Free Press.

Ozer, E. M., & Bandura, A. (1990). Mechanisms governing empowerment effects: A self-efficacy analysis. *Journal of Personality and Social Psychology, 58*, 472-486.

Peck, M. S. (1979). *The different drum: Community making and peace.* New York: Simon and Schuster.

Peterson, C., & Barrett, L. C. (1987). Explanatory style and academic performance among university freshmen. *Journal of Personality and Social Psychology, 53*, 603-607.

Phares, E. J. (1976). *Locus of control in personality.* Morristown, NJ: General Learning Press.

Phares, E. S. (1991). *Introduction to personality* (Third Edition). New York: Harper Collins.

Ridgeway, C. L., (1983). *The dynamics of small groups.* New York: St. Martin's Press.

The Roanoke Times (1995). Pair breathes life into driver, September 8, pp. A1, A2.

Roberts, D. S., & Geller, E. S. (1995). An "actively caring" model for occupational safety: A field test. *Applied & Preventive Psychology, 4*, 53-59.

Rogers, C. (1957). The necessary and sufficient conditions of therapeutic personality change. *Journal of Consulting Psychology, 21*, 95-103.

Rogers, C. (1977). *Carl Rogers on personal power: Inner strength and its revolutionary impact.* New York: Delacorte.

Rosenhan, D. L. (1970). The natural socialization of altruistic autonomy. In J. Macaulay & L. Berkowitz (Eds.), *Altruism and helping behavior.* New York: Academic Press.

Rosenberg, M. (1965). *Society and the adolescent self-image.* Princeton, NJ: Princeton University Press.

Rotter, J. B. (1966). Generalized expectancies for internal versus external control of reinforcement. *Psychological Monographs, 80*, No. 1.

Rushton, J. P. (1984). The altruistic personality: Evidence from laboratory, naturalistic and self-report perspectives. In E. Staub, D. Bar-Tal, J. Karylowski, & J. Reykowski (Eds.), *Development and maintenance of prosocial behavior.* New York: Plenum.

Rutkowski, G. K., Gruder, C. L., & Romer, D. (1983). Group cohesiveness, social norms, and bystander intervention. *Journal of Personality and Social Psychology, 44*, 545-552.

Scheier, M. F., & Carver, C. S. (1985). Optimism, coping and health: Assessment and implications of generalized outcome expectancies. *Health Psychology, 4*, 219-247.

——— & ———. (1993). On the power of positive thinking: The benefits of being optimistic. *Current Directions in Psychological Sciences, 2*, 26-30.

Scheier, M. F., Weintraub, J. K., & Carver, C. S. (1986). Coping with stress: Divergent strategies of optimists and pessimists. *Journal of Personality and Social Psychology, 51*, 1257-1264.

Schwartz, S. H. (1977). Normative influences on altruism. In L. Berkowitz (Ed.), *Advances in experimental social psychology* (Vol. 10). New York: Academic Press.

Seligman, M. E. P. (1991). *Learned optimism.* New York: Alfred A. Knopf.

Sherer, M., Maddox, J. E., Mercandante, B., Prentice-Dunn, S., Jacobs, B., & Rogers, R. W. (1982). The self-efficacy scale: Construction and validation. *Psychological Reports, 51*, 663-671.

Sherrod, D. R., & Downs, R. (1974). Environmental determinants of altruism: The effects of stimulus overload and perceived control on helping. *Journal of Experimental Social Psychology, 10*, 468-479.

Simon, S., & Simon, R. (1992, October). *Improving safety through innovative behavioral and cultural approaches.* Presentation at the National Safety Council Congress and Exposition, Orlando, Fl.

Staub, E. (1974). Helping a distressed person: Social, personality, and stimulus determinants. In L. Berkowitz (Ed.), *Advances in experimental social psychology* (Vol. 7). New York: Academic Press.

———. (1978). *Positive social behaviors and morality: Social and personal influence.* New York: Academic Press.

Strickland, B. R. (1989). Internal-external control expectancies: From contingency to creativity. *American Psychologist,* 44, 1-12.

Straumann, T. J., & Higgins, E. G. (1988). Self-discrepancies as predictors of vulnerability to distinct syndromes of chronic emotional distress. *Journal of Personality,* 56, 685-707.

Tavris, C., & Wade, C. (1995). *Psychology in perspective.* New York: Harper Collins College Publishers.

Taylor, S. E. (1991). *Health psychology* (2nd Edition). New York: McGraw-Hill.

Tucker, R. K., & Byers, P. Y. (1987). Factorial validity of Mertz's psychological reactance scale. *Psychological Reports,* 61, 811-815.

Vollmer, T. R., & Iwata, B. A. (1991). Establishing operations and reinforcement effects. *Journal of Applied Behavior Analysis,* 24, 279-291.

Waterman, A.S. (1988). On the uses of psychological theory and research in the process of ethical inquiry. *Psychological Bulletin,* 103, 283-298.

Weiss, H. M., & Knight, P A. (1980). The utility of humility: Self-esteem, information search and problem solving efficiency. *Organizational Behavior and Human Performance,* 25, 216-223.

Wells, L. E., & Marwell, G. (1976). *Self-esteem.* Beverly Hills, CA: Sage.

Wheeless, L. R., Wheeless, V. E., & Dickson-Markman, F. (1982). The relations among social and task perceptions in small groups. *Small Group Behavior,* 13, 373-384.

Wilson, J. P. (1976). Motivation, modeling, and altruism: A person X situation analysis. *Journal of Personality and Social Psychology,* 34, 1078-1086.

Wylie, C. (1974). *The self-concept* (Vol. 1). Lincoln: University of Nebraska Press.

Zuckerman, M. (1979). *Sensation seeking: Beyond the optimal level of arousal.* Hillsdale, NJ: Lawrence Erlbaum Associates.

Chapter 14

Increasing Actively Caring Behaviors

This chapter integrates principles and procedures from previous chapters to address the most critical question regarding the achievement of a Total Safety Culture. Namely, how can we increase actively caring behavior throughout a work culture? Some conditions and interpersonal techniques facilitate this behavior indirectly by benefiting self-esteem, empowerment, or belongingness. Other procedures can directly boost actively caring with certain activators and consequences. In addition, the social influence principles of reciprocity and consistency can be applied to enhance the caring behavior we desire. In sum, this chapter shows you how to increase the likelihood that people will go beyond their normal routines to help keep people safe.

In a Total Safety Culture . . . people actively care *on a continuous basis for safety.*

This quotation from my 1994 article in *Professional Safety* (Geller, 1994, p.18) reflects both the ultimate goal and the fundamental challenge of a Total Safety Culture: Everyone must periodically go beyond his or her personal routine for the safety and health of others. To meet the goal, and the challenge, we need to find ways to increase actively caring behaviors.

So how do we do this at work, in our homes, and the community at large? I hope you can already provide some answers after reading Chapters 12 and 13, where I alluded to some ways to increase actively caring behaviors. Here I want to expand on earlier suggestions and add more. We can classify these approaches as indirect or direct.

Indirect strategies for facilitating actively caring behavior follow from the theory and principles discussed in Chapter 13. The actively caring model supported by research proposes that certain person states inside people increase their willingness to look beyond self-interests and consider the safety of others. Thus, conditions and procedures that increase these states will indirectly increase the probability of active caring. More direct ways to increase these behaviors can be derived from behavior-based principles of learning and social influence. These are discussed in the latter part of this chapter.

Enhancing Actively Caring Person States

Sometimes at seminars and workshops I hear participants express concern that the actively caring person-state model might not be practical. "The concepts are too soft or subjective," is a typical reaction. Employees accept the behavior-based approach because it is straightforward, objective, and clearly applicable to the workplace. But person-based concepts like self-esteem, empowerment, and belongingness appear ambiguous, "touchy-feely," and difficult to deal with. "The concepts sound good and certainly seem important, but how can we get our arms around these 'warm fuzzies' and use them to promote safety?"

These person states are more difficult to define, measure and manage than behaviors. That's why I said early on in this text that it's more cost-effective to work on behaviors first. Whenever the ultimate outcome is behavior change, it's usually most efficient to deal directly with behaviors. However, it's always important to consider people's feelings when designing behavior-change interventions. That's why I recommend against using negative consequences whenever possible. And why I offer ways to design intervention strategies that account for subjective person-states like commitment, ownership, and involvement.

You see, we just can't ignore the importance of how people feel about a behavior-change intervention. For people to accept a behavior-change process and sustain the target behaviors long-term, we must consider internal person states while designing and implementing an intervention. We can measure and evaluate person states through one-on-one interviews, group discussions, or questionnaires. The twenty-item questionnaire in Figure 13.8, for example, illustrates how the actively caring person states can be measured systematically. Using more extensive surveys than this, researchers have demonstrated significant positive relationships between these person states and actively caring behavior. So, while measures of what occurs inside people—those person states—are less objective and precise than systematic observations of behavior, such measurement is possible. And it can help gauge the impact of strategies to increase these states and actively caring behavior.

After introducing the actively caring person-state model at my workshops, I often divide participants into discussion groups. I ask group members to define events, situations, or contingencies that decrease and increase the person state assigned to their group. Then I ask the groups to derive simple and feasible action plans to increase their assigned state. This promotes personal and practical understanding of the concept.

Feedback from these workshop tells me the concept may be soft, but it's not too hard to grasp. Action plans have been practical and quite consistent with techniques used by researchers. Also, there has been substantial overlap of practical recommendations—workshop groups dealing with different person states have come up with similar contributory factors and action plans.

Some techniques suggested to increase self-esteem, for example, have been offered by groups assigned to increase self-efficacy, personal control and optimism. Such interconnectedness of the person states is consistent with our research (Geller & Roberts, 1993; Geller *et al*, 1996; Roberts & Geller, 1994). Distinct action plans, however, have emerged from discussions of each particular person state, verifying the utility of teaching a five-state actively caring model.

Let's take a look at what workshop participants have come up with for factors and strategies regarding each of these person states:

SELF-ESTEEM

Factors consistently mentioned as shaping self-esteem include communication techniques, reinforcement and punishment contingencies, and leadership styles. Participants suggest a number of ways to build self-esteem, including:

- Provide opportunities for personal learning and peer mentoring
- Increase recognition for desirable behaviors and individual accomplishments
- Solicit and follow up on a person's suggestions

COMMUNICATION STRATEGIES. Figure 14.1 lists eleven words beginning with the letter "A" that imply a specific verbal technique for increasing a person's self-esteem. Each "A" word suggests a slightly different communication approach, from stating simple words of agreement, admiration, appreciation, and approval to acknowledging the achievement and individual creativity of others through active listening and praise.

- ***Accept:*** appreciate diversity
- ***Actively listen:*** with verbal & nonverbal behavior
- ***Agree:*** with verbal & nonverbal behavior
- ***Admire:*** "attractive dress," "nice tie"
- ***Appreciate:*** "please " (activator) & "thank you" (consequence)
- ***Acknowledge:*** the achievements of others
- ***Approve:*** praise for good behavior
- ***Ask:*** for feedback, advice, opinions
- ***Attend:*** lend a helping hand
- ***Avoid criticizing:*** it won't be accepted anyway
- ***Argue less:*** arguments are win/lose situations

Figure 14.1. Apply "A" strategies to increase others' self-esteem.

It's also a good idea to argue less and avoid criticizing. Arguments waste time and usually promote a win/lose perspective, and criticism always does more harm than good.

No one likes to be criticized, but some people are more resilient to negative feedback than others. People with a high and stable self-esteem have developed mechanisms to protect themselves. Hence, these people take criticism in stride, protecting their self-esteem, for example, by just denying a mistake, rationalizing that the flack was biased and unwarranted, or optimistically focusing on positive aspects of the verbal exchange. On the other hand, some people are overly sensitive to critical comments. These folks usually have a relatively low or unstable self-esteem, and perceived daggers can devastate them.

CORRECT WITH CARE. Figure 14.2 depicts a supervisor giving feedback to a worker quite appropriately. He notes achievements before pointing out a flaw. This represents a commendable strategy of making several deposits before making a withdrawal from that emotional bank account (Covey, 1989). Still, this person is overwhelmed by the corrective feedback and perceives it as an attack, possibly because his current self-esteem is low. He could be "emotionally bankrupt." Perhaps prior exchanges did not focus on the positive.

So despite an effective feedback technique, the net effect in this particular situation is still negative. What can you do to avoid or fix these situations?

First, recognize that the situation depicted in Figure 14.2 is possible. Observe body language carefully to assess the impact of your words. If you note potential blows to self-esteem, then do some damage control. You might reemphasize the positive: "I see much more good stuff here than bad." You could indicate that such errors are not uncommon and, in fact, you've made them yourself. Above all, focus on the act, not the actor. Stress that the error only reflects behavior that can be corrected, not some deeper character flaw. The worst thing you can do is be judgmental, implying that a mistake suggests some subjective personal attribute like "carelessness," "apathy," "bad attitude," or "poor motivation."

Figure 14.2. We selectively hear more negatives than positives.

When offering corrective feedback, it's critical to be a patient, active listener. Allow the person to make excuses, and don't argue about these. Resist the temptation. Giving excuses is just a way to protect self-esteem, and it's generally a healthy response. Remember, you already made your point by showing the error and suggesting ways to avoid the mistake in the future (as discussed in Chapter 11 on coaching). Leave it at that.

If a person doesn't react to corrective feedback, it might help to explore feelings. "How does this make you feel?" you might ask. Then listen empathetically to assess whether self-esteem has taken a hit. You'll learn whether some additional communication is needed to place the focus squarely on what is external and objective, rather than on very subjective, internal states.

SELF-EFFICACY

The feedback situation depicted in Figure 14.2 is clearly job related, so it's likely to have greater impact on self-efficacy than self-esteem. As discussed in Chapter 13, self-efficacy is more situation specific than self-esteem, and so it fluctuates more readily. Job-specific feedback should actually affect only one's perception of what's needed to complete a particular task successfully. It should not influence feelings of general self-worth. Keep in mind, though, that repeated negative feedback can have a cumulative effect, chipping away at an individual's self-worth. Then it takes only one remark, perhaps what you would think is innocuous and job-specific, to trigger what seems like an overreaction. The internal reaction "He hates me" in Figure 14.2 suggests the negative statement influenced more than self-efficacy—probably fragile self-esteem.

Hence, it's important to recognize that our communication may not be received as intended. We might do our best to come across positively and constructively, but because of factors beyond our control, the communication might be misperceived. One's inner state can dramatically bias the impact of feedback. I'm referring, of course, to variation in personal perception, or selective sensation, as discussed earlier in Chapter 5.

The communication process. To try to avoid these problems, let's deconstruct the process of communication. As shown in Figure 14.3, a one-to-one exchange consists essentially of six components or steps—three per individual. There is plenty of room for bias or misperception. First, the sender's idea (or intention) is filtered through psychological mechanisms influenced by personal history and person states. The message is then affected by the sender's verbal and nonverbal behavior during the exchange. I'm sure you've heard more than once about the critical importance of tone and body language. These are affected by the sender's prior experience and current person states.

The sender's imperfect message, biased by perceptions and presentation mannerisms, is transmitted to the receiver's imperfect system. (Let's face it, none of us are perfect.) As discussed in Chapter 5, we selectively receive information, sometimes filtering out messages we don't want to hear, and then we exert personal bias when comprehending the distorted message. Thus, the last stage of the communication cycle depicted in Figure 14.3 is critical. Listeners need to check their understanding of the message, and senders need to encourage clarification. In this way, constructive feedback allows for continuous improvement of speaking and listening. Again, it's often important to check for changes in feeling states, and make needed repairs.

Achievable tasks. What makes for a "can do" attitude? Personal perception is the key. A supervisor, parent, or teacher might believe he or she has provided everything needed to complete a task successfully. However, the employee, child, or student

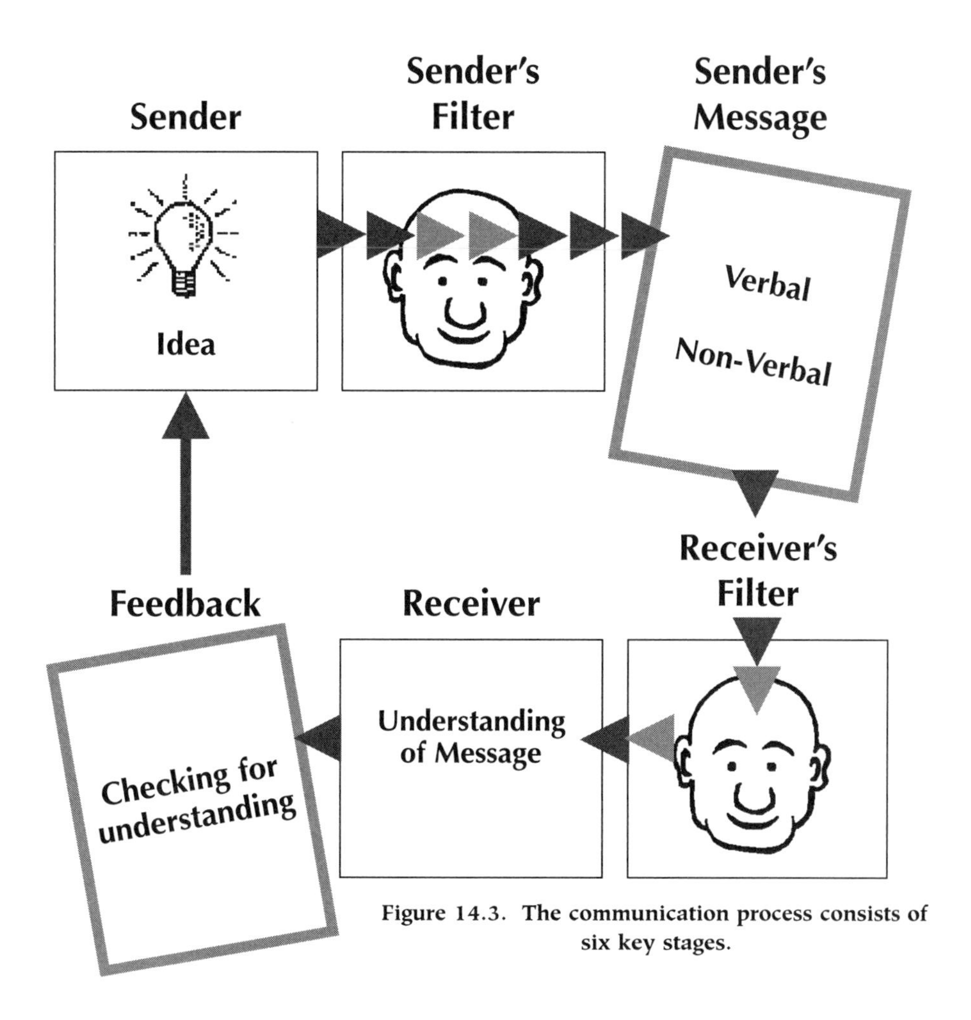

Figure 14.3. The communication process consists of six key stages.

might not think so. Hence, the importance of asking, "Do you have what you need?" We're checking for feelings of self-efficacy. This is easier said than done, because people often hesistate to admit they are incompetent. Really, who likes to say, "I can't do it?" Instead, we want to maintain the appearance of self-efficacy.

I have often found it necessary to ask open-ended questions of students to whom I give assignments, in order to assess whether they are prepared to get the job done. In large classes, however, such probing for feelings of self-efficacy is impossible. As a result, many students get left behind in the learning process (frequently because they skipped classes or an important reading assignment). As they get farther and farther behind in my class, their low self-efficacy is supported by the self-fulfilling prophecy and diminished optimism. Sometimes this leads to "give-up behavior" and feelings of helplessness (Peterson et al, 1993; Seligman, 1975). All too often, these students withdraw from my class or resign themselves to receiving a low grade.

Figure 14.4 reflects the need to focus on "small wins" (Weick, 1984) when assigning tasks and communicating performance feedback. Of course, the kind of situation depicted in Figure 14.4 requires one-on-one observation and feedback in which the individual's initial competencies can be assessed. Then successively more difficult performance steps can be designed for the learner. The key is to reduce the probability that the learner will make an error and feel lowered effectiveness or self-efficacy. Celebrating small-win accomplishments builds self-efficacy and enables support from the self-fulfilling prophecy.

Suppose you were teaching a young child to put together a puzzle. How would you apply the principle of reducing errors and celebrating small wins to build self-efficacy? Would you lay the puzzle pieces on a table and then encourage each attempt to find the right piece and put it in the right location? This might work if the child were experienced with the puzzle. But, this approach could be perceived as overwhelming by the child, and a lack of initial success and perception of effectiveness could lead to frustration and give-up behavior. In other words, the first performance step could be too large.

Figure 14.4. Starting small reduces errors and builds confidence.

Analogous to the scenario in Figure 14.4, a puzzle-learner could experience initial success and self-efficacy if he or she watched the teacher put all of the pieces of the puzzle together except for the last piece. Then the child has a relatively easy task to do, and the teacher and child can celebrate the completed picture. Note the role of observation learning when the child watches the teacher pick and place the various puzzle pieces. A good teacher would include appropriate verbal description with each selection and placement of the puzzle pieces, and ask the child for suggestions along the way.

On the next trial, the teacher would complete the puzzle again with the exception of two or three puzzle pieces. The child would handle the next step successfully with no frustration, and thus would feel more effective at the task. Eventually, with a patient and actively caring teacher, the child would put the entire puzzle together without experiencing significant errors or debilitation to self-efficacy. The result is a "can do" attitude in this situation, as well as perceptions of personal control and optimism. Hopefully, readers can relate this story to situations in their workplace and consider ways to apply the "reduced errors and small win" approach with their colleagues.

PERSONAL STRATEGIES. Professors David Watson and Roland Tharp (1993) suggest the following five steps to increase perceptions of self-efficacy. First, select a task at which you expect to succeed, not one you expect to fail. Then, as your feelings of self-efficacy increase, you can tackle more challenging projects. A cigarette smoker who wants to stop smoking, for example, might focus on smoking 50 percent fewer cigarettes per week rather than attempting to quit "cold turkey." With early success at reducing the number of cigarettes smoked, the individual could make the criterion more stringent (like smoking no cigarettes on alternate days). Continued success would lead to more self-efficacy.

Second, it's important to distinguish between the past and the present. Don't dwell on past failures. Instead, focus on a renewed sense of self-confidence and self-efficacy. Past failures are history—today is the first day of the rest of your life. The cigarette smoker might review past attempts to quit smoking, for example, in order to decide on initial goals. (Recall the discussion of SMART goals in Chapter 9). Choosing achievable goals gives self-efficacy a chance to build when the goal is reached. Third, it's important to keep good records of your progress toward reaching your goal. Our cigarette smoker should record the number of cigarettes smoked each day, and note when the rate of smoking is 50 percent less for a week. This should be noted as an achievement, and then a new goal should be set. Focusing on your successes (rather than failures) represents the fourth step in building self-efficacy.

The fifth step is to develop a list of tasks or projects you'd like to accomplish and rank them from easiest to most difficult to accomplish. Then, whenever possible start with the easier tasks. The self-efficacy and self-confidence developed from accomplishing the less demanding tasks will help you tackle the more challenging situations on your list.

FOCUS ON THE POSITIVE. Many of the strategies I've presented for improving behaviors and person-states include a basic principle—focus on the positive. Whether attempting to build our own self-efficacy or that of others, success needs to be emphasized over failure. Thus, whenever we have the opportunity to teach others or give them feedback, we must look for small-win accomplishments and give genuine approval before commenting on ways to improve. This approach is easier said than done.

Failures are easier to spot than successes. They stick out and disrupt the flow. That's why most teachers give rather consistent negative attention to students who disrupt the classroom, while giving only limited positive attention to students who remain on task and go with the flow. Furthermore, many of us have been conditioned (unknowingly) to believe negative consequences (penalties) work better than positive consequences (rewards) to influence behavior change (Notz *et al,* 1987).

Figure 14.5 illustrates how natural variation in behavior can lead to a belief that penalties have more impact on behavior than rewards. If people receive rewards for their superior performance and penalties for their inferior performance, natural variation in performance (or regression to average performance) can give the impression that penalties are more powerful than rewards. In other words, the natural tendency to do less than superior work over time can mask the behavioral impact of a reward given for superior performance. On the other hand, the natural tendency to improve following inferior performance can give the impression that prior punishment improved the performance.

Imagine you see an employee demonstrating peak performance at a particular task. This behavior deserves special commendation, and so you offer genuine words of appreciation. The next time you see the individual you rewarded, you note less than optimal performance. There are many reasons for the noticeable drop in performance, including the fact that few people can perform at peak levels all of the time. There is natural regression to average performance. You recall giving this person spe-

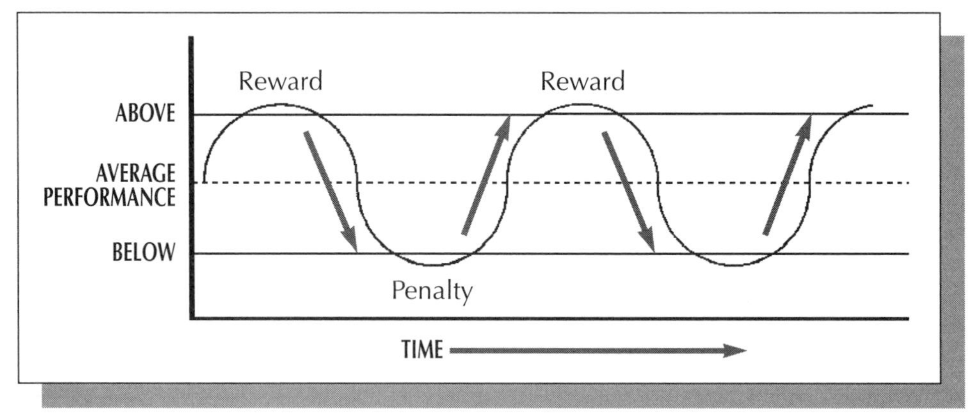

Figure 14.5. Natural performance variations make it easier to see a presumed behavioral impact of negative rather than positive consequences.

cial recognition (a reward) earlier for superior performance, and now you notice a decrement. Might you conclude that rewards don't work to improve performance?

Now imagine you observe an employee doing below-average work, and you decide to intervene to improve performance. You issue a progressive discipline warning citation which goes in the employee's file. Sometime later you observe this person performing notably better. Although there are many reasons for the improvement, you naturally presume your punishment procedure the other day was responsible for this beneficial change. Consequently, you develop the inaccurate belief that negative consequences work better than positive consequences to change behavior.

In summary, normal circumstances make it relatively difficult to focus on the positive. Mistakes are more noticeable than "go-with-the-flow" successes, and natural regression to average performance can develop a faulty belief that negative consequences have more behavioral impact than positive consequences. However, considering the impact consequences have on internal person states (especially self-efficacy and self-esteem), a positive consequence (like praise and social approval) is always preferable to a negative consequence (like criticism or ridicule). Indeed, as illustrated in Figure 14.6, it's often critical to search for admirable aspects of a situation and emphasize these in verbal interaction.

Figure 14.6. Look for the silver lining to build self-efficacy and self-esteem.

PERSONAL CONTROL

Most people find a healthy imbalance between internal (personal) and external control. They essentially believe, "I'm responsible for the good things that happen to me, but bad luck or uncontrollable factors are responsible for the bad things" (Beck, 1991; Taylor, 1989). Thus, people are apt to attribute injuries to rotten luck and beyond their control, especially when they happen to them. (Recall our discussion of attributional bias in Chapter 6.) This is the prime reason industrial safety must focus on process achievements rather than failure (injury) outcomes. To achieve a Total Safety Culture, people must believe they have personal control over the safety of their organization.

Employees at my seminars on active caring have listed a number of ways to increase perceptions of personal control, including:

1. Setting short-term goals and tracking progress toward long-term accomplishment
2. Offering frequent rewarding and correcting feedback for process activities rather than only for outcomes
3. Providing opportunities to set personal goals, teach others, and chart "small wins" (Weick, 1984)
4. Teaching employees basic behavior-change intervention strategies (especially feedback and recognition procedures)
5. Providing people time and resources to develop, implement and evaluate intervention programs
6. Showing employees how to graph daily records of baseline, intervention, and follow-up data
7. Posting response feedback graphs of group performance

Figure 14.7 illustrates humorously a personal control perspective. Obviously, this is an extreme and unrealistic scenario. But wouldn't it be nice if people would attempt to take personal control of safety issues at their industrial sites with the same passion and commitment some individuals have for their golf game? I believe differences in perceived personal control for safety versus golf are largely due to contrasting scoring procedures.

Figure 14.7. Sometimes we try extra hard to exert personal control.

Suppose you couldn't receive direct and immediate feedback about your golf game. That is, each time you hit a golf ball you wore a blindfold and couldn't see where the ball landed. Even when putting on the greens, you are blindfolded and can't tell whether your ball goes into the cup. Imagine also that you don't receive a score per hole or per game. However, you do receive negative feedback whenever your ball lands in a sandtrap. Under these circumstances, would you feel "in control" of your golf game? Would you attribute balls hit into sandtraps to personal control or just bad luck? Would you continue playing golf, or give it up for an activity in which you can experience greater personal control?

Of course, the golf scenario I've asked you to imagine is far-fetched. But, isn't this the way it is for safety at many industrial sites? The primary evaluation tool used to rank companies and determine performance appraisals and bonuses is an outcome number (such as total recordable injury rate) which is quite remote from the daily plant processes people have control over. Without a scoring system that focuses on controllable processes (as discussed earlier in this text), safety will be viewed as beyond personal control. An injury is just "bad luck," analogous to hitting a golf ball in a sandtrap while blindfolded.

Now imagine you're playing golf without a blindfold and you're really playing well. In fact, your scores indicate peak performance. Suddenly, it begins to storm. If you (unwisely) continue to play, your golf game would deteriorate, and your score could well exceed par. How would you evaluate this situation? Would you continue to feel a high degree of personal control and self-efficacy, or would you give up some control to uncontrollable factors?

My point here is that people need to distinguish between factors they can control on a personal level and factors beyond their domain of influence. Similarly, Covey (1989) recommends we distinguish between our "Circle of Concern" and "Circle of Influence," and focus our efforts in the Circle of Influence. Thus, it's healthy to admit there are things we are concerned about but have little influence over. Then, when negative consequences occur outside our domain of personal influence, we will not attribute personal blame and reduce our sense of self-efficacy, personal control, or optimism.

Obviously, we cannot have complete control over all factors contributing to an injury. That's why I think it's wrong to say "all injuries are preventable." However, there is much we can do within our own domain of influence, and we can prepare for factors outside our personal control. Thus, we take an umbrella to the golf course in case it rains, and we wear personal protective equipment in case we are exposed to risks beyond our domain of personal control. Likewise we protect our children for events beyond their control, as illustrated in Figure 14.8.

Figure 14.8. We can't control everything!

THE POWER OF CHOICE. More than 25 years ago when I was conducting research in cognitive science, I carried out a very simple experiment and obtained very simple results. The implications of the findings, however, were immense and relevant to this discussion. Half the 40 subjects in this experiment were shown a list of five three-letter words (cat, hat, mat, rat, bat) and asked to select one. Then, after a warning tone, the selected word was presented on a screen in front of the subject, and he or she pressed a lever as fast as possible after seeing the word. The delay in milliseconds between the presentation of the word and the subject's response was a measure of simple reaction time. This sequence of warning signal, word presentation, and subject reaction occurred for 25 trials. If a subject reacted before the stimulus word was presented, the reaction time was not counted, and the trial was repeated. The experimental session took less than 15 minutes per subject.

The word selected by a particular subject was used as the presentation stimulus for the next subject. Thus, this subject did not have the opportunity to choose the stimulus word. As a result, the word choices of 20 subjects were assigned (without choice) to 20 other subjects. Therefore, this simple experiment had two conditions—a "choice" condition (in which subjects chose a three-letter word for their stimulus) and an "assigned" condition (in which subjects were assigned the stimulus word selected by the previous subject). The mean reactions of subjects in the "choice" group were significantly faster than those of subjects in the "assigned" group.

I explained these results by presuming the opportunity to choose a stimulus word increased motivation to perform in the reaction time experiment. I must admit, however, I didn't expect the differences to be as large as they were. How could the simple choice of a three-letter word motivate faster response in a simple reaction time experiment? I frankly didn't feel confident in my explanation for these results, and thus I didn't pursue publication of these data in a professional research journal. However, subsequent laboratory studies verified these findings and were published (Monty *et al*, 1979; Monty & Perlmuter 1975; Perlmuter *et al*, 1971).

FROM LABORATORY TO CLASSROOM. About a year after the simple reaction time experiment described above, I tested the theory of "choice" as a motivator in the college classroom. I was teaching two sections of social psychology; one at 8:00 a.m. Monday, Wednesday, and Friday, and the other at 11:00 a.m. on the same days. There were about 75 students in each class. Instead of distributing a prepared syllabus with weekly assignments on the first day of classes, I distributed only a general outline of the course. This outline introduced the textbook, the course objectives, and the basic criteria for assigning grades (a quiz on each textbook chapter and a comprehensive final exam on classroom lectures, discussions, and demonstrations).

I told the 8:00 class they could choose the order in which the ten textbook chapters would be presented, they could submit multiple choice questions for me to consider using for the chapter quizzes, and they could hand in short answer and discussion questions for possible use on the final exam. The 11:00 class received the order of textbook chapters selected previously by the 8:00 class in an open discussion and voting process. Also, the 11:00 class was not given an opportunity to submit quiz or exam questions.

Thus, I derived "choice" and "assigned" classroom conditions analogous to the two reaction-time groups I had studied earlier. Two of my undergraduate research assistants attended each of these classes, posing as regular students, and systematically counted amount of class participation. These observers did not know about my intentional choice-assigned manipulations. From the day the students in my 8:00 class voted on the textbook assignments, this class seemed more lively than the later 11:00 class. And, my perception was verified by the participation records of the two classroom observers. Furthermore, the quiz grades, final exam scores, and my teaching evaluation scores from standard forms distributed during the last class period were significantly higher in the "choice" class than the "assigned" class. Although several students from the 8:00 class submitted potential quiz and final exam questions, I didn't use any of these questions per se. Each class received the same quizzes and final exam, and the final grades were significantly higher for the 8:00 class than the 11:00 class.

There are several possible reasons for the group differences, but I'm convinced the "choice" versus "assigned" manipulation was a critical factor. I believe the initial opportunity to choose reading assignments increased class participation. This increased involvement fed on itself and led to more involvement, choice, and learning. The students' attitudes toward the class improved as a result of feeling more "in control" of the situation rather than "controlled." And it's likely the "choice" opportunities in the 8:00 class were especially powerful because they were so different than the traditional top-down classroom atmosphere experienced in other classes (and typified by my 11:00 class). In other words, the contrast with the students' other courses made the choice opportunities especially salient and powerful.

The implications of these "choice" versus "assigned" findings in laboratory and classroom investigations are far reaching. Indeed, the notion that "choice increases involvement" relates to a number of motivational theories supported by psychological research (Steiner, 1970; White, 1959). Essentially, when people believe they have personal control over a situation they are generally more motivated to achieve and get more involved. And, how can this sense of personal control be increased? You guessed it. We can increase a belief in personal control by increasing the number of "choice" opportunities in a situation. Opportunities to choose lead to involvement, and more involvement leads to increased perceptions of personal control. More personal control leads to more choice and more involvement—and this continuous involvement cycle continues. This is a primary route to feeling empowered.

Some people's past experiences have made them less likely to develop a sense of personal control when "choice" is offered. They may mistrust the "choice" situation, or lack confidence in their ability to make something positive come out of a "choice" opportunity. They might not feel comfortable with the added responsibility of "choice," and thus resist the change implied by new "choice" potentials.

Usually the best way to deal with this resistance is to not confront it directly. Hopefully, enough other people will take personal control of their choice opportunities and eventually convince the resisters to get involved. (I address the topic of resistance more completely in Chapter 16.) It is important, however, that people with "choice" feel competent to make appropriate decisions, and therefore education and training might be needed.

The consequences of choice are critically important. If the subjects in my simple reaction time experiment or my social psychology class did not believe their choices made a difference in the situation, the choice opportunities would not have made a difference in their motivation. The reaction-time subjects saw me use their choice in the stimulus presentations, and the psychology students observed me change the class structure and process as a function of their choices. When we see a consequence consistent with decisions from our choice opportunities, we increase our trust of the people who gave us the power to choose. We gain confidence in our abilities to take personal control of the situation.

FROM RESEARCH TO REAL WORLD. An Exxon Chemical facility with 350 employees exemplifies the power of choice to make a difference for safety. The employees initiated an actively caring observation, feedback, and coaching process in 1992, reaping amazing benefits for their efforts. In 1994, for example, 98 percent of the workforce had participated in behavioral observation and feedback sessions, documenting a total of 3,350 coaching sessions for the year. A total of 51,408 behaviors were systematically documented on critical behavior checklists, of which 46,659 were safe and 4,389 were at risk. Such comprehensive employee involvement led to remarkable outcomes. At the start of their process in 1992, the plant safety record was quite good (13 OSHA recordables for a Total Recordable Incidence Rate of 4.11). They improved to 5 OSHA recordables in 1993 (TRIR=1.60), and in 1994 they had the best safety performance in Exxon Chemical with only one OSHA recordable (TRIR=0.35).

I've seen many companies subtantially improve safety performance with processes based on the principles of behavior-based safety, but this plant holds the record for efficiency in getting *everyone* involved and in obtaining exceptional results. I'm convinced a key factor was the employees' "choice" in developing, implementing, and maintaining the process. Choice has led to ownership.

Here's what I mean: Each month, employees schedule a behavioral observation and feedback session with two other employees, who are safety observers. They select the task, day and time for the coaching session, as well as two individuals to be observers. Employees choose their observers—and coaches—from *anyone* in the plant. At the start of their process the number of volunteer safety coaches was limited to about 30 percent of the workforce, but today everyone is a potential coach.

At first, some employees did not completely trust the process and resisted active participation. Some tried to beat the system by scheduling observation and feedback sessions at slow times when the chance of an at-risk behavior was minimal—such as when watching a monitor or completing paperwork. Today, most employees choose to schedule their coaching sessions during active times when the probability of an at-risk behavior is highest. Frequently, the observed individual uses the opportunity to point out an at-risk behavior necessitated by the work environment or procedure, such as a difficult-to-reach valve, a hose-checking procedure too cumbersome for one auditor, a walking surface made slippery by an equipment leak, or a difficult-to-adjust machine guard. This often leads to improved environmental conditions or operating procedures. Another benefit is an increased perception of personal control for safety.

OPTIMISM

As discussed in Chapter 13, optimism results from thinking positively, avoiding negative thoughts, and expecting the best to happen. Anything that increases our self-

I was in New York the other day and rode with a friend in a taxi. When we got out my friend said to the driver, "Thank you for the ride. You did a superb job of driving."

The taxi driver was stunned for a second. Then he said: "Are you a wise guy or something?"

"No my dear man, and I'm not putting you on. I admire the way you keep cool in heavy traffic."

"Yeh," the driver said and drove off.

"What was that all about?" I asked.

"I am trying to bring love back to New York," he said. "I believe it's the only thing that can save the city."

"How can one man save New York?"

"It's not one man. I believe I have made the taxi driver's day. Suppose he has twenty fares. He's going to be nice to those twenty fares because someone was nice to him. Those fares in turn will be kinder to their employees or shopkeepers or waiters or even their own families. Eventually the goodwill could spread to at least 1,000 people. Now that isn't bad, is it?"

"But you're depending on that taxi driver to pass your goodwill to others."

"I'm not depending on it," my friend said. "I'm aware that the system isn't foolproof so I might deal with 10 different people today. If, out of 10, I can make three happy, then eventually I can indirectly influence the attitudes of 3,000 more."

"It sounds good on paper," I admitted, "but I'm not sure it works in practice."

"Nothing is lost if it doesn't. It didn't take any of my time to tell that man he was doing a good job. He neither received a larger tip nor a smaller tip. If it fell on deaf ears, so what? Tomorrow there will be another taxi driver whom I can try to make happy."

"You're some kind of nut," I said.

"That shows you how cynical you have become. I have made a study of this. The thing that seems to be lacking—besides money, of course—for our postal employees, is that no one tells people who work for the post office what a good job they're doing."

"But they're not doing a good job."

"They're not doing a good job because they feel no one cares if they do or not. Why shouldn't someone say a kind word to them?"

We were walking past a structure in the process of being built and passed five workmen eating their lunch. My friend stopped. "That's a magnificent job you men have done. It must be difficult and dangerous work."

The five men eyed my friend suspiciously.

"When will it be finished?"

"June," a man grunted.

"Ah. That really is impressive. You must all be very proud."

We walked away. I said to him, "I haven't seen anyone like you since *The Man from La Mancha.*"

"When those men digest my words, they will feel better for it. Somehow the city will benefit from their happiness."

"But you can't do this all alone!" I protested. "You're just one man."

"The most important thing is not to get discouraged. Making people in the city become kind again is not an easy job, but if I can enlist other people in my campaign..."

"You just winked at a very plain looking woman," I said.

"Yes, I know," he replied. "And if she's a schoolteacher, her class will be in for a fantastic day."

Figure 14.9. A kind word can go a long way. (From Adler & Towne, 1990, p. 47)

efficacy should increase optimism. Also, if our personal control is strengthened we perceive more influence over our consequences. This gives us more reason to expect the best. Again, we see how the person states of self-efficacy, personal control and optimism are clearly intertwined. A change in one will likely influence the other two. Recall from Chapter 13 that simple events like finding a dime in a coin return, receiving a cookie, listening to soothing music, and being on a winning football team are sufficient to boost optimism and willingness to actively care. It's not necessary for a person to perceive personal control over a pleasant consequence for that consequence to build optimism and possible actively caring behavior. I can think of no better reason for offering words of appreciation and approval to others. As told by Art Buchwald in Figure 14.9, a kind word can go a long way.

BELONGINGNESS

Here are some common proposals given by my seminar discussion groups for creating and sustaining an atmosphere of belongingness among employees:

- Decrease the frequency of top-down directives and "quick-fix" programs
- Increase team-building discussions, group goal-setting and feedback, as well as group celebrations for both process and outcome achievements
- Use self-managed or self-directed work teams

When groups are given control over important matters like developing a safety observation and feedback process or a behavior-based incentive program, feelings of both empowerment and belongingness can be enhanced. When resources, opportunities, and talents enable team members to assert, "We can make a difference," feelings of belongingness occur naturally. This leads to synergy, with the group achieving more than could be possible from members working independently (see Figure 14.10).

Figure 14.10. Higher goals can be reached through synergy.

Social psychologists have studied conditions that influence the productivity of people working alone versus in groups. Results of this research are relevant to increasing actively caring behavior, because conditions that spur on group productivity and synergy also boost feelings of empowerment and belongingness. When group members experience synergy, they especially appreciate the group process and project pride in their output. This feeds a win/win perspective and more perceptions of group empowerment and belongingness. Obviously, a Total Safety Culture requires teamwork, belongingness, and synergy. Yet, as you'll see below, traditional approaches to safety management can throw a wet blanket on this process.

SOCIAL LOAFING: The whole can be less than the sum of its parts. First, I need to digress for a moment to describe a phenomenon known as social loafing. In the 1930s, social psychologists measured the effort people exerted when pulling a rope in a simulated tug-of-war contest (Dashiell, 1935). Researchers measured the force exerted by each of eight subjects when pulling on the rope alone and when pulling on the rope as a team. If the group effort was greater than the sum of the individual efforts, synergy would have been shown. What do you think happened?

The eight subjects worked harder alone than as a team. In fact, the total pulling force of the group was only about half the total of the eight individual efforts. This finding was not a fluke. This phenomenon, termed *social loafing,* has been demonstrated in more than 50 experiments conducted in the United States, India, Thailand, Japan, and Taiwan (Gabrenya *et al,* 1983). For example, blindfolded subjects were asked to shout or clap their hands as loud as they could while listening through headphones to the sound of loud shouting or hand clapping. Subjects told they were doing the clapping or shouting with others produced about one-third less noise than when they thought they were performing alone (Latané *et al,* 1979). So what contributed to the social loafing?

IDENTIFIABILITY: CAN I BE EVALUATED? When subjects were told the sound equipment could measure their individual clapping as well as the team effort, social loafing disappeared (Williams *et al,* 1981). Thus, when subjects believed their personal effort could be identified, they worked just as hard in the group as alone. When are individual efforts identifiable in the workplace? Are there more opportunities to recognize individual effort in production and quality aspects of the job than for safety?

One thing is certain. The typical approach to measuring safety achievement—by counting number of hours or days without an OSHA recordable or lost-time injury—does not provide much opportunity for individual recognition, unless a worker is injured. And in this case, the recognition is failure-based, which motivates behavior through fear rather than desire. So what happens? The injury is often covered up to avoid individual embarrassment or punishment. That's not the behavior we want to see.

RESPONSIBILITY: AM I NEEDED ON THIS JOB? Social loafing has been shown to increase when a person's efforts are duplicated by another person. In one experiment, for example, subjects worked in groups of four to report whenever a dot appeared in a particular quadrant of a video screen (Harkins & Petty, 1982). Social loafing did not occur when subjects were assigned their own quadrant to watch. However, social loafing was found when all four subjects were asked to observe all four quadrants. Social loafing can be expected to increase as people's sense of personal responsibility decreases. When people believe their individual contribution is important, perhaps indispensable, for the team effort, synergy is more likely than social loafing.

Do your safety processes allow individuals to feel a sense of personal responsibility for safety achievements? A safety system that is evaluated only by numbers of injuries or workers' compensation costs does not provide individuals with feedback regarding their contributions. But a safety monitoring process that tallies the safe versus at-risk observations of team members can build a sense of individual responsibility for a group's safety record. Likewise, when teams develop interventions to reduce hazards and increase safe behaviors, personal responsibility for team progress is realized and social loafing reduced. Tracking the success of intervention efforts verifies perceptions of empowerment and promotes feelings of group belongingness.

INTERDEPENDENCE: WE NEED EACH OTHER. Stephen Covey (1989) reminds us that we all come into this world completely dependent on others to survive. As we mature and learn, we reach a level of independence—we strive to achieve success or avoid failure on our own. But higher levels of achievement and quality of life are usually reached after we develop a perspective of interdependence, and act accordingly. According to Dr. Covey, acting to achieve interdependence means we actively listen (Habit 5: "Seek first to understand, then to be understood") and develop relationships or contingencies with people that reflect positive outcomes for everyone involved (Habit 4: "Think win/win"). Consistently practicing these habits leads to synergy (Habit 6).

Research on social loafing supports Covey's idea, and suggests approaches for facilitating synergistic teamwork. Social loafing is reduced when group members know each other well and agree on common goals (Williams *et al*, 1981). Assessing personal performance against an objective standard or the performance of other team members also diminishes social loafing. This happens even when evaluations are not publicized and there are no external consequences (Harkins & Szymanski, 1989; Szymanski & Harkins, 1987).

In safety management this can be achieved by tracking individual contributions to safety observations and intervention processes. For example, team members can conduct regular audits of safe versus at-risk conditions and behaviors, providing invaluable data for calculating group percentages of safe conditions and behaviors. Work teams can also develop intervention strategies to correct environmental hazards or motivate safe work practice, and then devise a process to monitor individual contributions to the process.

HOW LARGE IS YOUR GROUP? Most of us feel a sense of belongingness with our immediate family, and as discussed in Chapter 13, we usually don't hesitate to actively care for the safety and health of family members (see Figure 13.7). In the same vein, the more interdependence and belongingness we experience at work, the more likely we are to look out for coworkers' safety. But how large is your work group?

Obviously, a large "family" means more people come under the protective wing of actively caring behavior. So it makes sense to have an extended family at work. In a Total Safety Culture, everyone actively cares for everyone else's safety. In effect, everyone belongs to one group. Now what are the barriers to an extended work family?

Figure 14.11 depicts an obstacle, or win/lose perspective, that I come across all too often. The "we-they" attitude spun off by traditional management-labor differences often makes for a dysfunctional "family." It seems as if some unions attempt to justify their existence by focusing on disagreement, conflict, and mistrust between management and labor. For its part, management supports this "we-they" split with an alienating communications style that asserts its ultimate power and control. I've seen management memos, for example, that might have been well-intentioned, but were written in top-down, control language that sounded like an adult talking to a child.

Figure 14.11. Win/lose inhibits teamwork and synergy.

Through actively caring, and enhancing self-esteem, empowerment, and belongingness, we can bring down the "us versus them" walls that entrap a work culture. Active caring spreads mutual trust and interdependence throughout the culture. In a Total Safety Culture, everyone benefits from each individual's efforts.

Does actively caring imply the elimination of unions? No. But it might suggest altered visions and mission statements for organized labor groups. Labor unions can certainly help enhance the five person states that facilitate actively caring behavior. To do this, they need to work with management from a win/win perspective that appreciates interdependence and the power of synergy.

At the same time, managers need to relinquish their hold on the "control buttons" of operations and processes that workers can manage themselves, perhaps through self-directed work teams. A truly "empowered work force" is one trusted by managers and supervisors to get the job done without direct supervision. Obviously, this can't happen overnight, but a solid foundation is cemented when the five actively caring personal states are strengthened.

Directly Increasing Actively Caring Behaviors

You can treat actively caring behavior just like any other target behavior. Many interventions that increase the occurrence of safe work behaviors can be used to boost the frequency of actively caring behaviors. The four chapters in Part Three covered principles and procedures for directly influencing behavior. You'll recall that the techniques were classified as activators and consequences, with activators considered directive or instructional, and consequences being motivational. Let's take up that discussion again, as it applies to actively caring behavior.

EDUCATION AND TRAINING

There is some evidence that educating people about the barriers inhibiting actively caring behavior, as detailed in Chapter 12, will increase acts of caring. Professor Arthur Beaman and colleagues (1978) randomly assigned students to listen to either a lecture about the bystander intervention research conducted by Latané and Darley (1970) or a lecture on unrelated topics.

Two weeks later, the students participated in a presumably unrelated sociology experiment. They encountered a student lying on the floor. He could have been hurt, of course, but then he could have only been resting after studying all night for an exam. An accomplice of the researchers posed as another participant and acted unconcerned. Most of the students followed the lead of the accomplice and did not stop to inquire whether the student needed help. But the lecture about bystander intervention appeared to make a difference. Of those who heard this lecture, 43 percent stopped to help the victim, compared to only 25 percent of the students who heard an unrelated lecture.

HIGHER EDUCATION. The impact of education should be even more dramatic if some of the concepts discussed in Chapter 12 are taught. It seems particularly useful to explain that actively caring behavior can be proactive or reactive, direct or indirect, and focused on the environment, internal person-states, and behavior. Also, it should be taught that direct, proactive, and behavior-focused active caring is most challenging—and most useful to prevent injuries.

Professor Joseph R. Ferrari and I have advocated getting college students actively involved in research "projects focusing on improving the quality of life of others...(thereby) nurturing a sense of personal growth, and belonging to a larger community" (Ferrari & Geller, 1994, p. 12). At junior and community colleges and a large university, Dr. Ferrari has involved undergraduate students in evaluating interventions to increase blood donations, promote the use of child safety-belt use in shopping carts, encourage young adults to lessen their risk of hearing problems from high-volume headsets, and assess the rewarding consequences from helping AIDS sufferers. As a faculty member at Virginia Tech for more than 25 years, I have involved numerous graduate and undergraduate students in community and organizational projects to increase recycling, vehicle safety-belt and child safety-seat use, and safe work practices; and to decrease littering, shoplifting, energy and water consumption, alcohol and other drug abuse, and alcohol-impaired driving.

Professor Ferrari and I are certainly not unique in involving university students in field studies designed to help others. Many of our colleagues have conducted applied research that involves students as intervention agents to promote proactively caring behavior (see, for example, numerous studies published in Greene *et al,* 1987). My point is we need more of this kind of instruction, not only at colleges and universities, but at elementary and high schools; also within civic and social organizations such as Girl Scouts, Boy Scouts, 4-H clubs, and church groups. Of course, we need similar education and training in industry.

Promoting actively caring education and participation among children and adults can dramatically increase the number of caring people in our society. Participative education is particularly powerful because it follows the classic Confucian principle:

Tell them and they'll forget...
Show them and they'll remember...
Involve them and they'll understand...

EDUCATION AT HOME. Parents have profound impact on their children's current and future acts of caring. As discussed in Chapter 8, children learn continuously and indirectly by watching their parents (observational learning), and their behavior is directly influenced by the consequences they receive (operant learning). Children are more apt to actively care for others when their mothers are warm, sympathetic, and empathic; and when their fathers are perceived as generous and compassionate (Rutherford & Mussen, 1968; Zahn-Waxler *et al,* 1979).

Other studies find that people are more actively caring as adults when their parents have been open and tolerant of other people, teaching them that they are part of humanity in general (collectivism) rather than some elite or special group (Staub, 1990, 1992). Moreover, children were observed to perform kind acts when they saw an adult set an example two months earlier (Rushton, 1975).

Parents should promote active caring among their children. Society will reap rewards for the seeds you plant now. Professor Ervin Staub (1975, 1979) found that children who made toys for poor hospitalized children or taught skills to younger children were later more likely to exhibit actively caring behavior. Keep in mind, though, it's important to provide a rationale for active caring. Stress the importance of empathy and caring (Eisenberg, 1992). In other words, give children internal attributions for their behavior. When you see them performing acts of caring, praise them as helpful and generous individuals.

Grusec and Redler (1980) conducted an interesting and instructive experiment to compare the impact of internal attribution and external praise on children's short-term and long-term actively caring behavior. First, children were persuaded to share some tokens with poor children; these tokens had been won previously in a game, and could be exchanged for prizes. Following this induced actively caring behavior, the experimenter either verbally praised the child—"It was good that you gave your tokens to those poor children . . . that was a nice and helpful thing to do"—or offered an internal attribution about the child's behavior—"I guess you're the kind of person who likes to help others whenever you can. You are a helpful person."

The researchers assessed the amount of actively caring behavior from the two groups of children immediately after these two verbal consequences, and then after one week and three weeks. Initially there were no differences between groups in amount of actively caring behavior, measured by the number of additional tokens given to poor children. But one week later, the group given an internal attribution for their earlier behavior made more playhouse roofs for the teacher than the group given only praise for their token contributions. Three weeks later, subjects in the internal attribution group collected significantly more craft material for sick children than did the children in the praise-only group. Helping children write an internal script that they are an actively caring person will likely increase their future behavior in the absence of external prompts, models, or rewards.

Researchers have also shown that actively caring behavior can be inhibited by negative consequences (Hoffman, 1975; Zahn-Waxler *et al,* 1983). When children act in noncaring or selfish ways, it's important to explain why this behavior is antisocial and inappropriate. Don't spank a selfish child for not sharing, but urge him or her on with words like "Please share your toys with Sherry. You made her cry and now she's unhappy. You can be a helper and make her feel better."

CONSEQUENCES FOR ACTIVELY CARING

Rewards should increase actively caring behavior. This comes from the basic operant learning principle that behavior increases following positive consequences. Indeed, researchers have demonstrated increases in active caring following monetary rewards (Wilson & Kahn, 1975) and social approval (Deutsch & Lamberti, 1986). Even a simple "thank you" can be effective.

In a study by McGovern *et al* (1975), a female research accomplice asked male subjects to endure a brief electric shock for her. For one group, she responded with a "thank you" if the student agreed. For a control group, she said nothing if they agreed. After the initial shock trial, subjects who received a thank you showed more actively caring behavior throughout the experiment than did subjects in the control group.

POSITIVE VERSUS NEGATIVE CONSEQUENCES. In a field study by Moss and Page (1972) individuals on a busy street were approached by a researcher and asked for directions to a local department store. When the individual agreed to give directions, the researcher varied the consequences for this act of caring. For one group of subjects, the researcher smiled and said, "Thank you very much, I really appreciate this." For another group of subjects, the researcher gave a negative consequence by rudely interrupting and remarking, "I can't understand what you're saying. Never mind, I'll ask someone else." For a control condition, the researcher listened to the directions and gave neither a positive nor negative consequence.

A short time after giving directions, the subjects encountered another person, a research accomplice, who "accidentally" dropped a small bag. Of those subjects rewarded with a "thank you" for their previous actively caring behavior, 93 percent stopped to help this person. In contrast, only 40 percent of those subjects who received a negative consequence for giving directions earlier offered help. Of those in the control group, 85 percent stopped to help.

You can see the importance of responding positively to an individual displaying actively caring behavior, but it's perhaps even more important not to respond negatively when observing a caring act. A negative reaction could make that person avoid a subsequent opportunity to actively care for safety. I made a similar point when discussing safety coaching in Chapter 11. A person's reaction to a safety coach could determine whether or not that person goes out of their way to coach again for safety.

The "Actively Caring Thank-You Cards" described in Chapter 10 (see Figure 10.14) are a rather generic technique to boost caring in the workplace. You'll recall that employees used these cards to thank their colleagues for going beyond the call of duty for the safety of others (Roberts & Geller, 1995). Some cards included a space for the intervention agent to define the act of caring. Some cards could be exchanged for inexpensive rewards, like a beverage in the company cafeteria. Other cards had peel-off stickers for public display. In some applications, each card was worth 25 cents or 50 cents toward corporate contributions to local charities.

The contributions to charity idea promotes active caring in more than one way. First, those more ready to actively care, perhaps because of high levels of the internal person states, will find these "charity thank-you cards" more rewarding (Roberts & Geller, 1995). These people not only set admirable examples for others, they're also likely to become champions of the process and the actively caring concept. Also, public displays of the charitable contributions can help get others involved in the process.

A CHILD RESEARCHER. Besides our own research, I could find only one other study that used the opportunity to help other people as the reinforcing consequence for an actively caring behavior. The research was conducted by the 10-year-old grandson of the eminent behavioral scientist and teacher, Dr. Fred S. Keller. Jacob Keller (1991) wanted to increase participation in the curbside recycling program in his neighborhood. After counting the percentage of homes putting out recyclables along two streets, Jacob intervened on one street by delivering weekly handwritten notes (activators) to each of the 44 homes.

Jacob's first note notified residents that he was monitoring recycling participation, and if recycling improved, a local grocery store "has offered to give two $10 gift certificates to a homeless shelter" (Keller, 1991, p.618). The next three weekly notes specified the amount of improvement and thanked the residents for their behavior. The improvement was significant, increasing from 34 percent participation before the intervention to 53 percent during the last two intervention weeks. Participation on the street that did not receive the intervention (40 homes) remained quite stable at about 35 percent throughout the study.

This community research project is noteworthy on two accounts. Sure, it illustrates the impact of rewards on actively caring behavior with an environment focus. But more importantly, it was conducted competently by a 10-year-old, who also documented it for professional publication. Children throughout communities should be empowered to conduct such projects. Their "small wins" will add up to substantial benefits to our environment and its inhabitants. Even more significantly, these kids will be introduced to the actively caring concept, and perhaps experience the self-rewarding power of this behavior. This will have a multiplying effect, leading to involvement in other actively caring projects.

The Reciprocity Principle

Some sociologists, anthropologists, and moral philosophers consider reciprocity a universal norm that motivates a good deal of interpersonal behavior (Cialdini, 1993; Gouldner, 1960). Simply put, people are expected to help those who have helped them. You can expect people to comply with your request if you have done a favor for them. This is the principle behind Covey's (1989) claim that we need to make deposits in another person's emotional bank account before making a withdrawal.

I witnessed the reciprocity norm rather dramatically when I worked in the Virginia prison system in the mid-1970s. Several inmates used gifts and other forms of personal assistance to influence other inmates. I knew an inmate, for example, who went to great lengths to place a gift in the cell of a new inmate he wanted to influence. Accepting such a gift meant that the target inmate, actually a victim, was now obliged to return a favor in order to "save face."

In a similar vein, my father-in-law had a favorite expression whenever someone offered to do him a favor. He would say, "I don't want to be obligated." Without knowing the principle, he was saying that he didn't want the favor because he would feel a duty to reciprocate.

Have you ever felt a little uncomfortable after someone did you a favor? I certainly have. I interpret my discomfort as the reciprocity principle in action. Another person's kind act makes me feel pressured to respond in kind. What does this mean for safety? I think it means we should look for opportunities to go out of our way for another person's safety. Doing this, we increase the likelihood they will help when we need them.

The workplace setting might enhance chances for reciprocity. Research shows that people are most likely to pay back individuals they expect to see again (Carnevale *et al,* 1982). Also, the more actively caring behavior one receives, the more such behavior they feel obligated to return (Kahn & Tice, 1973). So when we actively care frequently for the safety of coworkers we see often, we'll obligate them to return in kind. This actually has far-reaching benefits.

Suppose you "do the right thing" for a coworker's safety, but you're not available to receive his or her reciprocal act of caring. Will that person be more likely to act on behalf of another individual? Yes, according to informative research by Berkowitz and Daniels (1964). Subjects in one experimental group received favors, while other subjects did not. Later, the person who did the favor was unavailable, but another individual (a research accomplice) was in apparent need of help. Subjects who had received the earlier favor were significantly more likely to assist than those who had not.

The reciprocity principle also works among strangers, or when there is no expectation of future interaction (Goranson & Berkowitz, 1966). Two researchers (Kung & Woolcott, 1976) demonstrated this by mailing Christmas cards to a sample of total strangers. To the researchers' surprise, many of these strangers (about 20 percent) responded by sending holiday greeting cards to the return address. They went beyond the call of duty to return a favor, even though they didn't know who they were sending the card to.

GIFTS AREN'T FREE. Has someone snared your attention to hear a sales pitch after giving you a free gift? Have you ever felt obligated to contribute to a charity after receiving gummed individualized address labels and a stamped envelope for your check? Ever purchase food in a supermarket after eating a free sample? Do you feel obliged to buy something after using it for a 10-day "free" trial period? As illustrated in Figure 14.12, how about feeling obligated to do the dishes after someone else fixes the meal?

If you answer "yes" to any of these questions, it's likely you've been influenced by the reciprocity principle. Many marketing or sales-promotion efforts count on this "free sample" gimmick to influence purchasing behavior. A classic experiment by

Figure 14.12. The dating game is frequently influenced by reciprocity.

Dennis Regan (1971) gives credibility to this ploy. During a break in an "art appreciation" experiment, in which pairs of individuals rated paintings, one subject (actually the experimenter's accomplice) performed an unexpected favor—returning from the break with a Coke for the other subject. For a control condition, the accomplice returned from the break empty handed.

At the end of the experiment, the accomplice asked the subject for a favor. He was selling raffle tickets for a new car and would win $50 if he sold the most tickets. The subjects who had received the unexpected favor purchased twice as many raffle tickets as those in the control group. This sounds fair because of the reciprocity norm. But soft drinks cost only 10 cents at the time of this study, and the raffle tickets were 25 cents. Attempting to reciprocate, the students purchased an average of two tickets each. The free drink was not "free." It cost them five times as much as if they had paid for it themselves.

Does this justify distributing free safety gifts, such as pens, tee-shirts, caps, cups, and other trinkets? Yes, to some extent, but you have to take into account perceptions. How special is the gift? Was the gift given to a select group of people, or was it distributed to everyone? Does the gift or its delivery represent significant sacrifice in money, time, or effort? Can it be purchased elsewhere, or does its safety slogan make it special?

A "special" safety gift—as perceived by the recipient—will trigger more acts of caring in response. Remember, too, that the way a gift is bestowed can make all the difference in the world. The labels and slogans linked with it can influence the amount and kind of responsive action. If the gift is presented to represent the actively caring safety leadership expected from a "special" group of workers, a certain type of reciprocity is activated. People will tell themselves that they are considered leaders, and they need to justify this label by going beyond the call of duty for others. And if they have learned about the various categories of actively caring behavior, they know the best way to lead is by taking action that is direct, proactive, and focused on supporting safe behavior or correcting at-risk behavior.

DOOR-IN-THE-FACE: START BIG AND RETREAT. Suppose the plant safety director pulls you aside and asks you to chair the safety steering committee for the next two years. This request seems outrageous, given your other commitments and the fact you never even served on the committee. You say, "Thanks, but no thanks!" The safety director says he understands, and then asks if you'd be willing to serve on the committee. Research shows that because the safety director "backed down" from his first request, you'll feel subtle pressure to make a similar concession—to reciprocate—and agree to the second, less demanding assignment.

Professor Robert Cialdini and his associates (1975) posed as representatives of the "County Youth Counseling Program" and asked students if they would be willing to chaperone juveniles on a trip to the zoo. Most of the students (16.7 percent) refused. But the researchers got 50 percent compliance when they preceded this minor request with a major one—to serve as counselors for juvenile delinquents for two hours a week over a two-year period. When the researchers accepted the subjects' refusal of their extreme request and retreated to a more reasonable one, half of the subjects apparently felt obligated to reciprocate and agree to the less demanding favor.

Commitment and Consistency

Robert Cialdini (1993) refers to commitment and consistency as an influence mechanism lying deep within us, directing many of our actions. It reflects our motivation to be, and appear to be, consistent. "Once we make a choice or take a stand, we will encounter personal and interpersonal pressures to behave consistently with that commitment" (Cialdini, 1993, p. 51). Professor Cialdini suggests that the pressure comes from three basic sources:

- Society values consistency within people
- Consistent conduct benefits daily existence
- A consistent orientation allows us to take shortcuts when processing information and making decisions: We don't have to stop to consider everything involved; instead, we fall back on our prior commitment or decision and act accordingly to remain true to ourselves

PUBLIC AND VOLUNTARY COMMITMENT. The "Safe Behavior Promise Card" described in Chapter 9 derives its power to influence from the commitment and consistency principle. When people sign their name to a promise card they commit to behaving in a certain way. And then they act in a way that is consistent with their commitment.

Commitments are most effective, or influential, when they are visible, require some amount of effort, and are perceived to be voluntary, not coerced (Cialdini, 1993). It makes sense, then, to have employees state a public rather than private commitment to actively care for safety, and to have them sign their name to a promise card rather than merely raise their hand. It's critically important for those making a pledge to believe they did it voluntarily. In reality, decisions to make a public commitment are dramatically influenced by external activators and consequences, including peer pressure. But if people sell themselves on the idea that they made a personal choice, consistency is likely to follow the commitment.

FOOT-IN-THE-DOOR: START SMALL AND BUILD. This strategy follows directly from the commitment and consistency principle. To be consistent, a person who follows a small request will likely comply with a larger request later. During the Korean War, the Chinese communists used this technique on American prisoners by gradually escalating their demands, which started with a few harmless requests (Schein, 1956). First, prisoners were persuaded to speak or write trivial statements. Then they were urged to copy or create statements that criticized American capitalism. Eventually, the prisoners participated in group discussions of the advantages of communism, wrote self-criticisms, and gave public confessions of their wrong-doing.

Research has found this "start small and build" strategy succeeds in boosting product sales, monetary contributions to charities, and blood donations. In a pioneering study, researchers posed as volunteers in a local traffic safety campaign and went door-to-door to ask residents' permission to install a large ugly sign in their front yards with the message "Drive Carefully." Only about 17 percent consented. However, of those residents who two weeks earlier had signed a safety legislation petition or had agreed to display a 3″-square "Be a Safe Driver" sign in their home, 76 percent allowed the large sign to be installed (Freedman & Fraser, 1966).

The "Safe Behavior Promise Card" uses this principle. After people sign on to perform a certain behavior for a specified period of time, such as "Buckle vehicle safety belts for one month," "Use particular personal protective equipment for two months," or "Walk behind yellow lines for the rest of the year," they are more likely to actually do it.

This foot-in-the-door technique only works when people go along with the first small request. If a person says "No" right away, he or she might find it even easier to resist subsequent, more important requests. So if your first call for actively caring behavior is shot down, you didn't start small enough. Be prepared to retreat to something less demanding and build reciprocity from there.

THROWING A CURVE BALL: RAISING THE STAKES LATER. This technique of "throwing a curve" occurs when you persuade someone to make a decision or commitment because of the relatively low stakes involved. Then you raise the level of involvement required. For example, being a safety steering committee member isn't asking too much if meetings are held only once a month. But after attending the first two safety meetings, the stakes are raised. More meetings are requested for a special project. Because of the commitment and consistency principle, the individual will likely stick with the original decision and remain an active member of the committee.

Professor Cialdini and colleagues (1978) used this technique to get college students to sign up for an early morning experiment on "thinking processes." During solicitation phone calls, the 7:00 a.m. start time was mentioned up-front for half the subjects. Only 24 percent agreed to participate. For the others, the caller first asked if they wanted to participate in the study. Then, after 56 percent agreed, the caller threw them a "curve" and said the experiment started at 7:00 a.m. The subjects had the chance to change their minds, but none did. Plus, 95 percent showed up at the 7:00 a.m. appointment time. Practically every one of them showed consistency and kept their commitment—in spite of being thrown a "curve." The effectiveness of this technique has been shown in several other studies (see, for example, Brownstein & Katzev, 1985; Burger & Petty, 1981; Joule, 1987).

This procedure is similar to the "foot-in-the-door" technique: a larger request occurs after you get agreement with a smaller one. A key difference, though, is that only one basic decision is made in the curve ball procedure, with costs or stakes raised after that initial commitment.

This compliance tactic is almost legendary among car dealers. A customer agrees to a special purchase price, say $800 below all other competitors. Then the price is raised. A number of reasons—we should say excuses—are given. The sales manager won't approve the deal. Certain options were not included in the special offer. The manager decreased the value of the customer's trade-in. But the ploy usually works. Customers agreeing to the special price usually don't flinch when thrown the curve because reneging might suggest a lack of consistency or failure to fulfill an obligation, even though the obligation is only imaginary. Often customers develop their own reasons—or excuses—to justify their initial choice and the additional costs (Teger, 1980).

SOME INFLUENCE TECHNIQUES CAN STIFLE TRUST

Throwing the curve ball raises a critical issue when trying to increase actively caring behavior. How do you feel when someone throws you a curve? Let's talk about trust for a moment.

Do you trust the waiter who brings you an expensive wine list—only after you've been seated and made selections from the food menu? It probably depends on whether you believe this sequence of events was done intentionally to get you to buy more. Similarly, you might not dislike or mistrust the car salesman who jacks up the price unless you suspect that his advertised price was just a lure to get you in the showroom. In other words, our trust, appreciation, or respect for people might fall off considerably if we believe they intentionally tricked or deceived us into modifying our attitude or behavior. Of course, there may be no harm done if the result is clearly for our own good, as for our health or safety, *and* we realize this.

Some of the influence strategies reviewed here are more likely to raise suspiciousness and reduce interpersonal trust. The curve ball and door-in-the-face techniques are usually most difficult to pull off. But reciprocity might do the most harm if the recipient believed the kindness or favor was a self-serving manipulation done *only* to force feelings of indebtedness, as is often the case in prison cultures.

Trying to increase actively caring behavior is not self-serving. We're trying to build a safety culture that benefits everyone. This must be made clear. Once the purpose of using the influence techniques described here is understood, interpersonal trust or mutual respect between agents and recipients of interventions will not diminish but may actually grow.

In Conclusion

There is a reason why this is the longest chapter in the book. It's probably the most important in terms of practical application. Integrating information from the prior chapters, I've tried to address this critical challenge: Continuous safety improvement leading to a Total Safety Culture requires people *to actively care for others as well as themselves.* Research-derived procedures to increase the frequency of actively caring behavior throughout a culture have been discussed. Some of these influence techniques indirectly increase actively caring behavior by benefiting the person states that facilitate one's willingness to care. Other influence strategies target behaviors directly.

Indirect strategies are deduced from the actively caring model explained in Chapter 13. Any procedure that increases a person's self-esteem, perception of empowerment—including self-efficacy, personal control, and optimism—or sense of belongingness or group cohesion will indirectly benefit active caring. A number of communication techniques enhance more than one of these states simultaneously, particularly actively listening to others for feelings and giving genuine praise for accomplishments. There are barriers to focusing on the positive, and we discussed them in the hope that awareness will help overcome the obstacles.

We need only reflect on our own lives to appreciate the power of choice, and how the perception of choice and personal control makes us more motivated, involved, and commited. Choice activates and sustains actively caring behavior.

Perceptions of belongingness are important, too. They increase when groups are given control over important decisions and receive genuine recognition for accomplishments. Synergy is the ultimate outcome of belongingness and win/win group involvement. It occurs when group interdependence produces more than what's possible from going it alone. We reviewed barriers to belongingness and synergy, specifically, social loafing.

The behavior-change techniques detailed in Chapters 9 and 10—from setting SMART goals and signing promise cards to offering soon, certain, and positive consequences for the right behavior—can be used to enhance those actively caring person states. They can also directly increase actively caring behavior. Education and direct participation in actively caring projects foster the behavior we're seeking, as well.

The interpersonal influence principles of reciprocity and commitment/consistency as they apply to our everyday decisions and behaviors were discussed. The commitment and consistency principle is behind the success of safe behavior promise cards, the foot-in-the-door technique ("start small and build"), and throwing a curve ball ("raise the stakes later"). These principles and specific strategies can be applied to directly boost actively caring behaviors.

Finally, it's critical to realize that social influence strategies can reduce respect or trust between people if they are applied within a context of win/lose or top-down control. On the other hand, when the purpose of the influence technique is clearly to increase actively caring behavior and improve the safety and health of everyone involved, a win/win climate is evident and mutual respect and trust is nurtured.

REFERENCES

Adler, R. B., & Towne, N. (1990). *Looking out/Looking in.* (Sixth Edition). Philadelphia, PA: Holt, Rinehart, & Winston, Inc.

Beaman, A. L., Barnes, P. J., Klentz, B., & McQuirk, B. (1978). Increasing helping rates through information dissemination: Teaching pays. *Personality and Social Psychology Bulletin,* 4, 406-411.

Beck, A. T. (1991). Cognitive therapy: A 30-year retrospective. *American Psychologist,* 46, 368-375.

Berkowitz, L., & Daniels, L. R. (1963). Responsibility and dependency. *Journal of Abnormal and Social Psychology,* 66, 429-436.

Brownstein, R., & Katzev, R. (1985). The relative effectiveness of three compliance techniques in eliciting donations to a cultural organization. *Journal of Applied Social Psychology,* 15, 564-575.

Burger, J. M., & Petty, R. E. (1981). The low-ball compliance technique: Task or person commitment? *Journal of Personality and Social Psychology,* 40, 492-500.

Carnevale, P. J., Pruitt, D. G., & Carrington, P. I. (1982). Effects of future dependence, liking, and repeated requests for help on helping behaviors. *Social Psychology Quarterly,* 45, 9-14.

Cialdini, R. B. (1993). *Influence: Science and practice* (Third Edition). New York: Harper Collins College Publishers.

Cialdini, R. B., Cacioppo, J. T., Bassett, R., & Miller, J. A. (1978). Low-ball procedure for producing compliance: Commitment then cost. *Journal of Applied Social Psychology,* 15, 492-500.

Cialdini, R. B., Vincent, J. E., Lewis, S. K., Catalan, J., Wheeler, D., & Darby, B. L. (1975). Reciprocal concessions procedure for inducing compliance: The door-in-the-face technique. *Journal of Personality and Social Psychology,* 1, 206-215.

Covey, S. R. (1989). *The seven habits of highly effective people.* New York: Simon and Schuster.

Dashiell, J. F. (1935). Experimental studies of the influence of social situations on the behavior of individual human adults. In C. Murcheson (Ed.). *A handbook of social psychology.* Worchester, MA: Clark University Press.

Deutsch, F. M., & Lamberti, D. M. (1986). Does social approval increase helping? *Personality and Social Psychology Bulletin,* 12, 148-157.

Eisenberg, N. (1992). *The caring child.* Cambridge, MA: Harvard University Press.

Ferrari, J. R., & Geller, E. S. (1994). Developing future caregivers by integrating research and community service. *The Community Psychologist,* 27(2), 12-13.

Freedman, J. L., & Fraser, S. C. (1966). Compliance without pressure: the foot-in-the-door technique. *Journal of Personality and Social Psychology,* 4, 195-203.

Gabrenya, W. K., Jr., Latané, B., & Wang, Y. E. (1983). Social loafing in cross cultural perspective. *Journal of Cross-Cultural Psychology,* 14, 368-384.

Geller, E. S. (1994). Ten principles for achieving a Total Safety Culture. *Professional Safety,* 39(9), 18-24.

Geller, E. S., & Roberts, D. S. (1993, January). *Beyond behavior modification for continuous improvement in occupational safety.* Paper presented at the FABA/OBM Network Conference, St. Petersburg, FL.

Geller, E. S., Roberts, D. S., & Gilmore, M. R. (1996). Predicting propensity to actively care for occupational safety. *Journal of Safety Research,* 27, 1–8.

Goranson, R., & Berkowitz, L. (1966). Reciprocity and responsibility reactions to prior help. *Journal of Personality and Social Psychology,* 18, 227-232.

Gouldner, A. W. (1960). The norm of reciprocity: A preliminary statement. *American Sociological Review,* 25, 161-178.

Greene, B. F., Winett, R. A., Van Houten, R., Geller, E. S., & Iwata, B. A., eds. (1987). *Behavior analysis in the community 1968-1986 from the Journal of Applied Behavior Analysis* (Reprint Series, Volume 2). Lawrence, KS: Society for the Experimental Analysis of Behavior, Inc.

Grusec, J. E., & Redler, E. (1980). Attribution, reinforcement, and altruism: A developmental analysis. *Developmental Psychology,* 16, 525-534.

Harkins, S.G., & Szymanski, K. (1984). Social loafing and group evaluation. *Journal of Personality and Social Psychology,* 56,934-941.

Harkins, S. G., & Petty, R. E. (1982). Effects of task difficulty and task uniqueness on social loafing. *Journal of Personality and Social Psychology,* 43, 1214-1229.

Hoffman, M. L. (1975). Altruistic behavior and the parent/child relationship. *Journal of Personality and Social Psychology,* 31, 937-943.

Joule, R. V. (1987). Tobacco deprivation: The foot-in-the-door technique versus the low-ball technique. *European Journal of Social Psychology,* 17, 361-365.

Kahn, A., & Tice, T. (1973). Returning a favor and retaliating harm: The effects of stated initiation and actual behavior. *Journal of Experimental Social Psychology,* 9, 43-56.

Keller, J. J. (1991). The recycling solution: How I increased recycling on Dilworth Road. *Journal of Applied Behavior Analysis,* 24, 617-619.

Kunz, P. R., & Woolcott, M. (1976). Season's greetings: From my status to yours. *Social Science Research*, 5, 269-278.

Latané, B., & Darley, J. M. (1970). *The unresponsive bystander: Why doesn't he help?* New York: Appleton-Century-Crofts.

Latané, B., Williams, K., & Harkins, S. (1979). Many heads make light the work: The causes and consequences of social loafing. *Journal of Personality and Social Psychology*, 37, 823-832.

McGovern, L. P., Ditzian, J. L., & Taylor, S. P. (1975). The effect of one positive reinforcement on helping behavior. *Bulletin of the Psychonomics Society*, 5, 421-423.

Monty, R. A., & Perlmuter, L. C. (1975). Persistence of the effect of choice on paired-associate learning. *Memory and Cognition*, 3, 183-187.

Monty, R. A., Geller, E. S., Savage, R. E., & Perlmuter, L. C. (1979). The freedom to choose is not always so choice. *Journal of Experimental Psychology: Human Learning and Memory*, 37, 170-178.

Moss, M. K., & Page, R. S. (1972). Reinforcement and helping behavior. *Journal of Applied Social Psychology*, 2, 360-371.

Notz, W. W., Boschman, I., & Tax, S. S. (1987). Reinforcing punishment and extinguishing reward: On the folly of OBM without SPC. *Journal of Organizational Behavior Management*, 9(1), 33-46.

Perlmuter, L. C., Monty, R. A., & Kimble, G. A. (1971). Effect of choice on paired-associate learning. *Journal of Experimental Psychology*, 91, 47-53.

Peterson, C., Maier, S. F., & Seligman, M. E. P. (1993). *Learned helplessness: A theory for the age of personal control*. New York: Oxford University Press.

Regan, D. T. (1971). Effect of a favor on liking and compliance. *Journal of Experimental Social Psychology*, 7, 627-639.

Roberts, D. S., & Geller, E. S. (1995). An "actively caring" model for occupational safety: A field test. *Applied & Preventive Psychology*, 4, 53-59.

Rushton, J. P. (1975). Generosity in children: Immediate and long term effects of modeling, preaching, and moral judgment. *Journal of Personality and Social Psychology*, 31, 459-466.

Rutherford, E., & Mussen, P. (1968). Generosity in nursery school boys. *Child Development*, 39, 755-765.

Schein, E. (1956). The Chinese indoctrination program for prisoners of war: A study of "brainwashing." *Psychiatry*, 19, 149-172.

Seligman, M. E. P. (1975). *Helplessness: On depression development and death*. San Francisco: Freeman.

Staub, E. (1975). To rear a prosocial child: Reasoning, learning by doing, and learning by teaching others. In D. DePalma & J. Folley (Eds.). *Moral development: Current theory and research*. Hillsdale, NJ: Lawrence Erlbaum Associates.

———. (1979). *Positive social behavior and morality: Socialization and development* (Vol. 2). New York: Academic Press.

———. (1990). Moral exclusion, personal goal theory and extreme destructiveness. *Journal of Social Issues*, 46, 47-65.

———. (1992). The origins of aggression and the creation of positive relations among groups. In S. Staub & P. Green (Eds.). *Psychology and social responsibility: Facing global challenges*. New York: New York University Press.

Steiner, I. D. (1970). Perceived freedom. In L. Berkowitz (Ed.), *Advances in experimental social psychology* (Vol. 5). New York: Academic Press.

Szymanski, K., & Harkins, S. G. (1987). Social loafing and self-evaluation with a social standard. *Journal of Personality and Social Psychology*, 53, 891-897.

Taylor, S. E. (1989). *Positive illusions: Creative self-deception and the healthy mind*. New York: Basic Books.

Teger, A. I. (1980). *Too much invested to quit*. Elmsford, NJ: Pergamon Press.

Watson, D. C., & Tharp, R. G. (1989). *Self-directed behavior: Self-modification for personal adjustment* (Sixth Edition), Pacific Grove, CA: Brooks/Cole Publishing Company.

Weick, K. E. (1984). Small wins: Redefining the scale of social problems. *American Psychologist, 39,* 40-44.

White, R. W. (1959). Motivation reconsidered: The concept of competence. *Psychological Review, 66,* 297-333.

Williams, K., Harkins, S., & Latané, B. (1981). Identifiability as a deterrent to social loafing: Two cheering experiments. *Journal of Personality and Social Psychology, 40,* 303-311.

Wilson, D. W., & Kahn, A. (1975). Rewards, costs, and sex differences in helping behavior. *Psychological Reports, 36,* 31-34.

Zahn-Waxler, C., Radke-Yarrow, M., & King, R. (1979). Child rearing and children's prosocial initiations toward victims of distress. *Child Development, 50,* 319-330.

Zahn-Waxler, C., Radke-Yarrow, M., & Chapman, M. (1983). Prosocial dispositions and behavior. In E. M. Hetherington (Ed.), *Manual of child psychology: Socialization, personality, and social development* (Vol. 4). New York: Wiley.

Part Five

Putting It All Together

Chapter 15
Evaluation for Continuous Improvement

Continuous improvement demands proper evaluation. This chapter explains how to evaluate the impact of safety interventions. The principles described here will make you a smarter consumer of marketed safety programs, and help you evaluate a customized intervention process.

What gets measured gets done; what gets measured and rewarded gets done well.

Larry Hansen (1994) used these words in his *Professional Safety* article on managing occupational safety (p. 41). You've probably heard words to this effect, indeed they are key to any continuous improvement effort. But there's a problem with how workplace safety is traditionally measured. As I indicated earlier in Chapter 3, too much weight is given to outcome numbers that people cannot control directly. People must be held accountable for results they can control. Yet corporations, divisions, plants, and departments are often ranked according to abstract outcome numbers like the total recordable injury rate. And these rankings often determine bonus rewards or penalties.

What behavior improves when safety awards are based only on an injury rate? If employees can link their daily activities to safety results, then celebrating reduced injury rates can be useful, even motivating. But it's critical to recognize the behaviors, procedures, and processes that led to fewer injuries or lower workers' compensation costs.

If you don't focus on the real causes of improvement, you run the risk of actually demotivating the folks deserving recognition. Employees might think continuous improvement is caused by luck or chance—events beyond personal control. This can lead to feelings of apathy or learned helplessness (Seligman, 1975), as I discussed earlier (see Chapters 12, 13, and 14). If we want employees to work for continuous improvement, we need to recognize and reward the "right stuff." And this requires the right kind of measurement procedures.

Measuring the Right Stuff

Deming (1991) admonished his audiences for ranking people, departments, and organizations. In fact, he recommended that grades and performance appraisals be abolished completely from education and business. In his words, "The fact is that performance appraisal, management by the numbers, M.B.O., and work standards have already devastated Western industry...the annual rating of performance has devastated Western industry...Western management has for too long focused on the end product (Deming, 1986, p.1).

Part of Dr. Deming's rationale comes from the fact that standard approaches to measuring academic and work performance are often subjective, relative, and not clearly related to individual behavior. Teachers and professors, for example, use contrived distribution curves and cutoffs to assure only designated percentages of students can attain certain grades. Knowledge tests are necessarily biased and imperfect assessment devices, and are often only remotely linked to specific behaviors within a student's control—including attending class, taking notes, reading the textbook, and studying material on a regular basis. I've known many demotivated students who felt their daily efforts were overshadowed by the emphasis on exams.

LIMITATIONS OF PERFORMANCE APPRAISALS

I've met very few employees who respect and appreciate annual performance appraisals. They don't see the evaluations as being fair, objective, and motivational. I'm sure you know department heads and supervisors who complete performance appraisals only because it's mandatory. They're not interested in providing constructive, performance-focused feedback for continuous improvement. They don't maintain ongoing records of employees' accomplishments and less-than-adequate performance. Instead, they typically wait until a few days before appraisals are due and then make their best-guess estimate of an individual's ranking, using nebulous performance dimensions like "competent," "enthusiastic," "self-motivated," "cooperative," "responsive to feedback," and "willing to improve."

Employees give these appraisals the attention they deserve. They simply ignore them, and try not to get too "stressed out" when appraisal time comes around. Some employees try to gain control of the situation—and turn distress into stress—by "performing" for the boss. They put on last-minute "exhibitions" in order to improve their "scores." All too often this strategy works. As a result, these short-term exhibitions are the only behavior reinforced as a result of the evaluation.

The manager in Figure 15.1. is obviously soliciting biased performance feedback. Given the principle of reciprocity discussed in Chapter 14, the suggestion boxes might contain a number of signed comments near the time he must write performance appraisals.

Actually, Figure 15.1 does depict a way for employees to deliver and receive timely performance-appraisal feedback whenever it's warranted. If the suggestions are based on the principles for giving rewarding and correcting feedback reviewed in Chapter 11, then performance appraisals would indeed facilitate continuous improvement.

Figure 15.1. For continuous improvement, performance appraisals need to be frequent, frank, and followed.

Some managers get proactive and behavior-based with their appraisal responsibilities, and measure and reward the right stuff. The process is not easy and takes extra time, but it's well worth the effort. Here's how it works: The manager meets with employees individually at the start of the evaluation period. The pair set operational (behavioral) and customized definitions for performance criteria. The process is very much interactive and includes setting SMART goals (see Chapter 9), as well as developing a way to report progress toward goal attainment and to receive feedback. The manager then keeps continuous records of the employee's behaviors related to those specific goals. The evaluation process is indeed an ongoing process—not some last-minute rush to judgment.

Performance appraisals should be reciprocal. Effective managers ask employees for input on their own management-related goals, and encourage feedback on how well they're progressing toward these goals. Employees should be required to complete periodic evaluations of supervisors, reflecting expected and desired performance. This increases their perceptions of empowerment and belongingness, and facilitates continuous performance improvement on the part of their supervisors.

WHAT IS PERFORMANCE IMPROVEMENT?

If "what gets measured gets done," what are we measuring when we measure performance? From a systems standpoint, performance means output. In the context of psychology, performance means behavior, or output from a human system. We can't discuss measurement and evaluation without an operational definition of performance. Let's examine different definitions more closely.

INDIVIDUAL PERFORMANCE. The first fact I learned in my first psychology course was that psychology is the study of the individual. Later, each psychology course I took in graduate school examined some aspect of individual performance (the dependent measure) as it was influenced by particular environmental or experimenter manipulations (the independent variables). For example, in my courses on learning I studied the effects of prior experiences and reinforcement history on individual performance. In social psychology, I learned how other people influence an individual's performance; and in physiological psychology, I learned how specific changes to one's nervous system through drugs, electrical stimulation, or surgery influenced individual performance. Today, in my management and organizational psychology courses I teach students how an individual's work performance can be affected by training techniques, the three-term contingency (activator-behavior-consequence), and differences in a person's knowledge, skills, ability, attitude, and personality.

SYSTEM PERFORMANCE. At a four-day Deming seminar in 1991, I learned that many corporate leaders and experts in the field of organizational management—including safety management—define performance from an industrial systems perspective. The second afternoon was devoted to the famous red/white bead game, a demonstration to convince the audience that individual performance is a relatively insignificant determinant of organizational, or system, performance. Dr. Deming asked for ten volunteers from the more than 600 attendees. He assigned six volunteers the job of "willing worker," two other volunteers became "inspectors" of each worker's output, another volunteer played the role of "quality control supervisor" or inspector of the inspectors' work, and the tenth volunteer was the "recorder."

After assigning work duties, Dr. Deming mixed 800 small red beads with 3,200 white beads—all the same size—in a box. He showed the "willing workers" how to scoop out beads from this box using a beveled paddle that had indentations for 50 beads. Each scoop of the paddle was considered a day's production. The inspectors counted the numbers of red and white beads on each worker's paddle, and after a reliability check by the bead inspector, the number of red beads (considered mistakes or defects) was recorded and displayed publicly on an overhead projector by the recorder.

Dr. Deming played the role of the corporate executive officer, and he urged the workers to produce no defects (red beads). When workers scooped fewer than 10 red beads, they were praised by the CEO. Scooping more than 15 red beads, however, resulted in severe criticism and an exhortation to do better, otherwise "we will go out of business." Dr. Deming reacted to each scoop of beads as a product of individual performance; if performance outcome was poor, he again demonstrated the correct bead-sampling procedure, or gave corrective feedback to a worker during the sampling process. It's noteworthy that regression to the mean resulted in good individual performance getting worse and poor individual performance getting better. As discussed in Chapter 14, this can give the impression that correcting feedback works better than rewarding feedback.

The sampling, inspecting, reliability checking, recording, public graphing of results, and corrective action for defects continued for ten samples (representing ten days) per worker. This portion of the demonstration lasted more than an hour, and led logically to Deming's concluding statements that "the actual number of red beads scooped by each worker was out of that worker's control. The workers only *delivered* defects. Management, which controls the system, caused the defects through system design."

This demonstration has profound lessons for achieving a Total Safety Culture. As I've discussed earlier, the typical evaluation procedure used by both the government and private sectors to judge the safety record of companies is based on organizational or system performance—"the total recordable injury rate" or number of "OSHA recordables"—uncontrollable by most individual workers. Individuals are not only held accountable, perhaps even disciplined, for their own injuries but also for the injuries of others, even employees outside their immediate work group. They are often blamed for injuries caused by a number of factors outside their control, such as at-risk environmental conditions or equipment, excessive workloads, system contingencies causing a fast work pace, and a culture that supports "rugged individualism" and thwarts group cohesion and a "brothers/sisters keeper" perspective.

In July, 1992, Dr. Deming wrote me this explanation to clarify his teaching intentions with his red/white bead demonstration.

> *Currently, management works under the assumption that people and not the systems they work in are responsible for performance. We therefore reward and punish people but the system they work in remains unchanged. The point is not that differences between individuals are unimportant in and of themselves (but) that focus on individual differences alone yields possibilities for improvement that are trivial compared with transformation of the entire system that they work in. In fact, only individuals can change a given system to improve performance. (W. E. Deming, personal communication, July 11, 1992)*

SYSTEM VS. INDIVIDUAL PERFORMANCE. In the Fall of 1991, a demonstration in my graduate course in behavior and systems management illustrated profoundly the role of individual performance in organizational output. It revealed the need to consider both individual and system-level factors when evaluating safety achievement. One of my students, Mario Beruvides, directed a probability "game" that, to his surprise and my delight, illustrated the significant and unexpected impact of individual performance factors. Playing cards were used instead of beads to illustrate the importance of system-level performance.

Mario asked four volunteers and myself to play the role of workers for a hypothetical company, "Geller Inc., Non-Face Card Manufacturers." He gave us each a sealed deck of cards, and instructed us to open them, remove the two jokers and instruction cards and shuffle them thoroughly. Then we placed our decks face down on the seminar table, and per Mario's instructions, drew five cards from the top of our decks to represent a day's work. Face cards were counted as defects or errors, and the number of face cards per "willing worker" was recorded and displayed to the class on a "production log."

As in Deming's bead game, the CEO, played by Mario, urged the workers to care for the welfare of the company and produce as few defects as possible. Defects from each worker were recorded, summarized, and displayed by a student assigned as "supervisor." The five-card drawing and recording continued for five trials, representing a work week. After each trial, the CEO reprimanded the workers whose rate of defects exceeded the average and praised those who had just one or zero defects. To the delight of the class and surprise of the CEO, my drawing always included the most face cards, or defects, and I always received the most criticism.

Before the sixth round of drawings, marking the second work week, the CEO unveiled a company-wide incentive plan: Whenever the face cards from a group of five "hands" are combined within 15 seconds to form a pair or three or more of a kind, as in a poker hand, the face cards will not be counted as defects. On each trial, I continued to draw the most face cards, usually three or more, and I immediately took charge of organizing the face cards from the other "hands" into poker combinations. This leadership and teamwork decreased the number of defects per individual dramatically. Now, my defect rate was at least as low as any other worker's.

During the lively classroom discussion that ensued, Mario admitted that our results were remarkably different than those obtained by the management training and development specialist who invented the card-game analog of Deming's bead demonstration (Storey, 1989). In a completely random system, one "worker" is not supposed to always produce the most defects. Plus, Chris Storey reported that the incentive condition should not reduce defects, even over five draws, presumably because some "workers" are reluctant to trade cards due to lack of experience, knowledge, or leadership.

While discussing reasons for the unexpected results from "Geller, Inc." I made a provocative discovery—I had been drawing cards from a pinochle deck containing 50 percent face cards. Mario had inadvertently purchased one pinochle deck, and happened to give me that deck by chance. This serendipitous innovation in Mario Beruvides' demonstration resulted in profound awareness of the power in individual differences.

Mario admitted his preset expectation that the game would show random, uncontrollable variation. His anticipation to find only "common cause" prevented him from recognizing "special cause" contributions from individuals. And he was caught completely off guard by the success of the incentive program. The students identified individual performance factors that made the incentive program effective, particularly the special effects of the pinochle deck and its chance assignment to the assertive class instructor. This led us to discuss how companies can benefit from appropriate employee selection and placement, education and training, interpersonal communication, and individual and group recognition processes. We also saw the need to shift from our usual preoccupation with individual performance and attend to group and system-wide performance as well. This requires an ecological analysis (Willems, 1974) of multiple factors that can affect individual, group, and organizational performance.

A Total Safety Culture, then, requires us to balance how we measure and manage both organizational performance and individual behavior. The challenge is to accurately attribute change in safety performance. Is it due to individual behavior, groups, or the system? Individuals should only be held directly responsible for their own safety performance; teams should be held accountable for outcomes directly related to their team performance.

Continuous improvement in organizational safety performance, using the yardstick of OSHA recordables, requires improvements in the system as a whole, of which individual employees play an integral part—both as individuals and members of work teams. As I've discussed earlier, holding individuals responsible for safety performance outside their perceived control develops attitudes and perceptions, such as apathy, helplessness, and pessimism, that interfere with both personal and organizational safety improvement.

Developing a Comprehensive Evaluation Process

In April and August of 1995, I had the pleasure of working with a panel of evaluation experts to develop a set of measurement guidelines for the National Safety Council and the Centers for Disease Control and Prevention.[4] Our mission was to develop a handbook of practical guidelines that field personnel can use to evaluate the impact of an intervention to improve safety on family and commercial farms. The results of our four days of deliberations are documented in Steel (1996), which includes useful insight and direction for the design of evaluation procedures.

We agreed that evaluation was essential to hold people accountable for achieving program objectives. This particularly pertains to those developing and implementing the intervention. The evaluation process should measure whether intervention procedures: a) are consistent with relevant principles and mission statements of the organization; b) reach the desired audience and are implemented as planned; and c) are efficient and effective (cf. Vojtecky & Schmitz, 1986). In sum, a complete evaluation process should assess how an intervention was conceived, designed, and implemented, and how efficient and effective it was.

Our panel discussed a number of approaches to assess these three basic areas—conceptualization, implementation, and impact. Each suggests that certain records be kept and examined systematically. For example, the rationale, goals, action plan, and techniques of the intervention must be documented in order to check if the intervention design is consistent with relevant corporate principles and aims of a safety effort. When a safety steering committee deliberates to decide whether to implement a specific training process or to hire a particular team of consultants, it essentially makes this kind of evaluation. The committee decides whether the goals, objectives, and procedures of the training or intervention program fit its philosophy, purpose, and mission. Obviously, this evaluation is typically made before training or intervention begins.

The actual intervention process must be documented to decide if the procedures are being implemented as planned, and whether the intervention is reaching appropriate numbers of people. Attendance records, for example, provide an efficient means of measuring the coverage of training programs. Participation in an intervention can be readily measured if the procedures include completion or delivery of materials with the participant's name, such as recognition thank-you cards, critical behavior checklists, or safe behavior promise cards.

Often it's useful to record reactions from participants and nonparticipants. What did they think of the intervention principles and procedures? This feedback can help find ways to increase program involvement. Thus, it's useful to conduct this kind of

4. These four days of meetings were organized by Dr. Sam Steel of the National Safety Council and facilitated by Dr. Jerry Burk of the BURK group (Bakersville, CA). Other panelists included Professor Joe Miller (Penn State), Dr. Doreen Greenstein (Cornell University), Professor Stephen J. Guastello (Marquette University), Professor Midge Smith (University of Maryland), Jim Williams (Country Companies Bureau Insurance), and David Hard (National Institute of Occupational Safety and Health).

Figure 15.2. Measures of intervention impact vary according to remoteness from immediate injury causation.

evaluation during the early stages of an intervention process. Personal interviews or questionnaires can rate levels of satisfaction or dissatisfaction, and lead to "course corrections" if needed.

Assessing intervention effectiveness is most difficult, but most critical. Our lengthy panel discussions on this issue revealed the different levels of performance discussed above—individual, group, and system. We also debated process versus outcome issues, and the most objective and efficient ways to measure safety processes and outcomes. We concluded that behavior was the optimal process measure and injury reduction the ultimate measure of intervention success. Attitudes, perceptions, and beliefs were presumed to influence whether the intervention is accepted and has potential for long-term success. We all agreed, though, that survey techniques to estimate these subjective person states are relatively difficult to develop and evaluate.

Issues of questionnaire reliability and validity need to be addressed, as I discuss later in this chapter.

Figure 15.2 summarizes our panel's deliberations on how to evaluate intervention impact. It integrates the primary issues discussed so far in this chapter. Lower levels of the hierarchy represent process activities needed to improve the higher-level outcomes of a safer environment and ultimate injury reduction. Immediate causes of injury reduction are changes in environment or behavior—or both.

Let's pause for a moment to consider the cause-and-effect connection between process and outcome. Behavior can be viewed as an outcome, shaped by a process that pursues changes in employee knowledge, perceptions, or attitudes. In this case, the behavior-change goal depends on employee participation in an intervention aimed at influencing a person state. Of course, intervention processes can be designed to circumvent person states and directly change behavior to reduce injuries, as discussed in Part Three of this text.

Figure 15.2 points out the relativity of process and outcome. An education process, for example, can lead to behavior change (an outcome), while a process to change behaviors might result in outcome changes to the environment, say improved housekeeping. And completing a work process in a safer environment can affect the ultimate outcome—fewer injuries.

Figure 15.2 also reflects the three basic areas requiring attention for injury prevention: environment, behavior, and person. This is, of course, the Safety Triad (Geller, 1989b) introduced in Chapter 2 to categorize intervention strategies, and referred to later in Chapter 12 to classify different types of actively caring behaviors. The hierarchical levels of intervention impact in Figure 15.2 reflect one or more of these three domains and suggest a particular approach to measurement. Changes in environments and behaviors can be assessed directly through systematic observation, but changes in knowledge, perceptions, beliefs, attitudes, and intentions are only accessible indirectly through survey techniques, usually questionnaires. Let's review these three basic areas of evaluation.

WHAT TO MEASURE?

Most safety interventions focus on either environmental conditions—including engineering controls—or human conditions, as reflected in employees' perceptions, attitudes, or behaviors. It might seem reasonable to us to evaluate change only in the area we've targeted—environment, behavior, or person state. When the target is corporate "culture," employee perceptions or attitudes are typically evaluated. If behavior change is the focus, then behaviors are observed and analyzed in terms of their frequency, rate, duration, or percentage of occurrence, as reviewed in Chapter 8. When environments or engineering technologies are evaluated, mechanical, electrical, chemical, or structural measurements are taken (Geller, 1992).

I've heard culture-change consultants advocate perception surveys in place of environmental audits. And presentations on behavior-based safety emphasize direct observations of work practices, often in lieu of the subjective evaluation of personal perceptions and attitudes. But given the need for employees to "feel good" about a behavior-based safety process, and their need to participate continuously in managing and monitoring this process, I think it's obvious that we need to check perceptions and attitudes and ongoing behaviors. For example, a comprehensive evaluation of a simple change in equipment design should probably include an assessment of relevant human factors, like employees' work behaviors around the new equipment and their attitudes and perceptions regarding the equipment change.

I hope you can see that a comprehensive evaluation for safety requires a three-way audit process covering environmental conditions, safety-related behaviors, and person states such as perceptions, attitudes, beliefs, and intentions.

EVALUATING ENVIRONMENTAL CONDITIONS. In many ways, environmental audits are the easiest and most acceptable type of evaluation. In fact, regular environmental or housekeeping audits are already standard practice at most companies. But these evaluations can often be improved by involving more employees in designing audit forms, conducting systematic and regular assessments, and posting the results in relevant work areas. A safety incentive program established in 1992 at a Hoechst Celanese plant in Narrows, Virginia, awarded employees weekly "credits" for accomplishing these components of environmental evaluation. At the end of the year, the credits could be exchanged for various commodities containing a special safety logo. This company recognized the need to involve as many employees as possible in the regular auditing of environmental conditions, including tools, equipment, and operating conditions. Its incentive/reward process motivated involvement.

Figure 15.3 depicts a generic environmental checklist for safety that can be used to graph for public display the percentage of safe conditions and the percentage of potential corrective actions taken for at-risk tools, equipment, or operating conditions. My associates at Safety Performance Solutions typically teach work teams the rationale behind the environmental checklist, and then assist them in applying the checklist in their plant. Employees then customize the checklist and graphing procedures for their particular work areas. Regular audits and feedback sessions increase accountability for environmental factors that can be changed to prevent an injury.

EVALUATING WORK PRACTICES. The systematic auditing of work practices was the theme of Chapters 8 and 11. In Chapter 8, I introduced the overall ***DO IT*** process: **D**efine target behaviors, **O**bserve target behaviors, ***I***ntervene to increase safe behavior or decrease at-risk behavior, and ***T***est (or evaluate) the impact of your intervention. Also in Chapter 8, we discussed how to develop two types of observation checklists—a generic version for basic work practices applicable anywhere, such as prescribed lifting techniques and use of certain personal protective equipment (see Figures 8.9 and 8.10); and a job-specific checklist for particular tasks, like the safe driving checklist my daughter and I developed (see Figure 8.7).

The coaching process detailed in Chapter 11 also discussed how to develop and apply both generic and job-specific checklists for one-on-one observation and feedback sessions with coworkers (see Figures 11.9 and 11.10). As mentioned above, using a behavioral checklist to observe and evaluate ongoing work practices is the type of performance appraisal that can lead to continuous improvement.

My experience has been that group auditing—of people using or not using personal protective equipment, for example—is readily accepted by most employees and relatively easy to implement. But the one-on-one audits in coaching (see Chapter 11), with employees observing other employees who volunteer to be monitored, are not

Observer:______________________ **Date:**___________ **Time:**_____________
Department:_________ **Building:**_________ **Floor:**_________ **Area:**_________

OPERATING CONDITIONS/TOOLS & EQUIPMENT	SAFE	AT-RISK	CORRECTIVE ACTIONS TAKEN*
Electrical wiring (properly enclosed)			
Air nozzles (limited to 30 psi)			
Chemicals (exposure concern)			
Eyewash station			
Emergency shower			
Barricades (in place where necessary)			
Storage of materials (neat/safe)			
Hazard communication labels (appropriate)			
Floors (dry)			
Exits, aisles, sidewalks and walkways (clear of debris)			
Lighting (adequate)			
Housekeeping (satisfactory)			
Tools (safe operating condition)			
Guards (adequate and in place)			
Fire extinguisher (in appropriate location)			
Forklift (safe operation)			
Forklift driver (license in possession)			
Tow truck (safe operation)			
Tow truck driver (license in possession)			
Chairs (safe condition)			
Totals			

Percent Safe Conditions: $\frac{\text{Total Safe Observations}}{\text{Total Safe Observations + At-Risk Observations}} \times 100 =$ _____%

Percent Corrective Actions Taken: $\frac{\text{Total Corrective Actions Taken}}{\text{Total At-Risk Observations}} \times 100 =$ _____%

*Please list and define the corrective actions taken on back of this sheet.

Figure 15.3. An environmental checklist to evaluate the safety of tools, equipment, and operating conditions.

readily accepted in some corporate cultures. A plant-wide education and training intervention is often necessary to teach the rationale and procedures for this evaluation process and to develop the necessary interpersonal understanding, empathy, and trust. My associates and I at Safety Performance Solutions have developed a Safety

Culture Survey to assess if a corporate culture is ready for one-on-one coaching. This involves evaluating person factors related to safety.

EVALUATING PERSON FACTORS

As discussed earlier in this text, person factors refer to subjective or internal aspects of people. They are reflected in commonly used terms like attitude, perception, feeling, intention, value, intelligence, cognitive style, and personality differences. You can find many surveys that measure specific person factors of target populations ranging from children to adults. Some of these factors are presumed to be *traits*, others are considered *states*. It's important to understand the difference when you consider the evaluation potential of a particular survey.

PERSON TRAITS. Theoretically, traits are relatively permanent characteristics of people; they don't vary much over time or across situations. The popular Myers-Briggs Type Indicator, for example, was designed to measure where individuals fall along four dichotomous personality dimensions: extroversion versus introversion, sensing versus intuition, thinking versus feeling, and judgment versus perception (Myers & McCaulley, 1985). The various combinations of these attributes allow for sixteen different personality types, each with special characteristics. These traits are presumed permanent and unchangeable, determined largely by physiological or biological factors.

Since traits are relatively permanent, questionnaires that measure them cannot gauge the impact or progress of a culture-change intervention. Trait measures serve as a tool to teach individual differences, but in safety management their application is limited to selecting people for certain job assignments (Geller, 1994). This is very risky, though, and should not be done without understanding the validity limitations described later in this chapter.

PERSON STATES. Person states are characteristics that can change from moment to moment, depending on situations and personal interactions (as discussed in Chapter 13). When our goals are thwarted, for example, we can be in a state of frustration. When experiences lead us to believe we have little control over events around us, we can be in a state of apathy or helplessness. Person states can influence behaviors. Frustration, for example, often provokes aggressive behavior (Dollard *et al,* 1939); and perceptions of helplessness inhibit constructive behavior or facilitate inactivity (Abramson *et al,* 1989).

In contrast, certain life experiences can affect positive person states, such as optimism, personal control, self-confidence, and belongingness. These, in turn, boost constructive behavior. This was the indirect approach to increasing actively caring behavior discussed in Chapter 14.

Measures of person states can be used to evaluate perceptions of culture change, and to pinpoint areas of a culture that need special intervention attention. Like most culture surveys, our Safety Culture Survey asks participants to answer questions on a five-point continuum (from highly disagree to highly agree) about their perceptions of the safety culture. Issues include the perceived amount of management support for safety, the willingness of employees to correct at-risk situations and look out for the safety of coworkers, the perceived risk level of the participant's job, and the nature of interpersonal consequences following an injury.

Our survey also measures factors that increase one's willingness to actively care for another person's safety. These include self-esteem, belongingness, and empowerment, as detailed in Chapter 13. Sample items from our survey that measure the actively caring person states are given in Chapter 13. They were adapted from professional measures of these characteristics, and have been evaluated for reliability and validity, as discussed below. Figure 15.4 contains twenty items from the safety perception and attitude portion of our survey. You'll note nothing very special about the items in this scale. They ask employees to react to straightforward statements about safety management and improvement.

You could compare employees' reactions to the items in Figure 15.4 before and after implementing a safety improvement process. Studying reactions prior to an intervention helps identify issues or work areas needing special attention. This information can lead you to choose a particular intervention approach, or customize one. Data from a baseline perception survey might even indicate that a culture is not ready for a given intervention process, suggesting the need for more education and discussion to get employees to "buy in."

The impact of an intervention can be measured by comparing perception surveys given before and after implementation. At one plant, our baseline Safety Culture Survey indicated that secretaries had below-average levels of perceived empowerment, as assessed by the measures of self-efficacy, personal control, and learned optimism described earlier in Chapter 13. A special recognition intervention was devised, and later the survey was administered again to measure changes in the five actively caring person states, as well as safety perceptions and attitudes.

Reviewing the results of this survey helped employees understand the relationship between work practices, perceptions, and attitudes. It also revealed that the recognition program improved some person states both for employees who received and administered it.

LIMITATIONS OF QUESTIONNAIRES. Although measures of person states are more useful for safety management than measures of person traits, there are critical limitations to both. Given the increasing popularity of these evaluation tools among safety professionals, I urge you to give these limitations serious thought. First, surveys of person factors, traits or states, are neither as objective nor as reliable as audits of behaviors. Second, results are not as straightforward and easy to analyze and interpret as information from behavioral observations. And third, developing, administering, and interpreting surveys designed to evaluate person factors relevant for safety requires a basic understanding of reliability and validity.

RELIABILITY AND VALIDITY

What is the practical value of a questionnaire or survey? This can be assessed with a variety of research methods and statistical tools. Many are beyond the scope of this text, but a few basic concepts are pertinent. First, questionnaires to measure person factors can be reliable, though not valid. But to be valid, they must be reliable. A reliable survey gives consistent results. You assess this by comparing answers

	HIGHLY DISAGREE	DISAGREE	NOT SURE	AGREE	HIGHLY AGREE
1. The risk level of my job concerns me quite a bit.	1	2	3	4	5
2. When told about safety hazards, supervisors are appreciative and try to correct them quickly.	1	2	3	4	5
3. My immediate supervisor is well informed about relevant safety issues.	1	2	3	4	5
4. It is the responsibility of each employee to seek out opportunities to prevent injury.	1	2	3	4	5
5. At my plant, work productivity and quality usually have a higher priority than work safety.	1	2	3	4	5
6. The managers in my plant really care about safety and try to reduce risk levels as much as possible.	1	2	3	4	5
7. When I see a potential safety hazard (e.g., oil spill), I am willing to correct it myself if possible.	1	2	3	4	5
8. Management places most of the blame for an accident on the injured employee.	1	2	3	4	5
9. "Near misses" are consistently reported and investigated at our plant.	1	2	3	4	5
10. I am willing to warn my coworkers about working unsafely.	1	2	3	4	5
11. Employees seen behaving unsafely in my department are usually given corrective feedback by their coworkers.	1	2	3	4	5
12. Compared to other plants, I think mine is rather risky.	1	2	3	4	5
13. Working safely is the Number One priority in my plant.	1	2	3	4	5
14. I have received adequate job safety training.	1	2	3	4	5
15. Many first-aid cases in my plant go unreported.	1	2	3	4	5
16. Information needed to work safely is made available to all employees.	1	2	3	4	5
17. Management here seems genuinely interested in reducing injury rates.	1	2	3	4	5
18. Safety audits are conducted regularly in my department to check the use of personal protective equipment.	1	2	3	4	5
19. I know how to do my job safely.	1	2	3	4	5
20. Most employees in my group would not feel comfortable if their work practices were observed and recorded by a coworker.	1	2	3	4	5

Figure 15.4. Questionnaire items designed to measure personal perceptions of organizational safety. (Selected from the Safety Culture Survey developed by Safety Performance Solutions, Inc.)

across different survey items that supposedly measure the same factor, or by comparing two different administrations of the same survey.

If a scale indicated that I weighed 250 pounds on Monday, 249 pounds on Wednesday, and 251 pounds on Saturday of a given week, the scale would get a high reliability rating, even though I really weigh in at about 180 pounds. This scale gave consistent results; it's reliable, but the numbers are invalid.

Validity refers to whether the survey instrument measures what it claims to measure. There are three basic types of validity for a measurement scale, each with particular experimental and statistical methodologies for evaluation. These are *content validity* (Do relevant experts agree that the survey appears to measure what it is supposed to measure?), *criterion validity* (Can scores from the survey be used to predict individual behavior or performance?), and *construct validity* (Are the relationships found with the survey consistent with relevant theory and research?).

In the weight example, the scale looks like it measures the correct weight of a person standing on it (content validity), but if results were compared with readings of other scales, or with results of another estimate of weight, the numbers would not correspond. Construct validity would be questionable. Plus, this weight scale could not predict other variations in individual performance, such as running speed or calories consumed per day, presumed to be influenced by weight. Criterion validity could not be demonstrated.

Criterion validity can be evaluated with two different validity-testing techniques: concurrent and predictive validity. *Concurrent validity* is most frequently used, and refers simply to the relationship between the scale results—in this case my weight in pounds—and another simultaneous assessment of the factor the scale is supposed to measure. This assessment could result from measuring my weight with another scale, or visually estimating my weight.

Predictive validity is a more rigorous test, much more difficult to administer. It refers to the ability of an evaluation tool to predict future behavior. In our example, testing predictive validity requires that the scale results (how much someone weighs) are compared with a future outcome that the scale is purported to predict, such as a person's quickness, general health, or diet. Determining if the results of a perception survey predict the degree of employee involvement in a safety-improvement effort is another example of testing for predictive validity.

The construct validity of a scale is usually evaluated with tests of convergent and divergent validity. *Convergent validity* refers to the extent that other measures of the same construct (for example, a visual estimate of weight) relate to each other. *Divergent validity* indicates the extent that scores from surveys unrelated to the construct *do not* correlate with survey scores related to the construct. In other words, divergent validity implies the extent a particular questionnaire measures special characteristics not measured by other scales.

Regardless of how a person scale is used, whether for teaching, pinpointing problems, or measuring trends or change, it's important to use measurement tools with acceptable levels of reliability and validity. However, if results of a person scale are only used to teach diversity or to measure group change, statistically unacceptable levels of reliability or validity will not cause harm or injustice to someone. This is obviously not the case, however, when a person scale is used to select individuals for a particular job, as some evaluation tools on the safety market purport to do (Burke, 1994; Job Safety Consultant, 1995; Krause & Kamp, 1994). When using a questionnaire to identify individuals, it's critical that prior research with the scale has demonstrated acceptable levels of criterion and construct validity.

WHAT IS ACCEPTABLE? The basic statistic used to measure reliability and validity is a correlation coefficient, which describes the relationship between two sets of survey scores with a number ranging from –1.0 to 1.0. The greater a positive number (between 0 and 1), the greater the direct relationship between measures (a high score on the predictor scale indicates a high score on the criterion). Negative correlations (between 0 and –1.0) indicate an indirect (or inverse) relationship (a high score on the predictor scale indicates a low score on the criterion), and the closer the correlation to –1.0, the greater the inverse relationship.

The closer the correlation to 1 or -1 and the larger the sample, the more confidence one can have that the relationship is true. However, it's important to realize the difference between statistical significance and practical significance. A correlation of .30, for example, would be statistically significant for most statistical tests and sample sizes. But in some cases, this number might not represent practical significance, given that the square of the correlation coefficient indicates the degree of variance overlap between the two measures.

For example, a correlation of .30 between the results of a safety perception survey and other measures of safety, such as employees' frequency of coaching sessions completed or percentages of at-risk behavior per observation period, sounds good until you realize that only nine percent of the variance in one measurement device could be accounted for by the other ($.30^2 = .09$ or 9 percent). In this case, 91 percent of the variance in people's safety perception scores could not be explained by the other estimate of a person's safety.

THE IMPORTANCE OF CONSTRUCT VALIDITY. It's possible a direct relationship between a predictor (such as a measure of accident proneness) and a criterion (such as number of at-risk behaviors or recordable injuries) can be found (predictive validity) without supporting the underlying principle(s) or theory. This would indicate the absence of construct validity. Suppose, for example, an individual could figure out how to answer the survey questions in order to receive a favorable score. Then construct validity would be questionable, even if criterion validity were high.

Every survey I've seen that attempts to assess injury proneness has items which are transparent and enable a respondent to "fake good." Faking good is called impression management in the research literature (Schlenker, 1980; Umstot, 1984), and leads to significant bias in many survey administrations. If a scale to measure injury proneness is used to select individuals for a job, and if the respondents know this, impression management could easily bias the test results. For example, honest risk seekers who confine their thrills to off-the-job free time may be rejected, while deceptive individuals covering up their risk-seeking tendencies may be selected.

The ethics of survey administration require that respondents give their informed consent to be tested, and that they know how their answers will be used. This second ethic, termed "demand characteristics" in the research literature (Orne, 1969), is problematic for an employee selection device with transparent items. A significant relationship between scores on such a survey and other indicators (convergent validity) could reflect principles and theory other than those presumed.

For example, the significant correlation could reflect motivation or intelligence factors rather than actual injury propensity, thus concurrent validity would be shown without construct validity. This kind of hiring survey might select individuals who are most skilled at impression management, rather than less injury prone.

Here's another important point to consider: Research done to test the validity of a survey must occur under the same demand characteristics as the proposed use of the survey. For example, if a safety survey is to be used to select individuals for a particular job, then tests of validity should occur with respondents knowing the survey will be used to select them for a job. This particular demand characteristic is often difficult to pull off in a testing situation; you should check for it when reading the technical manual accompanying a survey used to screen individuals.

I hope you can see that determining an acceptable level of validity is not a straightforward process. It requires you to distinguish between statistical and practical significance, and to carefully evaluate the experimental methodology used to assess validity. Unfortunately, this kind of evaluation requires special training and experience beyond the purview and expertise of most safety professionals. I think it's wise to seek advice from an appropriate consultant—one who has nothing to gain if the target survey is used or not used. A more cost-effective approach is to study the research literature associated with the survey. If there is research published in a peer-reviewed scientific journal, it's likely the survey has passed at least one rigorous test of validity.

If the survey has not been reviewed and accepted by the scientific community, it should not be used to select out individuals for any purpose. There are too many potential biases in using any survey of subjective human factors, and to use a survey without sufficient validity to identify individuals is just too risky, especially for a safety professional. It's far safer to use surveys of person factors to *identify* person characteristics contributing to the increase or decrease of injuries, and to *monitor* the impact of interventions designed to change attitudes, perceptions, intentions, or mood states. Although it might sound good to use a person scale to select safe employees, I urge extreme caution.

SOCIAL VALIDITY. We must consider one final type of validity when evaluating a safety intervention. It comes from researchers and practitioners in behavior-based psychology (Baer *et al*, 1968), and essentially refers to practical significance. It includes using rating scales, interviews, or focus-group discussions to assess: "1. The social significance of *goals*...2. The social appropriateness of the *procedures*...(and) 3. The social importance of the *effects*" (Wolf, 1978, p. 207).

This is more complex than it seems, as I reviewed in a lengthy monograph on social validity (Geller, 1991). There are many perspectives on what makes intervention goals socially significant, procedures socially appropriate, and results socially important. Plus, there are various ways to assess the social validity of an intervention, from unobtrusive behavioral observation to surveys of reactions from those involved in the process.

To understand various perspectives on social validity, I have found it useful to consider the four basic components of an intervention process: selection, implementation, evaluation, and dissemination (Geller, 1989). Selection refers to the importance or priority of the target problem and population addressed. The social validity of selecting workplace and community safety as an intervention target is obvious, given that unintentional injury is responsible for the greatest percentage of years of potential life lost before age 65 (Sleet, 1987).

Assessing the social validity of the implementation stage includes evaluating the goals and procedures of the program plan—how acceptable are they to potential participants and other parties, even those tangentially associated with the intervention (Schwartz & Baer, 1991)? In the case of a corporate safety program, this means obtaining acceptability ratings not only from employees, but also employees' family members and customers of the company. This assessment clearly relates to one of the evaluation recommendations from the panel of experts mentioned earlier in this chapter. That is, are intervention procedures consistent with an organization's values, and do they reach the appropriate audience? One difference is that Drs. Ilene Schwartz and Donald Baer recommend a broader assessment of acceptability from both direct and indirect consumers of the intervention process.

The social validity of the evaluation stage refers, of course, to the impact of the intervention process. This includes estimates of the costs and benefits of an intervention (discussed later in this chapter), as well as measurers of participant or consumer satisfaction. Figure 15.5 depicts the various ways to evaluate program impact. The far-left column of Figure 15.5 lists aspects of a work setting that can be measured before, during, and after implementation of a safety intervention. The order of these characteristics reflects the evaluation hierarchy presented earlier in Figure 15.2. The top items are directly measurable and relate most immediately to the ultimate purpose of a safety intervention—injury prevention. Therefore, improvements in injury-related incidents, behaviors, and environmental conditions would indicate more social validity for the evaluation phase than would beneficial changes in attitudes, perceptions, knowledge, opinions, or program participation.

The center column of Figure 15.5 includes examples of the type of measurement tool or index that can measure the dimensions in the left column. Each of these measurement devices can be classified according to three basic sources of data: a) direct observation, b) archival data (obtained from examining plant documents, memos, and government reports), and c) self report (such as verbal answers to interview questions or written reactions on questionnaires). If the direct observations and archival data are reliable, these measures have greater social validity than self-report measures (Hawkins, 1991).

The third column in Figure 15.5 reflects the scores or numbers obtained from various measurement devices. What are meaningful and useful numbers? Of course, they need to be reliable and valid. But they also need to be understood by the people who use them. If they are not, the evaluation scheme cannot lead to continuous improvement.

Meaningless numbers also limit the dissemination potential and large-scale applicability of an intervention. This is the social validity of the dissemination stage of the process. Next, I want to address confusing characteristics of statistical analysis

WHAT IS MEASURED	HOW IS IT MEASURED?	WHAT IS THE SCORE?
Injury-related Incidents	Near hit reports	Frequency & type of near hits
	Injury reports	Number & type of injury-producing incidents
	Worker compensation costs	Monetary expenditures
Environment	Observation of worksite	Percentage of safe conditions per opportunity
	Housekeeping audit	Percentage of items in proper location
Behavior	Direct observation	Percentage of safe behaviors per opportunity
	Corrective action survey	Number of items corrected for safety
Attitudes Perceptions Person States	Questionnaire	"Safety score" reflecting overall safety attitude, perception or person state
	Interview	Statements of specific and general attitudes about safety
Knowledge	Questionnaire	Percentage correct
	Interview	Statements indicating awareness of a hazard or a safety procedure
Opinions	Questionnaire	Opinion score
Participation	Direct observation Attendance records	Number of participants per opportunity

Figure 15.5. A summary of measurement devices to evaluate intervention impact.

that reduce social validity as it pertains to large-scale acceptance and application of an intervention process.

Cooking Numbers for Evaluation

The issue of using socially valid numbers reminds me of insightful lessons from a good friend and eminent behavioral scientist, Ogden R. Lindsley (Professor Emeritus, University of Kansas). Dr. Lindsley completed his graduate studies at Harvard University with B. F. Skinner and has dedicated most of his creative and prolific research and scholarship to applying the principles and procedures of behavior-based

Figure 15.6. Raw data is cooked before evaluation.

psychology to improving education (Lindsley, 1992). Now he spends considerable time and effort sharing his profound knowledge with corporations, especially regarding the use of behavior-based principles to evaluate organizational change at individual and group levels.

Figure 15.6 illustrates wisdom from Dr. Lindsley that is relevant to our discussion of evaluation. Program evaluators often lose important information from their observations and reduce social validity by "cooking" their "raw data" with complex statistical tests. That is, they transform the numbers from the field into composite scores and test results in order to determine whether the differences (or similarities) they found are statistically significant. Often the outcome of these statistical tests are meaningless to those responsible for improving the process. This is similar to using injury-rate data to try to change safety behaviors.

Computers and software programs make it easy to crunch numbers that might be statistically significant, but all too often they have limited practical utility or social validity. Many of my graduate students, for example, become remarkably skilled at running impressive statistical tests on raw data. They speak eloquently and persuasively about the step-by-step procedures required for a particular statistical technique. But many are stumped by my simple question, "What does it mean? How does your interesting and competent statistical evaluation apply in the real world?" Often the theory or rationale behind statistical results is lost when software programs mix numbers with a formula and churn out pages of computer output.

WHAT DO THE NUMBERS MEAN?

Years ago, my students and I posted our evaluation data daily on large graphs. Posted numbers represented the frequency or percentage of certain target behaviors observed. The baseline data, obtained from observations prior to implementing an intervention, told us the amount and variability of the desired or undesired behaviors we were targeting. This feedback was invaluable when deciding whether to intervene. If intervention was called for, the graphs helped us decide when to act. For example, it's easier to show the clear impact of an intervention if the baseline data is relatively stable at the start of the process.

Posting numbers from daily observations allowed us to monitor the progress of our attempts to improve performance. Sometimes we modified intervention procedures as a result. We all paid attention to these daily numbers, getting a surge of motivation whenever they improved.

When the numbers stabilized, we typically withdrew the behavior-change procedures and noted whether the target behavior(s) reverted to baseline levels. Obviously, many real-world applications of an intervention process do not include a Withdrawal Phase. We implemented this evaluation approach for research purposes. If target behaviors returned to near-baseline levels after the intervention process was removed—which was indeed the case for most of the behavior-change techniques described in Chapters 9 and 10—we had the most convincing demonstration of the intervention's impact (Hersen & Barlow, 1976; Kazdin, 1994).

I'm sure you've noted the similarity between this scheme for intervention evaluation and the DO IT process introduced in Chapter 8. The ***T*** or test phase, of DO IT implies a comparison of the target behaviors before and after the ***I*** or intervention phase, is initiated.

Note also that I've used the past tense to describe this evaluation process. Now my students punch numbers into high-tech computer programs to allow for multiple statistical transformations. Still, I recommend the earlier, straightforward and low-tech approach over our current protocol.

We still record behavioral frequencies or percentages on a session-by-session basis, but these graphs are not processed daily for ongoing feedback and evaluation. Instead, weeks after a study has been completed—and all data has been entered into the computer—we get a final printout. The figures are quite attractive and ready for publication, but I miss the frequent posting of daily observations and the personal reaction and interaction it stimulated. Obviously, we could still post the results of our daily observations and reap the benefits of both evaluation approaches. Several times my students and I have discussed the need to revive the low-tech approach, but other priorities take control over the people empowered to make this happen.

My students feel pressure, actually mandates, from other faculty to conduct sophisticated statistical conversions of research data, so computer processing takes priority. Hopefully, most of you feel no need to "cook" raw data, and will reap the benefits of the low-tech way. You'll be spared the interpretative expertise of a statistical consultant, as depicted in Figure 15.7, plus the raw numbers will be most useful for directing action and motivating continued involvement.

Figure 15.7. Cooked data is often confusing and requires special consultaton.

THE PROBLEM WITH PERCENTAGES. Ogden Lindsley (in press) even advises against one of the simplest calculations—percentages. Percentages are not problematic when comparing static levels of performance across different groups or conditions. But they can be very misleading when monitoring fluctuations in the performance of a single group over several observation sessions. This is because percent change is not symmetrical. If you add 20 percent to a number (like 100) and then subtract 20 percent from the result, you are not back where you started. You'll be below the start point (at 96). In fact, if you start at 100 and then add and subtract 20 percent on ten trials, you'll end up at 66.6—a result well below the starting point.

Another problem with using percentages, of course, is the disregard for sample size. The basketball player who goes to the foul line once and makes one of two free throws is at the same 50-percent effectiveness level as the player who makes 10 of 20 free-throw attempts. But it's impossible to evaluate obvious differences in performance. The player going to the line more often might be more aggressive and valuable to the team effort. The same holds true in safety. Performing one at-risk behavior out of two opportunities seems less problematic in terms of exposure and potential injury than performing 50 at-risk behaviors on 100 trials.

CHANGES IN PERCENTAGES. Another problem is that people can be confused by changes in percentages. Suppose during a coaching observation session you record 60-percent safe behavior on your critical behavior checklist. Then, during subsequent sessions, you observe 70-percent safe behavior. How do you report this increase in safe behavior percentage? Is this a 10 percent increase? I've seen many people report changes in percent this way, because it seems logical. By this logic, however, the percentage increase from 20 percent safe to 30 percent safe is the same as the percentage increase from 60 to 70 percent safe. There is a flaw in such logic, at least as conceptualized by a percent-change transformation. A percent increase from one number to another implies reference to the starting number. Thus, the percentage increase from 20 to 30 is 50 percent, calculated by subtracting 20 from 30 to determine amount of increase, and then dividing 20 (the beginning amount or reference number) into 10 (the amount of increase). By similar logic and calculations, the percentage increase from 60 to 70 is 16.7 percent (10 divided by 60).

Now watch what happens when calculating the percentage *decrease* from the second to first numbers given above. It can get quite confusing when realizing that the decrease percentages are not the same as the increase percentages. For example, the percentage decrease from 30 to 20 is 33.3 percent (10 divided by 30), and the percentage decrease from 70 to 60 is 14.3 percent (10 divided by 70). Of course, the root problem of this confusion is the change in the reference number when going up versus going down. That's why starting at 100 and adding and subtracting 20 percent results in 66.6 after ten trials (Lindsley, in press).

Calculations of percent change are logical and understandable when we keep the reference number (or baseline) in mind. Thus, a 50 percent increase from 100 is 150, and a 50 percent decrease from 100 is 50. Problems can occur if we lose sight of the starting point, or reference number. But some people find it disconcerting that adding 10 to 10 (as in 10 percent safe) comes across as 100 percent improvement, while adding 10 from a starting point of 80 percent safe might get reported as only 12.5 percent improvement. With this percent-change logic, safety excellence rewards are more easily won by organizations that start with the worst safety record.

One approach to handling the confusion of shifting the point of reference when calculating change percentages is to report change in *percentage points*. With this logic, the difference between percentages are reported as an increase or decrease in percentage points with no reference to starting (or baseline) levels. For example, increases in percentage safe behavior from 10 to 20, 50 to 60, and 80 to 90 would all be reported as increases of 10 percentage points. Likewise, changes from the second to first number in each pair would be refereed to as decreases of 10 percentage points.

Is it fair to determine safety improvement awards on the basis of change in percentage points? Or, do you believe it's critical to consider an organization's baseline level at the beginning of the evaluation period? Should the company with the lower baseline, and thus greater opportunity to improve, have an advantage? Actually, the better one's safety record, the more difficult it is to improve. Not only are the percent-change calculations biased against organizations with enviable baseline records, the reality of making a noticeable improvement stacks the deck for companies with the most improvement needed. Now, I hear Dr. Deming (1991, 1992) warning us again to "abolish the ranking of people, departments, and organizations."

AN EXEMPLAR

I think you can see how important it is to realize that even the simplest transformation of raw numbers can add confusion, eliminate instructive information, and detract from constructive feedback. While writing this chapter, I received the display depicted in Figure 15.8 from Kitty Morgan of Paxon Polymer Company in Baton Rouge, Louisiana. She was rightfully proud of the safety-belt use at her plant, and the benefits of unannounced buckle-up checks conducted by the "Paxon Family Safety Council." Indeed, this is an example of the DO IT process, and in this case it might have saved a life. Shortly after leaving the plant following his night shift, a Paxon employee was involved in a crash with an oncoming vehicle traveling without headlights. He sustained only minor injuries because he was buckled up, something he claimed always to do "strictly because of the seat belt inspections" (personal communication from Kitty Morgan, November 3, 1995).

The numbers in Figure 15.8 provide more information than percentages. Paxon employees look at such a display and know exactly where they stand regarding safety belt use. They see how many coworkers are like them, buckled or unbuckled; and they clearly see a dwindling of peer conformity in the at-risk category over four gate checks, from 120 to only 13. It's noteworthy that the first three data points were obtained before the recent safety-belt use law went into effect in Louisiana. Thus, the high rate of voluntary buckling up at Paxon is something to be proud of.

When I received this data, I immediately calculated the buckle-up percentages of 79, 91, 95, and 97 across the four check periods, respectively. My extensive personal history of examining safety-belt use percentages had conditioned me to do this. After completing these calculations and obtaining results I could compare with prior results, I realized the display of raw numbers Kitty Morgan sent me was in fact the most complete and clear way to share the results. Often the simplest approach to an evaluation process has the most social validity.

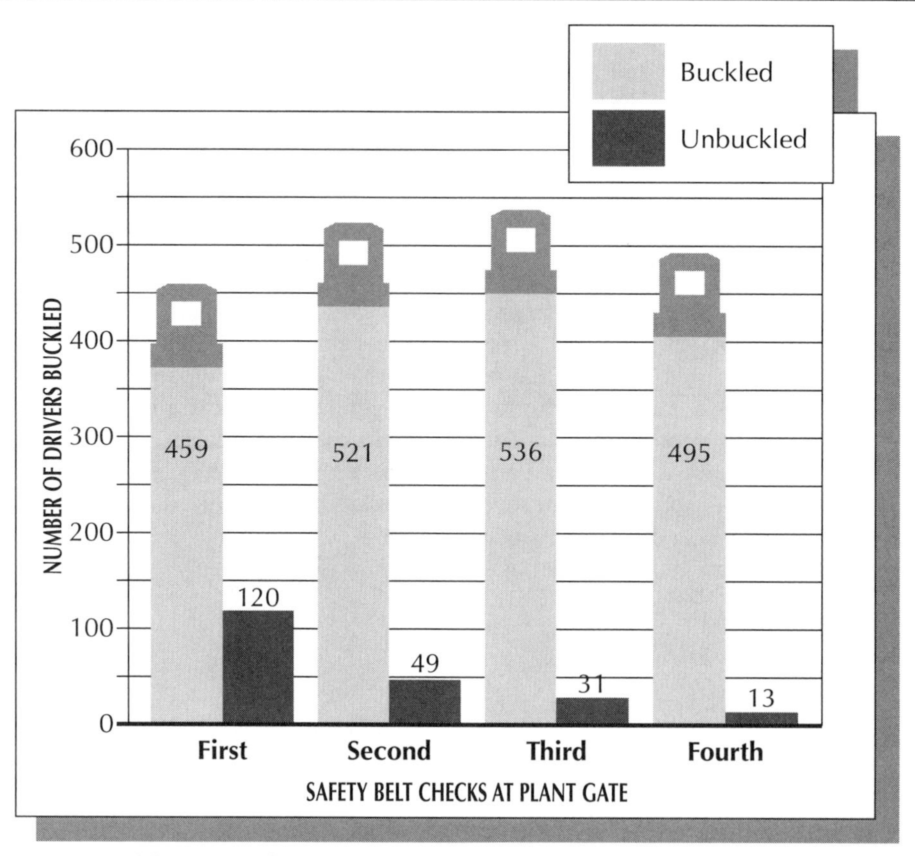

Figure 15.8. The 1995 safety belt use numbers from Paxon Polymer Company, Baton Rouge, Louisiana.

Evaluating Costs and Benefits

The Paxon Family Safety Council can readily justify its safety-belt promotion efforts. The benefits from protecting one employee from serious injury, perhaps a fatality, are clearly greater than the effort and financial costs of the program. For example, it has been conservatively estimated by the National Highway Traffic Safety Administration (1984) that every employee fatality costs industry more than $120,000 in direct payments, property damage, and medical care; and it would take $2,400,000 in sales at a five-percent margin to offset such a loss. Plus, the costs of a nonfatal employee accident can be much greater if, for example, the injury causes permanent disability and requires lengthy rehabilitation. Also, estimating direct costs does not include the expense of hiring and training replacements and associated productivity losses.

The benefits of an occupational safety program are illustrated dramatically by the cost-benefit evaluation of the company-wide safety-belt program initiated at Ford Motor Company (Chapter 8). This corporate-wide program was initiated in the spring of 1984, before any state passed a mandatory seat-belt use law. Safety committees at the Ford plants developed their own safety-belt programs, based on the behavior-change principles described in Part 3 of this text. The combination of activator interventions like awareness sessions, incentives, and commitment pledge cards with rewarding consequences more than tripled safety-belt use among Ford employees. After only one year, this increase saved the lives of at least eight employees, spared about 400 others from serious injury, and reduced corporate costs by ten million dollars. After the second year of increased belt use, the benefits more than doubled, amounting to direct monetary savings of approximately $22 million (Geller, 1985; Gray, 1988).

This cost-benefit analysis was invaluable in sustaining top-management support for the campaign at Ford. It also provided motivating feedback to the many employees who developed, managed or evaluated the plant-based interventions. It's important to maintain records of direct and indirect costs associated with injuries—and with injury prevention—even if these calculations are only estimates. Comparing estimates with the costs of implementing and maintaining particular safety programs illustrates specific benefits and justifies continued program support, especially if you can show that the program has substantially reduced injury frequency and costs.

When calculating program costs, you should document every expense, including: promotional materials, teaching aids, evaluation supplies, rewards, media expenditures, and wages paid for program assistance. Employees' time used to plan, present, evaluate, or participate in the program should be estimated, even if the time is voluntary—for example, on evenings or weekends. If you are comprehensive when calculating program costs, then you are justified in estimating the numerous direct and indirect costs resulting from a job-related injury.

Injury records should be consulted before and after an intervention process is implemented in order to show savings from fewer work-related injuries. Direct costs that should be calculated per injury include:

- Wages paid to absent employees (workers' compensation)
- Property damage
- Medical expenses
- Physical and vocational rehabilitation costs
- Survivor benefits

These direct costs may be the proverbial tip of the iceberg when considering the indirect or hidden costs of business disruptions caused by a lost-time injury. These costs might be difficult or impossible to calculate, but you should try to estimate costs in these categories:

- Overtime pay used to cover the work of an injured employee
- Scheduling work tasks to cover for an injured employee
- Additional administrative hassles, extra wages, training time, and inefficient work associated with temporary replacements
- Special costs of losing a skilled employee
- Extra time from work supervisors to schedule shift changes, temporary replacements, or employee training necessitated by the absence of an injured worker
- Retraining and readjustment for employees returning to work after an extended absence (a permanent disability from the injury may call for a new job assignment)

- Special costs for extensive recruitment procedures and on-the-job-training for permanent replacements for injured employees who don't return to work or who return with a permanent disability
- Special administrative costs to investigate and document the incident and medical treatments for compliance with state workers' compensation laws and with other state and federal regulations, such as OSHA standards

Considering the direct and indirect costs to a company from work-related injury is overwhelming in at least two respects. It's certainly an intimidating chore to estimate these costs, and it's stunning to think that the corporate losses from one employee's injury can be so great. These dollars clearly justify considerable intervention activity. Just anticipating the negative consequences from a work-related injury should motivate support and participation in proactive efforts. This is the first critical step in "selling" safety—the theme of the next chapter.

You Can't Measure Everything

Deming (1991, 1992) condemned grades and performance appraisals because they provide a limited picture of an individual's contributions and potential. And they might constrain the number and type of interventions used to improve the quality of a work culture. If, for example, the only procedures implemented to improve safety are those that allow for objective measurement, the number and quality of safety interventions is severely restricted. In Chapter 14, for example, I discussed a number of ways to increase actively caring behaviors directly, through applications of learning and social influence principles, and indirectly, through improving the five personal states that increase willingness to actively care. It's impractical and impossible to measure the impact of many of these interventions. Should we avoid doing so just because we can't measure their occurrence and impact?

Dr. Deming (1991, 1992) explained there are many things we should do for continuous improvement without attempting to measure their impact. We shouldn't do these things only to influence performance indicators, but because they are the right things to do for people. You might never be able to measure the impact of treating an employee with special respect and dignity, but you do it anyway. Such treatment may, in fact, contribute to achieving a Total Safety Culture but you'll never know it. Likewise, you'll never know how many injuries you prevent with proactively caring behaviors, and you'll never know how much actively caring behavior you'll promote by taking even small steps to increase coworkers' self-esteem, empowerment, and belongingness. You need to continue doing these things anyway. Many things that can't be measured and rewarded still need to get done.

I first realized the fallacy of the common management dictum, "You can't manage what you can't measure" in May, 1991, when attending my first Deming workshop. Each afternoon during the three-day seminar, the 600 participants split into small work groups to discuss various topics. On the second day, the groups were asked to "explain why it is wrong to suppose that if you cannot measure something, you cannot manage it effectively—a costly myth." Except for our facilitator, this statement caused mental conflict or cognitive dissonance (Festinger, 1957) for all of us.

Each of us had been living by the manage-by-measurement philosophy and believed in the opening quote of this chapter.

We entertained the points reviewed earlier in this chapter, that:

- Outcome measures are usually imperfect and deceptive estimates of critical process activities
- Outcome measures remotely connected to process activities provide minimal if any useful direction and motivation
- There are numerous immeasurable things we need to do on a regular basis to help people optimize their system
- We cannot expect external rewards and recognition for most of the important management-related behaviors we need to do: In other words, we cannot measure everything for which we need to be accountable; we need to start with the right vision, theory, and principles and hold ourselves accountable with internal consequences

Our discussion group did, however, agree that we should try to develop objective, process-based measures for our quality—or safety—objectives. Although we cannot measure every important process directly, defining and tracking desired actions or behaviors guide proper procedures and motivate continuous improvement. In other words, the quote from Hansen (1994) at the start of this chapter is accurate, but it doesn't say it all. Many factors affect performance. Not only is it impossible to monitor all of them, it's often impossible to identify the specific change in performance that led to an improved system.

In Conclusion

At the start of this text, I explained the fallacy of basing decisions on common sense. Rather than adopt intervention programs that sound good, we need to use procedures that work. But how do we know what works? Of course, you know the answer to this question. Only through rigorous program evaluation can we know whether an intervention is worth pursuing. Now comes the more difficult question. What kind of program evaluation is most appropriate for a particular situation?

Actually, every chapter of this text has addressed program evaluation in one way or another. Early on I explained the need for achievement-oriented methods to keep score of your safety efforts. This enables people to consider safety in the same work-to-achieve context as production and quality. This implies, of course, the need for program evaluation numbers people can understand and learn from. This is how evaluation leads to continuous improvement.

Throughout this text, I have referred to published research in order to justify psychological principles or recommendations for intervention procedures. Information presented is founded on rigorous evaluation, not common sense. Evaluation techniques used in published research are indeed more rigorous and complex in terms of reliability, validity, and statistical analyses than those needed for continuous improvement of real-world safety programs. But the basic principles and issues presented in this chapter are relevant to both researchers (seeking to contribute to professional scholarship) and practitioners (seeking continuous improvement of an intervention process).

To publish their findings, researchers need to demonstrate reliability and validity of their measures and find statistical significance. However, they can and do ignore several evaluation principles presented in this chapter. For example, their measures typically: a) target only one dimension (environment, behavior, or person factors); b) are short-term (applied for a limited number of experimental phases); c) are subjected to statistical transformations and analyses that take substantial time to complete and are not readily understood by the average person; and d) often do not include a cost-benefit analysis. You see, reports of their procedures and results only need to be understood and appreciated by a select, often esoteric, group of professionals who specialize in the particular issue or problem addressed by the research.

But you can't overlook most of the basic principles when evaluating practical interventions to achieve continuous improvement. Data collection procedures and statistical analyses often can be less rigorous. But safety practitioners need to address several important issues often bypassed by professional researchers. Specifically, they need to: a) define the level of performance targeted by the intervention, while appreciating limitations in attacking individual versus organizational performance; b) use measures for the three dimensions of safety improvement—environment, behavior, and person factors; c) apply process measures periodically over the long term, especially checks on environmental conditions and work practices; d) include a cost-benefit analysis to justify continued intervention and evaluation efforts; and e) keep score with numbers that are both meaningful to all program participants and provide direction to refine interventions. These last two principles are critical to meet the challenge addressed in the next chapter—obtaining and maintaining support for an effective intervention process.

REFERENCES

Abramson, L. V., Metalsky, G. I., & Alloy, L. B. (1989). Hopelessness depression: A theory-based subtype. *Psychological Review, 96*, 358-372.

Baer, D. M., Wolf, M. M., & Risley, T. (1968). Some current dimensions of applied behavior analysis. *Journal of Applied Behavior Analysis*, 1, 91-97.

Burke, A. (1994). Putting job candidates to the safety test. *Industrial Safety and Hygiene News*, 28(4), 19-20.

Deming, W. E. (1986, July). *Drastic changes for western management.* Abstract for the meeting of TIMS/ORS at Gold Coast City, Australia.

———. (1991, May). *Quality, productivity, and competitive position.* Four-day workshop presented in Cincinnati, Ohio by Quality Enhancement Seminars, Inc., Los Angeles, CA.

———. (1992, January). *Instituting Dr. Deming's methods for management of productivity and quality.* Two-day workshop presented in Washington, D. C. by Quality Enhancement Seminars, Inc.

Dollard, J., Doob, L., Miller, N., Mowrer, O. H., & Sears, R. R. (1939). *Frustration and aggression.* New Haven: Yale University Press.

Festinger, L. (1957). *A theory of cognitive dissonance.* Evanston, IL: Row, Peterson.

Geller, E. S. (1985). *Corporate safety belt programs.* Blacksburg, VA: Virginia Polytechnic Institute and State University.

———. (1989a). Applied behavior analysis and social marketing: An integration for environmental preservation. *Journal of Social Issues*, 45, 17-36.

———. (1989b). Managing occupational safety in the auto industry. *Journal of Organizational Behavior Management,* 10(1), 181-185.

———. (1991) (Eds.). *Social validity: Multiple perspectives,* Monograph Number 5. Lawrence, KS: Society for the Experimental Analysis of Behavior, Inc.

———. (1992). *A critical review of human dimension presentations given at the 1992 National Safety Council Congress and Exposition.* Technical Report for Exxon Chemical Americas. Blacksburg, VA: Virginia Polytechnic Institute and State University.

———. (1994). What's in a perception survey? *Industrial Safety and Hygiene News,* 28(11), 11-12.

Gray, D. A. (1988, October). Introduction to invited address by E. S. Geller at the annual National Safety Council Congress and Exposition, Orlando, FL.

Hansen, L. (1994). Rate your B.O.S.S.—Benchmarking organizational safety strategy. *Professional Safety,* 39(6), 37-43.

Hawkins, R. P. (1991). Is social validity what we are interested in? Argument for a functional approach. *Journal of Applied Behavior Analysis,* 24, 240-213.

Hersen, M., & Barlow, D. H. (1976). *Single case experimental designs: Strategies for studying behavior change.* New York: Pergamon Press.

Job Safety Consultant (1995). Hiring safety-conscious employees—Can it be done? Is it legal? August, Issue 267, pp. 1, 6, 7.

Kazdin, A. E. (1994). *Behavior modification in applied settings* (Fifth Edition). Pacific Grove, CA: Brooks/Cole.

Krause, T. R., & Kamp, J. (1994). Viewpoint: Selecting safe employees. *Industrial Safety and Hygiene News,* 28(10), p.25.

Lindsley, O. R. (1992). Precision teaching: Discoveries and effects. *Journal of Applied Behavior Analysis,* 25, 51-57.

———. (in press). Performance is easy to monitor and hard to measure. In R. Kaufman, S. Thiagarajan, & P. MacGillis (Eds.), *Handbook of human performance systems.* San Diego, CA: University Associates.

Myers, I. B., & McCaulley, M. H. (1985). *Manual: A guide to the development and use of the Myers-Briggs Type Indicator.* Palo Alto, CA: Consulting Psychologists Press.

National Highway Traffic Safety Administration (1984). *The profit in safety belts: A handbook for employees,* DOT HS 806 443, Washington, DC: The U.S. Department of Transportation.

Orne, M. T. (1969). Demand characteristics and the concept of quasi-controls. In R. Rosenthal & R. Rosnow (Eds.), *Artifact in behavior research.* New York: Academic Press.

Schlenker, B. R. (1980). *Impression management.* Monterey, CA: Brooks/Cole.

Schwartz, I. S., & Baer, D. M. (1991). Social validity assessments: Is current practice state of the art? *Journal of Applied Behavior Analysis,* 24, 189-204.

Seligman, M. E. P. (1975). *Helplessness: On depression, development, and death.* San Francisco: W. H. Freeman.

Sleet, D. A. (1987). Motor vehicle trauma and safety belt use in the context of public health priorities. *The Journal of Trauma,* 27, 695-702.

Steel, S., (1996) (Eds.). *Evaluating the effectiveness of agricultural safety and health initiatives.* Itasca, IL: The National Safety Council.

Storey, C. (1989). Excellence: It's in the cards. *Training & Development Journal,* September, 46-48.

Umstot, D. D. (1984). *Understanding organizational behavior.* St. Paul, MN: West Publishing Company.

Vojtecky, M. A., & Schmitz, M. F. (1986). Program evaluation and health and safety training. *Journal of Safety Research,* 17, 57-63.

Willems, E. P. (1974). Behavioral technology and behavioral ecology. *Journal of Applied Behavior Analysis,* 7, 151-165.

Wolf, M. M. (1978). Social validity: The case for subjective measurement or how behavior analysis is finding its heart. *Journal of Applied Behavior Analysis,* 11, 203-214.

Chapter 16
Obtaining and Maintaining Involvement

You can't effectively put to use the principles in this book without sustained support from both managers and employees. This chapter focuses on ways to initiate and maintain that support, including ways to promote leadership, build commitment and involvement, expand the scope of interventions, and reduce active resistance.

Culture change is never quick, never easy. The "quick-fix" illustrated in Figure 16.1 is clearly ridiculous. As absurd as this notion is, it comes to us naturally. We want speedy solutions to difficult challenges. And it's easy to lose patience, enthusiasm, and optimism along the way. After all, our society demands immediate gratification—just look at all the movies and television shows that begin with dramatic problems and come to happy endings within 30 to 90 minutes. Plus, the faster we solve any problem, the sooner we experience rewarding consequences.

Figure 16.1. There's no quick fix solution to culture change.

Of course, we know the insinuation depicted in Figure 16.1 is preposterous. Common sense tells us, correctly this time, that fairy-tale solutions are utopian and unrealistic. It makes much more sense to plan and work for the small wins that come from SMART goals (Chapter 10). We need to remember the well-known answer to the silly question, "How do you eat an elephant?" One bite at a time!

This chapter brings us to the point where we start pulling things together. I want to discuss the broad challenge of initiating a culture-change process aimed at achieving a Total Safety Culture. First come general guidelines for starting a process and maintaining support. Then, I address concepts of leadership, communication, and resistance. You won't find step-by-step cookbook procedures here; a generic recipe is just not available. Instead, take the principles and procedures presented in prior chapters, add the information found here, and you'll be well on your way to an innovative experience in safety improvement.

Starting the Process

MANAGEMENT SUPPORT

You can't do without it. How many times have you heard: "Whatever management really pushes and supports will happen." And, implicitly, "Whatever upper management does not push and does not support will fail." If managers emphasize housekeeping, quality control or cost-reduction, improvements in these areas are likely to follow. Strong top-down support, involvement, and commitment alone won't make a campaign succeed, but they are essential ingredients. Plus, management and labor must collaborate to make the process work.

CREATING A SAFETY STEERING COMMITTEE

A Safety Steering Committee (SSC) plays a critical role in developing a Total Safety Culture, providing policy-making, oversight, and general support. All this is simply more than any one person can handle. At the start, there are at least two functions that the commitee serves:

Content function: groups typically produce better ideas, and more creative problem solutions than individuals working alone

Process function: group-based solutions or decisions lead to more commitment and enthusiastic involvement than individual solutions or decisions handed down

Before creating a Safety Steering Committee, it's important to look at the existing committee structure in your organization. You don't want to duplicate the efforts of the safety department, or some other relevant standing committee. For example, a current employee committee might be able to take on the responsibility of coordinating efforts to achieve a Total Safety Culture.

Careful planning is needed to determine:

- What is the mission of the Safety Steering Committee?
- What is the vision for SSC operation?
- What are the group's limitations or restrictions?
- What are the priorities?
- Who should be on it?

DEVELOPING EVALUATION PROCEDURES

"Did it do any good?" This is the central question that the Safety Steering Committee must be prepared to answer about any intervention. Chapter 15 offered guidelines for evaluating impact. At the start of an evaluation process, these questions need to be answered:

- What indicators should we look at—behaviors, attitudes, opinions, or outcomes?
- When should we measure?
- What types of data should we collect and analyze?
- What is the cost of this evaluation process?
- How should we summarize and display results?

SETTING UP AN EDUCATION AND TRAINING PROCESS[5]

Ensuring that employees learn key principles and procedures to improve safety is a major responsibility for the committee. At minimum, the following elements should be incorporated into planning an effective education and training program:[6]

- Develop education content and procedures
- Plan the education and training process
- Plan for follow-up sessions
- Identify and prepare instructors
- Measure the impact of the program

Let's discuss each of these elements in a bit more detail, keeping in mind that my suggestions need to be customized at the plant level to get the most "bang for your buck."

IDENTIFYING AND PREPARING INSTRUCTORS. Selecting the "right" instructors is critical because teaching is at the heart of effective education and training. If the teachers do a poor job, they undermine other training tools such as videos and booklets. You should consider these factors when choosing instructors:

5. I distinguish between educating and training, with *educating* referring to teaching the principles or theory behind a process or set of procedures and training referring to teaching techniques or procedures with hands-on experience and behavioral feedback.

6. A process is a long-term continuous effort, mission, or set of procedures which might include specific programs with a beginning and end. Thus, a training process might include a particular training program that is refined or updated on a regular basis.

- Prior experience in educating or training
- Current level of teaching ability (aptitude and achievement)
- Credibility with employees to be educated
- Level of motivation and interest in doing the instruction
- Prior familiarity with psychology, especially behavior-based principles
- Belief that the principles and procedures can help achieve a Total Safety Culture

To help selected teachers prepare for their task, they will need to: a) understand the principles and relevant procedures in this text so they can represent them accurately to the participants; b) feel comfortable with the specific process of the plant-wide education and training which, ideally, they will help to develop; c) practice basic communication skills; d) demonstrate leadership at meetings; and e) learn how to facilitate discussions—particularly if they have little experience at stand-up training and small-group meetings.

In-house training staff can help volunteers build their teaching skills. These topics are often covered in various corporate training programs. Plus, instructional programs are readily available from outside vendors or consultants.[7]

DEVELOPING TRAINING CONTENT. When considering the specific topics to include in each instructional session, you need to address these points:

- What are the specific goals of a particular session—for example, to teach principles, train procedures, or build commitment and motivation?
- In addition to this text, what sources are available for relevant content and support, such as local case studies of behavior-based safety?
- What relevant films, videotapes, and other instructional materials are available?[8]
- To meet session goals, what key content points should be covered, and in what order?
- What specific plant facts, statistics, and case studies can be incorporated into the session?
- How much information can be covered effectively in one session?

Involve your selected instructors as much as possible in developing the specific education and training plan.

PLANNING THE INSTRUCTIONAL PROCESS. "Everything was covered but nobody paid much attention." This is a common complaint about education or training.

7. My associates at Safety Performance Solutions (SPS) provide training and guide organizations through the entire process reviewed here. For more information write Safety Performance Solutions, 1007 N. Main Street, Blacksburg, VA 24060 or call (540) 951-SAFE (7233).

8. A videotape training series (including four videotapes, workbooks, and facilitator guides) has recently been developed from the information presented in this text. The series is called "Actively Caring for Safety," with the four separate programs entitled: 1) Motivating Safe Behavior, 2) Implementing Behavior-Based Safety, 3) Coaching Safe Behavior, and 4) Making Safety Incentives Work. For more information write Tel-A-Train, Inc., PO Box 4752, Chattanooga, TN 37405 or call 1-800-251-6018.

Translation: Good content is important, but not sufficient. You've got to "package" your content and present it in a way that hits home with participants. Obviously, you want them to practice what is preached. Most instructors know their material; the challenge lies in conveying that knowledge. How do you get your message across? Here are some points to consider:

- An interactive/participative approach is typically more effective than a "top down" lecture coming from the podium
- Like a good pitcher, change speeds. Don't rely on one pitch, one way of presenting information
- It is easier to involve small groups of participants than large ones
- Regardless of the main objective of a session, some initial awareness raising makes participants more receptive to the content
- Integrate demonstrations into the program
- If possible, have participants practice the skill taught, with appropriate feedback in the classroom or on the job
- Resolve those administrative questions, including where and when the instructional sessions should be held, and how long they should take

EVALUATING EFFECTIVENESS. As covered in Chapter 15, you can't overlook measurement. At this point in the overall intervention process, you need to determine impact of education and training on participants' knowledge, skill, and attitude.

Knowledge of content can be assessed the old-fashioned way, through written tests given at the end of a session. Skills can be evaluated by systematic behavioral observation in the classroom or out in the workplace. Participants' reactions to the session can be measured with brief questionnaires. For this you should keep in mind:

- Brevity
- Choice and complexity of wording
- Combining objective rating and written comments

Figure 16.2 depicts a simple evaluation form to assess reactions to an instructional session. Distributing a form like this shows that teachers care about improving the sessions. Often you'll find that participants give useful information to develop action plans for implementing principles. The form also solicits suggestions for improving subsequent sessions. Consider using the form shown in Figure 16.2, or a refinement, to help along your education and training process.

SUSTAINING THE PROCESS

CONTINUED UPPER-MANAGEMENT SUPPORT. Many safety programs get a big send-off, only to drop off the radar screen. Then it's "out of sight, out of mind" as some new program is pushed. This is why safety is often derided for its "flavor of the month" approach. So how do we sustain a safety process? (Note that I prefer the concept of "process" to "program" here because processes flow on while programs begin and end.) First and foremost we need continued and visible support from top manage-

Training Evaluation

Please evaluate this training session along the characteristics listed below (circle the number corresponding to your answer).

Effectiveness of the presentations:

1	2	3	4
Ineffective	Somewhat Effective	Definitely Effective	Highly effective

Satisfaction with what you learned:

1	2	3	4
Not Satisfied	Somewhat Satisfied	Definitely Satisfied	Highly Satisfied

Usefulness of material presented:

1	2	3	4
Not Useful	Somewhat Useful	Definitely Useful	Highly Useful

The training session was:

1	2	3
Too Long	About Right	Too Short

If too long or short, what would you add, delete or change?

- From your viewpoint, what are the most significant principles or procedures you learned from this session?
- What needs to happen over the course of the next six months to implement these principles or procedures?
- What would you like to learn next about the "Psychology of Safety"?
- Is there anything else you would like to tell us?

Figure 16.2. Questionnaire to Evaluate Safety Training Session.

ment. If management endorses the process on an ongoing basis, it can become integrated into normal plant operations.

The Safety Steering Commitee needs to work hard to convince managers that their commitment is fundamental to the process—not only to get it going, but to keep it going. Here are some thoughts on maintaining that all-important backing of top managers:

First, you need to gain access to upper management. Identify a manager to champion your cause in the executive offices. Bring him or her into the loop, ask them to attend all committee meetings.

Keep managers informed. Submit committee reports to them on a regular basis. But don't overwhelm busy managers with minutiae.

Ask for input, keep managers involved. Solicit comments and "concerns" about the process you have underway. Of course you should be doing this with all levels of the organization to create a top-to-bottom sense of ownership.

Promote and market your efforts. Publish articles and announcements about the safety process in the employee newsletter on a regular basis.

Keep at it. Identify the benefits of your process and continue to "sell" them to upper management.

AWARENESS SUPPORTS: ACTIVATORS

Reminders of various kinds can be extremely helpful to keep managers, supervisors, and employees aware and involved in your safety process. Vary these support devices and by all means keep them coming. In addition to management commitment and activity, other sustaining process drivers are:

- Environmental supports, such as signs, slogans, newsletters, articles
- Incentive supports, especially publicity about reward strategies
- Promise cards to pledge behavioral commitment
- Social support at group meetings and social gatherings

PERFORMANCE FEEDBACK: CONSEQUENCES

As you know, a cornerstone of the DO IT approach to motivating behavior change is the principle that behavior is influenced by its consequences. But in most plants there are no clear-cut consequences for performing safe acts. Sometimes there are actually subtle rewards for *failing* to follow proper safety procedures. Employees might save time and avoid discomfort by not wearing appropriate protection, for instance. As detailed in Chapter 11, *rewarding* feedback should be given following safe behaviors, and *correcting* feedback should be given when at-risk behaviors are observed.

Effective performance feedback is, of course, closely linked to measurement and evaluation—the "test" phase of DO IT. As with measurement, performance feedback should be given systematically and consistently. Ideally, feedback regarding safe behavior should occur at several levels, including plant-wide (for overall organizational performance), across shifts or work groups (for group performance), and individually (for personal performance).

Finally, it's important to acknowledge supervisors and others for supporting a safety achievement effort. As discussed in Chapter 11, safety coaches must feel appreciated for their actively caring efforts. They need to be recognized for their competence at giving supportive and corrective feedback to others.

Without clear behavioral consequences, performance tends to drift (Hayes *et al*, 1980), usually in the direction of minimal effort. If you want to sustain the energy it takes to be a safety coach, you need to recognize their work. You have to convince top managers that coaches need their support, as well.

TANGIBLE CONSEQUENCES. Performance-contingent consequences can go beyond just saying "good job" or "you need to do better." The use of tangible rewards for desirable behaviors is fundamental to changing behaviors. The advantages of incentive strategies over the disincentive approach are detailed in Chapter 10. That chapter also describes procedures for implementing an incentive/reward program. At this point, it's important to consider these general questions about tangible reward interventions:

- Should rewards, penalties, or both be used?
- What specific events or items should be used as rewards or penalties?
- Should the rewards be individual or group-based, or both?
- Who should administer tangible consequences?
- How often should these consequences be available?

ONGOING MEASUREMENT AND EVALUATION

You can't maintain and improve a safety process unless you regularly measure its impact. As discussed in Chapters 10 and 11, even the mere act of tracking a given set of behaviors usually improves the performance of those behaviors. Performance tracking should be done continously, and questions to be resolved include:

- Who conducts ongoing measurement?
- What will be measured?
- How frequently should the program's impact be measured?
- What types of evaluation data should be taken (for example, frequency counts or percentages)?
- How will data be summarized and tabulated?
- What are the costs of the evaluation?
- What are the benefits—real and potential?

FOLLOW-UP INSTRUCTION/BOOSTER SESSIONS

Even with ongoing support, a comprehensive safety achievement process cannot succeed without carefully planned follow-up instruction. From time to time education and training content must be updated to reflect changes in plant conditions, the use of new machines or protective equipment, and the like. Don't delay in keeping pace with change; what is being taught should correspond exactly to current plant conditions. Follow-up education and training includes these issues:

- When and how should basic instruction be repeated for new employees?
- What are the objectives of follow-up instruction?
- How should new material be integrated?
- How often should follow-up sessions occur?
- What should be the content?
- How should the material be presented—videotapes, lecture, on-the-job training, discussion groups?
- How can monthly safety talks be used as boosters or activators?

INVOLVEMENT OF CONTRACTORS

Let's pause to consider the influence of those outside our process, specifically contractors. Noncompliance by outside contractors can undermine our efforts by demonstrating dangerous behavior to plant personnel—in addition to possibly compromising the safety of everyone on site.

Are outside personnel exempt from wearing safety glasses? Exempt from wearing hard hats and steel-toed shoes in construction areas? From following speed limits posted on plant property? If so, you've probably heard one of your employees complain: "If they don't follow safe work practices, why should we?"

Here are some strategies for getting outside personnel to understand and comply with your safety procedures:

- Request information pertinent to safety-related issues in the bidding process
- Conduct safety meetings and behavior-based instruction for contractors
- Obtain full cooperation/commitment of the contractor to follow local safety practices
- Provide verbal instructions and feedback (rewarding and correcting) to contractors
- Enable and support safety coaching of contractors by regular plant employees, and vice versa
- Include contractors in DO IT and reward/recognition programs.

TROUBLESHOOTING AND FINE-TUNING

Once a safety achievement process is up and running, the Safety Steering Commitee confronts the responsibility of fine-tuning the procedures. This is based on ongoing evaluations. If the process is going to be sustained, employees and managers must perceive it as current—state of the art—in terms of content, and also adaptable and responsive. It simply cannot be "frozen," nor left unattended. You can't "wind up" a process at the start and expect it to run forever like that battery-powered bunny. Some keys to fine-tuning include:

Discuss the impact: Is it working? What do the data from participants' reaction sheets, as well as other measures, tell us?

Identify strengths and weaknesses: Based on the data, which elements should be kept, changed (and how), or replaced?

Cope with change: Be sure all those affected by the process are fully informed about changes when they occur. Ideally, all participants should be actively involved in troubleshooting and fine-tuning interventions.

Cultivating Continuous Support

Starting a safety achievement process and maintaining it for long-term improvement requires the three essential support processes depicted in Figure 16.3. Leaders are needed to champion new principles and procedures. In fact, leadership makes the difference between a "flavor of the month" safety initiative and a long-term continuous improvement process.

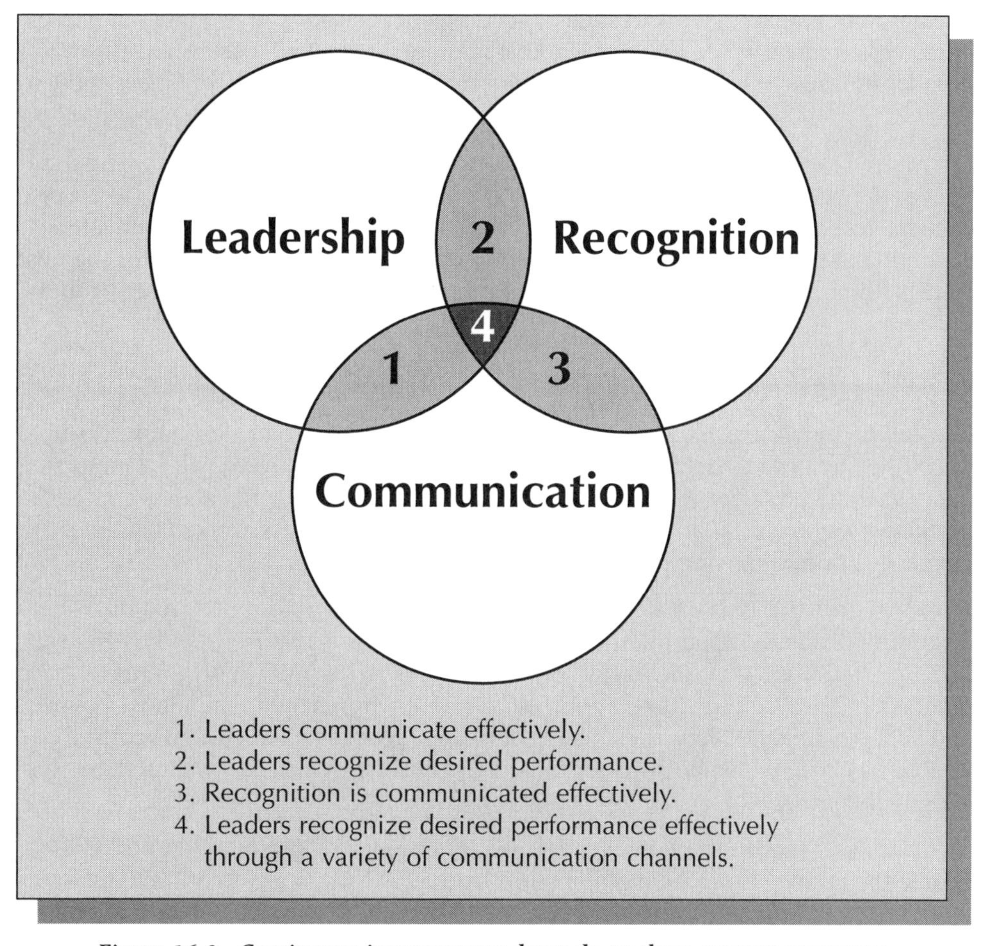

Figure 16.3. Continuous improvement depends on three support processes.

My colleagues and I at Safety Performance Solutions have seen the principles and procedures presented in this book lead to remarkable success and eventually a Total Safety Culture. But all too often we've seen good intentions and superb introductory instruction fizzle out and go nowhere. Why? It's a matter of leadership. You can launch a process with excellent education and training, but you can't keep the momentum going without individuals who provide energy, enthusiasm, and the right example. This section covers some essentials of effective leadership.

WHERE ARE THE SAFETY LEADERS?

First, we have to find the leaders. Who are they? The traditional definition of one person exerting influence over a group doesn't quite work for safety. Ask any safety manager who has been expected to do it all. To achieve a Total Safety Culture, everyone needs to accept a leadership role in reducing injuries. Everyone feels responsible for safety and goes beyond the call of duty to protect others. This requires leadership skills: giving supportive feedback for another person's safe behavior and constructive feedback for at-risk behavior.

Psychologists have studied leadership rigorously for over half a century in an attempt to define the traits and styles of good leaders (Yukl, 1989). Still, many questions remain unanswered, making leadership more an art than a science. But several decades of research has turned up some important answers, which we'll now apply to safety.

Many psychologists consider the characteristics that define leaders to be permanent and inborn personality traits (Kirkpatrick & Locke, 1991), but I prefer to consider them response styles or personality states that can be taught and cultivated. If action plans or interventions can be developed to promote styles typical of the best leaders, then the number of effective safety leaders in an organization can be increased.

PASSION. The most successful leaders show energy, desire, passion, enthusiasm, and constant ambition to achieve. Passion to achieve a Total Safety Culture can be fueled by clarifying goals and tracking progress. Put a positive spin on safety, make it something to be achieved—not losses to be controlled. Then employees will be motivated to achieve shared safety goals just like they work toward production and quality goals. Marking progress leads to the genuine belief that the process works. This fires up employees to continue the process.

HONESTY AND INTEGRITY. Effective leaders are open and trustworthy. A Total Safety Culture depends on open interpersonal feedback. This obviously requires honesty, integrity, and trust. It's quite useful for work groups to discuss ways to nurture these qualities in their culture. Take a look at certain environmental conditions, policies, and behaviors. Some arouse suspicions of hidden agendas, politics, and selfish aims. You can work to eliminate some of these trust-busters by first identifying them, discussing their purpose, and devising alternatives.

MOTIVATION. Since most people really care about reducing personal injuries, even to people they don't know, the motivation to lead others will spread naturally throughout a work culture when people believe they can have personal control over injuries. This occurs when they learn effective techniques to prevent injuries (as presented in Part Three of this text) and feel empowered to apply them (as covered in Part Four).

Self-confidence. Effective leaders trust in their own abilities to achieve (Baron, 1995; Kirkpatrick & Locke, 1991). Education helps convince people they can achieve, but they need ongoing support and recognition for their efforts. For example, the self-confidence needed to give safety feedback can be initiated with appropriate education and training, and can be maintained with coaching, communication, and recognition. Notice that enhancing the three empowerment factors—self-efficacy, personal control, and optimism—as discussed in Chapter 14, builds self-confidence.

Thinking skills. Successful leaders can integrate large amounts of information, interpret it objectively and coherently, and act decisively as a result (Baron, 1995; Kirkpatrick & Locke, 1991). Constructive thinking skills evolve among team members when objective data is collected on the progress of safety interventions and used to refine or expand these processes and develop new ones.

When teams work through the DO IT process (as presented in Part Three), participants develop skills to evaluate behavioral data and use the information to make intervention decisions. This is basic scientific thinking, the key to substituting profound knowledge for common sense. This critical thinking leads to special expertise.

Expertise. To achieve a Total Safety Culture, everyone needs to understand the principles behind policies, rules, and interventions to improve safety. When employees teach these principles to co-workers, they develop the level of profound knowledge and expertise needed for exemplary leadership.

Flexibility. Successful leaders size up a situation, and adjust their style accordingly (Hersey & Blanchard, 1982). At times some groups and circumstances call for firm direction—an autocratic style. At other times, the same people might work better under a nondirective, hands-off approach—a democratic style. The best leaders are good at assessing people and situations, and then matching their behavior to fit the need (Zaccaro *et al,* 1991).

Tuckman and Jensen (1977) noted that work groups or teams progress through four development stages: forming, storming, norming, and performing. During the early stages of forming and storming, there is a need for structure, clear vision, and a sense of mission. Autocratic or directive leadership is often most appropriate—though it's good to get input from group members before the first meeting. When group members become familiar with each other and start implementing their assignments, the norming and performing stages, democratic leadership is usually called for.

In summary. So here we have seven personal qualities that distinguish leaders from followers. It's common to consider these characteristics permanent inborn traits. But a more optimistic perspective, one helpful to achieving a Total Safety Culture, is that employees can nurture these ideals in themselves and others. There is no quick and easy way to to do this. But the principles reviewed here should help you discriminate between action plans that facilitate versus inhibit safety leadership. And people already possessing leadership qualities must help bring them out in others.

COMMUNICATE TO SELL THE PROCESS

How we talk about safety influences whether people will contribute their leadership skills to a safety process. Indeed our language can determine acceptance or rejection of the process.

> *Words are magical in the way they affect the minds of those who use them...words have power to mold men's thinking, to canalize their feelings, to direct their willing and acting.*

This quote from Aldous Huxley's "Words and Their Meanings" (Hayakawa, 1978, p.2), reflects the power of words to shape our feelings, expectancies, attitudes and behaviors. When people use expressions like, "Say that enough times and you'll start to believe it," "Can't I talk you into it," or "Do as I say, not as I do," they acknowledge the influence of words on behavior.

Figure 16.4. Language can influence attitude and behavior.

Words also affect feeling states. Years ago, when my two daughters discussed horse manure at the dinner table, I lost my appetite. Likewise, using negative, uninspiring words to describe ourselves or everyday events can contribute to losing our appetite or passion for daily life.

Figure 16.4 illustrates how a simple word change can influence a feeling state, and then more behavior. The scenario might seem far-fetched at first, but it's really not if you consider that the child is responding to the parents' reactions. Initially, mom and dad are unhappy with the word "crib" and their body language contributes to the child's negative reactions. However, the change of language made the parents happier. This is perceived by the child and leads to his positive reaction.

What does this have to do with safety? I think some words we use in the safety and health field are counter-productive. Let me point out a few that should be eliminated from our everyday language if we want to "sell" safety and increase involvement to prevent injury.

"ACCIDENT" IMPLIES "CHANCE." The first definition of "accident" in my *New Merriam-Webster Dictionary* (1989) is "an event occurring by chance or unintentionally" (p. 23). Workplace accidents are usually unintentional, of course, but are they truly chance occurrences? There are usually specific controllable factors, such as changes in the environment, behaviors, and/or attitudes, that can prevent "accidents."

We want to develop the belief and expectation in our work culture that injuries can be prevented by controlling certain factors. "Accident," then, is the wrong word to use when referring to unintentional injuries. It reduces the number of people who believe with true conviction that personal injuries can be prevented.

RESTRAINTS DON'T INVITE USE. For almost two decades I've been urging transportation and safety professionals to stop using the terms "occupant restraints" and "child restraints" for vehicle safety belts and child safety seats. These terms imply discomfort and lack of personal control, and fail to convey the devices' true function. "Seat belt" is better than "occupant restraint," but this popular term is not really adequate because it doesn't describe the function or appearance of today's lap-and-shoulder belts. We need to get into the habit of saying "safety belt" and "child safety device."

PRIORITY OR VALUE. Priority implies importance and a sense of urgency, and safety professionals are often quick to say "safety should be a priority." This seems appropriate, since my *New Merriam-Webster Dictionary* defines "priority" as "taking precedence logically or in importance" (p. 577). But everyday experience teaches us that priorities come and go. Depending upon the demands of the moment, one priority often gets shifted for another. Do we really want to put safety on such shifting ground?

I believe a Total Safety Culture requires safety to be accepted as a value. The relevant definition of "value" in my *New Merriam-Webster Dictionary* is "something (as a principle or ideal) intrinsically valuable or desirable" (p. 800). Safety should be a "value" that employees bring to every job, regardless of the ongoing priorities or task requirements.

DON'T SAY "BEHAVIOR MODIFICATION." Over recent years I've seen "behavior modification" used many times for titles of safety presentations at regional and national conferences. I've heard trainers, consultants, and employees use the term to describe behavior-based safety. In fact, I've often been introduced at conferences as a specialist in "behavior modification." This is the wrong choice of words to use if we want acceptance and involvement from the folks who are to be "modified." Who wants to be "modified"?

This lesson was learned the hard way more than 30 years ago by the behavioral scientists and therapists who developed the principles and techniques of "behavior modification." Whether it applied to teachers, students, employees, or prisoners, the term "behavior modification" was a real turn-off. It conveyed images of manipulation, top-down control, loss of personal control, and "Big Brother." For example, in 1974 my colleagues and I developed a behavior-change process and training program for the Virginia prison system (Geller *et al*, 1974). But our innovative and very effective plan was never fully implemented, partly because the inmates and guards associated "behavior modification" with brainwashing and lobotomies.

Actually, the term "behavior" alone carries negative associations for many—as in "Let's talk about your behavior last night"—but I can't see any way around using it. We need to teach and demonstrate the benefits of focusing on behaviors, especially desirable behaviors. But we don't have to link "modification" with behavior; it only adds to the negative feelings.

"Behavior analysis" is the term used by researchers and scholars in this area of applied psychology. This implies that behavior is analyzed first, and if change is called for, an intervention process is developed with input from the client(s). Given that "analysis" can sound cold or bring to mind Freud, I have recommended the label "behavior-based approach" for several years. This contrasts nicely with the "person-based approach" that focuses on attitudes, feelings, and expectancies. As I have repeatedly emphasized, a Total Safety Culture requires us to consider both behavior-based and person-based psychology.

I hope the basic message is clear. We need to understand that our language can activate feelings and even behaviors we don't want. Figure 16.5 illustrates what I mean. If we want to communicate in order to "sell" the process, we need to consider how our language will be perceived by the "customer."

Figure 16.5. The words we use can increase participation or resistance.

As we end this discussion on language, I'd like you to ponder the following word-attitude associations. What terms better contribute to employee involvement?

"Peer pressure" or "peer support"?
"Loss control managers" or "safety facilitators"?
"Compliance" or "accomplishment"?
"Meeting OSHA standards" or "fulfilling a corporate mission"?
"30 days without an injury" or "30 safe days"?
"I must meet this deadline" or "I choose to achieve another milestone"?
"I've got to do this" or "I get to do this"?

It's a good personal or group exercise to consider the ramifications of using these terms and phrases, and adding alternatives to this list is even more beneficial. But understanding the critical relationship between words, attitudes, and deeds is only half the battle. We need to change verbal habits, and this is easier said than done. When we communicate with greater passion and optimism about safety we'll attract more people to our safety mission. We'll also reduce resistance to change.

Overcoming Resistance to Change

"How do we deal with people who resist change?" "How do we get more people to participate?" I frequently hear these questions at training seminars and workshops. First, let's face reality: Change is unpleasant for many people, and some are apt to react poorly. Change often threatens our "comfort zones"—those predictable daily routines we like to control. In fact, it takes a certain amount of personal security and leadership to try something new. A certain kind of risk taking is needed to lead change, and some people want no part of exploring the unknown.

We've all been in unfamiliar situations where we're not sure how to act. We feel awkward and uncomfortable. If someone gives us direction, helps increase our sense of control, it's easier to adjust. We might even help others deal with the change. But without leaders and adequate tools to cope with change, we might retreat, withdraw from the situation, or even actively resist the change.

So how do we deal with resistance? Simply put, we should teach people the skills and give them the tools to handle change, plus support those who set the right examples. This seems logical and intuitive, but it doesn't always happen. Instead, managers too often try to identify the malcontents and discipline them for not participating.

Let's try to better understand resistance by considering one of the classic awkward situations thrust upon many of us—our first school dance. Remember it? For me, it was a high school homecoming dance in 1957.

Attending your first dance is like a rite of passage. If you were anything like me, you were a bit nervous about this change in your social world. You might have been prepared for it. Family, friends and teachers probably told you what to expect. Maybe you even had dance lessons. But these "tools" didn't make it any easier for some of us to participate. Not me, anyway. I didn't participate, but I wanted to. Before the dance, I practiced how to ask a girl to dance. I took four, two-hour dance lessons at an "Arthur Murray Dance Studio." I felt ready, but never once did I dance the entire night. I didn't feel too embarassed, though, because there were so many others not dancing. As is the custom, boys stood on one side of the gym and the girls on the other.

As illustrated in Figure 16.6, some kids were dancing and seemed to be having a great time. They danced almost every number, and tried to lure others out on the floor. I couldn't be enticed, though. I hung back in my comfort zone. But at least I was in the dance hall.

As illustrated in Figure 16.7, some students stood around in the parking lot, talking, drinking, and smoking cigarettes. These were the resisters. Some were active resisters. They stayed in their cars, never intending to enter the dance hall. Now and

Figure 16.6. Different reactions to change can be seen at the high school dance.

Figure 16.7. Some actively resist change and others follow.

then, these guys started up their cars and cruised around town for awhile, and then returned to the parking lot. They'd persuade others to hop in their car, try some beer, smoke a cigarette, fool around, or cruise.

LEVELS OF PARTICIPATION. There are essentially five ways of reacting to change—call them levels of participation—and they were all on display at the dance. First, there are the true leaders who get totally involved. They are the innovators—those who view change as necessary, and an opportunity to improve. At the dance, they were the teenagers on the floor for almost every number. They had the most fun. They didn't necessarily know what they were doing when it came to dancing, but they got out there and tried. They took a risk. They got totally involved and benefited most from the occasion. A dance might start with only a few of these "risk takers," but they often persuaded a number of others to get involved as the night wore on.

Some people want to change, but need direction and support. They're motivated to participate, but need models or leaders. At the dance, these were the kids who hung back at first. With a little encouragement they danced a few numbers. By the end of the evening you couldn't get them off the dance floor. They were now totally involved.

Most of us are at the third level of participation. We're ready to get involved, but we'll stay in our comfort zones until we're directed and motivated to participate. It might look like we're resisting change, but not really. Call us neutral when it comes to our attitude about change. We just aren't sure what to do. We need self-confidence that we can handle the change. And we need genuine support (positive recognition) when we try to participate. Once in a while, just getting started or "breaking the ice," is enough to turn a passive observer into an active participant. But for the most part, we stand on the sidelines and watch. This level of participation was represented by the boys and girls who lined each side of the gym.

TYPES OF RESISTANCE. The final two levels of participation are passive and active resistance. Passive resisters perceive change as a problem. They complain a lot. They're critical and untrusting of something new imposed on them. They seem to see only the negative side of a new program, policy, or challenge. They rationalize their position by gathering with others at their same level of nonparticipation, and they grumble and whine about proposed changes or about others who are participating in a change effort. But their whining and complaining usually stops when participation in a new process is clearly enjoyed by the majority. Passive resisters are followers, and they'll do what they see most people doing.

These are the teenagers who came to the dance because everyone else would be there, but they felt so insecure or anxious they didn't enter the building. They looked for others hanging around outside, and made fun of the silly dancing going on inside. Sometimes these nonparticipants ran into an active resister.

Fortunately, active resisters are few in number. But it doesn't take many of these characters to slow down a change process. These individuals view change as a threat or an opportunity to resist. They see any change effort that was not their idea as a potential loss of personal control, and they often exert countercontrol to assert their control or freedom. Psychologists call this "psychological reactance" (Brehm, 1966, 1972), a phenomenon most parents observe when their children reach the teens. Teenagers want to feel independent, and at times will disobey their parents' directions—break the rules—to gain a sense of independence or self-control.

We all feel overly controlled at times, and perhaps react to regain independence or assert personal freedom. Sometimes our reactions are not thoughtful, caring, or safe. But, active resisters feel the need to resist change, the status quo, or authority much of the time. This is partly because their contrary behavior brings them special attention—recognition for resisting.

WHO GETS THE ATTENTION? Active resisters stick out and attract attention. Nonparticipants use them to rationalize their own commitment to comfort zones. Managers monitoring the workplace often hit them with discipline. But this can backfire. This makes the top-down control more obvious for resisters. Discipline builds their resentment of the system and makes it even less likely they will join the change process. For some individuals, disciplinary attention only fuels their burning desire to exert independence and resist change. As a result, they might become more vigorous in recruiting others to oppose change. As I discussed in Chapter 10, top-down discipline should be used sparingly if the ultimate purpose is total participation in an improvement process.

How were resistant teenagers brought inside to the dance? The harsh warnings of the school principal shouting from the steps didn't work; neither did the one-on-one confrontation between one of the adult chaperones and the "leader of the pack." Whenever I saw a resister come inside, it was always the result of urging by another teenager. Peer pressure (or peer support) is still the most powerful motivator of human behavior. Therefore, the best way to deal with resistance is usually to arrange for situations that enable or facilitate peer influence. Eventually, some of the resistant teenagers came into the building so they wouldn't miss something. As depicted in Figure 16.8, they saw from their remote comfort zones that the people inside were really enjoying themselves, and they chose to participate.

The dance party will become more enticing as more and more teenagers dance. To increase such active involvement, the right kind of encouragement and support is needed. Will motivational lectures from a teacher, counselor, concerned parent, or

Figure 16.8. When a critical mass of the culture changes, others follow.

outside consultant make that happen? It might make a temporary difference, but not over the long haul. The best way to deal with nonparticipation is usually to set up situations that allow for peer influence. This could mean managers do nothing more than support the change process, and let peer pressure or support occur naturally.

POWER OF PEER INFLUENCE. So how do you facilitate peer influence? The best way is through empowerment, but this is easier said than done. You can give people more responsibility, such as the challenge to lead others in a change process, but they must *feel* responsible. As I discussed in Chapter 12, they need to have sufficient self-efficacy ("I can do it"), personal control ("I am in control"), and optimism ("I expect the best").

Some people already have sufficient self-efficacy, personal control, and optimism when assigned leadership responsibilities. As discussed in Chapter 12, they feel empowered ("I can make a difference"). Still, these individuals may need some basic training in communication, social recognition, and behavior-change principles. Others may lack one or more of the three person states that facilitate feelings of empowerment. So, in addition to education and training, they need a support system to build their sense of self-efficacy, personal control, and optimism.

As discussed in Chapter 14, there's no quick fix for increasing perceptions of empowerment. But if you ask people to define policies, settings, interactions, and contingencies that influence these three person states, you're on your way to developing action strategies to improve them. This in turn increases people's readiness to be empowered.

Remember the bottom line: Resistance to change is overcome most effectively through peer influence. This requires, of course, that people are willing to accept the leadership position of *change agent* (Chapter 11). Certain individuals naturally rise to the occasion and welcome opportunities to lead. Others may be committed to continuous improvement through change, but need some direction and encouragement from the natural leaders. Both of these groups need training and practice in behavior-change principles and social influence strategies. Then management needs to give these people direction and the opportunity to work with those folks standing on the sidelines in need of a nudge to leave their comfort zones.

It's usually best to ignore the resisters. Don't give them too many opportunities to say "no." The more often they publicly refuse, the more difficult it will be for them to change their minds and participate. So don't pressure these folks. Invite them to contribute whenever they feel ready to achieve success with those willing to try something new. When the majority buys in and eventually celebrates its accomplishments, resisters will choose to come on board. The key is for them to perceive that they chose to get involved rather than being forced by a top-down mandate.

My teenage daughter's recent experience at her first high-school dance allows me to use this analogy once more to make a final point. Karly was well prepared for the dance, but perhaps not for her first date. She was quite talented at performing the latest dances, she was wearing a new outfit, and she had planned to meet a number of her girlfriends who were also attending their first homecoming dance with a date.

After the football game, Karly was to meet her date at the dance. She arrived at the gym before him, and instead of waiting outside, she went on inside. She then got totally involved in the dance, heading out to the floor for just about every song with whomever was available and willing, boy or girl. In this situation, Karly was an innovator and a leader. She felt empowered enough to lure others onto the dance floor so they could join in the fun. When Karly's date finally arrived, he rushed up to her, led her off the dance floor, and admonished her for not waiting outside. Why should she have fun without him?

It's important to realize that even leaders need support for their leadership efforts. Those who benefit from a leader's inspiration or coaching should share their appreciation. Karly's date actually punished her for her initiative and total involvement.

Fortunately, Karly had sufficient self-efficacy, personal control, and available support from friends to ignore her date's reprimand. In fact, she went back to the dance floor and participated with the support group that had evolved before her date arrived. And she chose to make her first date with that boy her last. That empowerment was sure appreciated by her dad!

Plan for Safety Generalization

I'd like to discuss another factor that can help us build a Total Safety Culture—the heavily researched phenomenon of generalization. What is it? Generalization has to do with the spread of behaviors, and it occurs in two ways:

Stimulus generalization refers to the spread of influence from one setting or environment to another.

Response generalization refers to the occurrence of one behavior leading to another (Kimble, 1961).

For example, if a safety training process increases the use of personal protective equipment on the job and at home, stimulus generalization has occurred. On the other hand, response generalization occurs when an increase in one safety behavior, such as vehicle safety belt use, is accompanied by an increase in another task-related behavior, like turn-signal use.

Obviously, generalization is a desired goal of safety efforts. Safety leaders hope safe operating procedures used in one setting will spread to other situations, including employees' homes (stimulus generalization), and to other procedures or behaviors (response generalization). Psychological principles already discussed in this text (especially Chapter 14) suggest ways to increase both types of generalization. Interventions set up to improve the driving safety of pizza deliverers support these principles, and demonstrate practical ways to increase generalization.

CASE STUDY: GENERALIZING SAFETY. When behavioral scientists evaluate the impact of an intervention they typically measure the quantity and/or quality of a target behavior before, during, and after the intervention. As discussed earlier in Chapter 8, this is the basic research design used for more than 30 years to demonstrate the impressive impact of behavior-based psychology, and to refine intervention procedures and develop new ones. Unfortunately, this method cannot be used to evaluate generalization because usually only one situation is observed and only the behavior targeted by the intervention is measured. A few years ago, Tim Ludwig, a former graduate student, and I measured changes in behaviors that were not targeted by a safety intervention, and the results were informative (Ludwig & Geller, 1991, in press).

In the first project, our goal was to boost the safety-belt use of pizza deliverers working out of two stores in southwest Virginia. Before, during, and after a safety-belt campaign, we unobtrusively measured the deliverers' safety-belt use when entering and exiting the store parking lots. The results showed remarkable benefits of the safety-belt intervention. We also discovered response generalization. Observing the drivers' daily use of turn signals, we found that the use of both safety belts and turn signals increased after the safety-belt campaign, which did not include any mention of turn signal use.

Tim Ludwig and I were actually surprised to find the marked increase in turn-signal use (response generalization) after a campaign to increase only safety-belt use. We concluded that certain aspects of the intervention process promoted a sense of personal control, commitment, and group ownership, which theoretically should promote generalization. Tim's Ph.D. dissertation, conducted two years later, verified these conclusions.

Our safety-belt campaign included a one-hour group discussion on the value of vehicle safety belts and ways to support each other's use of safety belts. Then, buckle-up promise cards were distributed to each participant and signed as a personal commitment to buckle up consistently for two months. Signed promise cards were entered into a random drawing for a $20 sweatshirt. Everyone signed the pledge. In addition, the group decided to do a few things on the job to promote safety-belt use. Specifically, buckle-up reminder signs were posted in the two stores, and the dispatchers agreed to remind drivers to buckle up when giving them their pizzas to deliver. In an analogous study (Streff *et al*, 1993), employees signed a promise card to use safety glasses. Subsequently, these employees increased their use of safety glasses on the job and their use of vehicle safety belts when entering and exiting the plant parking lot.

For his Ph.D. dissertation research, Tim varied the safety intervention strategies between two pizza stores to test whether the choice, commitment, and involvement aspects of this intervention process were critical to obtaining generalization (Ludwig & Geller, in press). Goal-setting and feedback were employed at each store to increase the target behavior—complete vehicle stopping at intersections. During group discussion among the employees at one store about the benefits of always stopping completely at intersections, the manager noted that he had observed 55 percent complete stops during the previous week. Then, the group selected a goal of 75 percent for each of the following four weeks. The manager agreed to post biweekly percentages of safe intersection stopping obtained from his periodic observations of vehicles leaving the parking lot.

Goal setting and feedback strategies were also used at the second pizza store, but the process of implementing these behavior-change techniques was different. Instead of introducing the procedure through interactive discussion, Dr. Ludwig and the store manager lectured about the benefits of complete intersection stopping, emphasizing the same points brought out through group discussion at the other store. Then the manager assigned this group the same goal chosen by the employees at the other store. He also indicated he would post biweekly percentages of safe stopping from his daily observations. The same weekly feedback percentages were posted at each store, and were not calculated from actual observations. The results were provocative and instructive.

The pizza deliverers at both stores significantly increased their percentage of complete intersection stops during the intervention period, but response generalization was observed only among employees involved in the open discussion and group goal setting. In addition to systematically evaluating vehicle stopping, our field observers recorded unobtrusively (from store windows across the street) whether the pizza deliverers buckled up and used their turn signals when leaving the parking lot. As we expected, only the pizza deliverers who chose their own safe-stopping goal following interactive discussion significantly increased their use of *both* safety belts and turn signals.

INVOLVEMENT MADE THE DIFFERENCE. This study demonstrates that safety generalization is most likely to occur when people feel a sense of commitment and ownership for the goals of a behavior-based intervention process. When this happens, some people remind themselves to perform the target behavior in various situations (stimulus generalization) and to perform other safe behaviors related to the target behavior (response generalization). But individuals who are told they must comply with a safety policy or mandate might only do so because the consequences of not complying could mean a penalty, not because they believe and "own" the reason for the safe behavior. These people will probably not show generalization to other situations or behaviors.

The important lesson here is that people who believe in the mission of a safety intervention are usually willing to extend their commitment to safety across situations and behaviors. Buying into the mission, they recognize the inconsistency and futility of limiting safety to only certain conditions and behaviors. This leads to stimulus and response generalization. The key to gaining commitment and ownership is involvement. When we participate in discussing the rationale and goals of an intervention process, we're apt to develop internal justification and support for the intervention process—and beyond.

Slight differences in the way Dr. Ludwig and the store managers implemented the behavior-based intervention apparently resulted in different feelings of personal commitment and group ownership. Involving employees in a discussion of the intervention process, rather than lecturing information, and allowing them to choose a group safety goal, rather than assigning one, led to response generalization.

These findings bring to mind the discussion of "choice" in Chapter 14 and the results of my manipulations of students' opportunity to choose stimulus materials in the laboratory and reading assignments in the classroom. The implications are critical for obtaining and maintaining involvement in a safety process. For the benefits of an intervention process to generalize across situations and behaviors, participants need to do more than comply with the specifics of a mandate. They need to believe in the goals and the methods used to reach those goals. Theory and research indicate that a prime way to develop this personal commitment and ownership is to involve the participants in deciding goals and ways to attain them. Perceptions of choice and control conducive to personal commitment, ownership, and involvement can be increased by applying the three support processes: *leadership, communication,* and *recognition.*

In Conclusion

This chapter began with a list of guidelines to initiate and sustain a culture-change process aimed at achieving a Total Safety Culture. Critical challenges include:

- Gaining sufficient top management support
- Creating a Safety Steering Committee
- Developing valid evaluation procedures
- Establishing an education and training process
- Sustaining the culture change with activators, consequences, evaluation techniques, and follow-up training
- Dealing with outside contractors
- Troubleshooting, fine-tuning, and updating the various process procedures

Three support processes were identified to maintain employees' long-term commitment and involvement in a culture-change effort—leadership, communication, and recognition. Each of these processes were discussed in various forms throughout previous chapters. Aspects covered in this chapter were: a) characteristics of effective leaders; b) safety language that increases resistance and should be avoided; c) levels of resistance that can be influenced by leadership, communication, and recognition; and d) the concept of generalization as a desired outcome of an intervention process that draws on leadership, communication, and recognition strategies.

Psychologists find the best leaders are enthusiastic, honest, motivated, confident, analytical, informed, and flexible. Although it's common to see these characteristics described as permanent personality traits, it's certainly reasonable to assume that they can be increased through education, communication, recognition, and involvement in a behavior-based safety process. Thus, while it is useful to look for "natural" leaders when selecting members of a Safety Steering Committee, it's important to realize that leadership qualities could be suppressed in some people by their lack of empowerment or belongingness. New safety processes and eventual culture change might bring out leaders you didn't know existed in the workforce.

Involvement is key to so many aspects of building a Total Safety Culture, and it can be increased many ways. Start by getting into the habit of using more positive safety language. And focus less attention on the active resisters. There are five levels of involvement in any change effort that you should recognize: a) total involvement from innovators who see change as an opportunity to improve; b) individuals committed but not totally involved until direction and support are given; c) people, usually the majority, ready but on the sidelines until prodded and encouraged by others; d) doubters who see change as a problem and use learned helplessness and criticism as excuses to remain detached; and e) the active resisters who see change as an opportunity to resist, complain, and promote mistrust.

Active and passive resisters (categories d and e) should be ignored, if possible. Recognize and support those willing to try the new process. Employees totally involved in the process (category a) need to help individuals committed but not yet totally immersed (category b). Then these two groups can work with the majority (category c) who need examples to follow. You can see why it's important to cultivate leadership, communication, and recognition skills among the "true believers" in innovation. Turn these leaders loose and they'll be your best recruiters to build the base of support for a Total Safety Culture.

One of the best ways to develop champions of change is to give true believers opportunities to teach others the principles and procedures of a safety process. When people teach stuff they believe in, they increase their personal commitment and become continuous role models and visible leaders of the process. A "teach-the-teacher"[9] process is critically important to achieve a Total Safety Culture. First, identify your potential change agents. Then teach them the right principles and procedures, and how to teach coworkers. The final chapter reviews the principles presented in this text. They are the building blocks for a teach-the-teacher process that integrates behavior-based and person-based psychology.

REFERENCES

Baron, R. A. (1995). *Psychology* (Third Edition). Boston, MA: Allyn and Bacon.

Brehm, J. W. (1966). *A theory of psychological reactance*. New York: Academic Press.

———. (1972). *Responses to loss of freedom: A theory of psychological reactance*. New York: General Learning Press.

9. A more standard term is "train-the-trainer," but since both education and training are needed (as defined earlier), I believe "teach-the-teacher" is a more appropriate label for this process.

Geller, E. S., Johnson, D. F., Hamlin, P. H., & Kennedy, R. D. (1977). Behavior modification in a prison: Issues, problems, and compromises. *Criminal Justice and Behavior,* 4, 11-43.

Hayakawa, S. I. (1978). *Language in thought and action* (Fourth Edition). New York: Harcourt Brace Jovanovich, Publishers.

Hayes, S. C., Rincover, A., & Solnick, J. V. (1980). The technical drift of applied behavior analysis. *Journal of Applied Behavior Analysis,* 13, 275-285.

Hersey, P., & Blanchard, K. (1982). *Management of organizational behavior* (Fourth Edition). Englewood Cliffs, NJ: Prentice-Hall.

Kimble, G. A. (1961). *Hilgard and Marquis' conditioning and learning.* New York: Appleton-Century-Crofts.

Kirkpatrick, S. A., & Locke, E. A. (1991). Leadership: Do traits matter? *Academy of Management Executives,* 5(2), 48-60.

Ludwig, T. D., & Geller, E. S. (1991). Improving the driving practices of pizza deliverers: Response generalization and moderating effects of driving history. *Journal of Applied Behavior Analysis,* 24, 31-44.

——— & ———. (in press). Improving driving practices of professional pizza deliverers: Participative versus assigned goal setting and response generalization. *Journal of Applied Psychology.*

The New Merriam-Webster Dictionary (1989). Springfield, MA: Merriam-Webster Inc., Publishers.

Streff, F. M., Kalsher, M. J., & Geller, E. S. (1993). Developing efficient workplace safety programs: Observations of response covariation. *Journal of Organizational Behavior Management,* 13(2), 3-14.

Tuckman, B. W., & Jensen, M. A. C. (1977). Stages of small group development revisited. *Group and Organizational Studies,* 2, 419-427.

Yukl, G. (1989). *Leadership in organizations* (Second Edition). Englewood Cliffs, NJ: Prentice-Hall.

Zaccaro, S. J., Foti, R. J., & Kenny, D. A. (1991). Self-monitoring and trait-based variance in leadership: An investigation of leader flexibility across multiple group situations. *Journal of Applied Psychology,* 76, 308-315.

Chapter 17

Reviewing Fifty Key Principles of a Total Safety Culture

Fifty principles for understanding the human dynamics of safety summarize what this book is all about. When you use these principles to design, execute, evaluate and continuously improve interventions to change safety-related behaviors and attitudes, you're well on your way to achieving a Total Safety Culture.

"How should we translate these concepts into real-world application?"

"Would you please put your theory into procedures or practices we could follow in our plant?"

I heard questions like these at each of the Deming workshops I attended. They seemed to disappoint Dr. Deming (1991, 1992), who would assert that the purpose of the seminar was to teach theory and principles, not specific procedures. It was up to the participants to return to their own organizations and devise specific methods and procedures that fit their culture. Dr. Deming stressed the need to start with theory and then customize practices.

This text also downplays a "one-size-fits-all" solution. Packaged programs are not the answer to safety problems, though they can be found everywhere. Sure, the quick fix might work for a while. Short-term success follows a familiar pattern: Injuries reach unacceptable levels, management hires a consultant, employees react with interest, injuries go down, statistics improve. Improvement is not difficult if injury rates are bad to begin with. But as I discussed earlier, this superficial approach only improves the worst organizations for a limited period. For companies with a good safety record, it doesn't work even for the short term.

As Dr. Deming well knew, and as I've discussed throughout this text, lasting improvement is built on specific procedures that fit the culture of an organization. Outside consultants can be invaluable, teaching appropriate principles and facilitating the implementation process. But if most of the employees do not understand and believe in the principles to begin with, well-intentioned efforts never take root.

That's why throughout this book I've presented theory and principles from psychological research to help you design safety-improvement interventions. The successful applications of these principles that I've described in brief case studies are not intended as step-by-step procedures to follow, but rather examples to consider when customizing a process for your culture.

It all starts with theory. And in this final chapter I want to pull together 50 important principles that summarize the psychology of safety, and lay the groundwork for building a Total Safety Culture. I know 50 sounds like a long list, but don't worry, I'll be brief. And the principles will be familiar to you, having come from information already covered.

I hope you'll find the list useful as a review, and a starting point for developing your safety-enhancement process. Some of the principles focus on design and implementation. Others explain why we often fail in safety. Most can be used as guidelines for checking potential long-term benefits of a specific safety-improvement procedure. All help you appreciate the complex human dynamics of safety and health promotion.

This is not a priority list. Don't read anything into the order of principles. What I hope is that you will teach them to others. You *can* make a difference, and bring about constructive culture change.

PRINCIPLE 1: Safety should be internally—not externally—driven.

It's common to hear employees talk about safety in terms of OSHA—the Occupational Safety and Health Administration. It often seems they "do" safety more to satisfy the mandates of this outside regulatory agency than for themselves. This translates into perceptions of top-down control, and performing to avoid penalties rather than to achieve success.

Ownership, commitment, and proactive behaviors are more likely when we work toward our own goals, not the government's. As discussed in Chapter 16, how we define programs and activities can influence attitudes that shape involvement. It makes sense to talk about corporate safety as a mission owned and achieved by the very people it benefits.

PRINCIPLE 2: Culture change requires people to understand the principles and how to use them.

In Chapter 16 (footnote 5), I distinguished between education and training, and emphasized that long-term culture change requires both. Education focuses on theory or principles. Training gets into the specifics of how to turn principles into practical procedures. Role playing or one-to-one interaction is very important because participants get direct feedback on how they're executing procedures or processes.

PRINCIPLE 3: Champions of a Total Safety Culture will emanate from those who teach the principles and procedures.

When people teach, they "walk the talk" and become champions of change. After more than 20 years of safety consulting, it's clear to me that success depends on the presence of these leaders. I've seen no better way to develop champions of a campaign than to first teach relevant theory and method, then show how others can be instructed, and finally allow opportunities for colleagues and coworkers to teach each other.

PRINCIPLE 4: Leadership can be developed by teaching and demonstrating the characteristics of effective leaders.

Just because you believe in something doesn't guarantee you'll be an effective champion of the cause. Leaders have certain characteristics, as discussed in Chapter 16, which can be learned and cultivated in others. People need to understand the principles behind good leadership, and the behaviors that reflect good leadership qualities. You can also learn by observing the leadership skills of others. And when you see leaders in action, reward their exemplary behavior with feedback.

PRINCIPLE 5: Focus recognition, education, and training on people reluctant but willing, rather than on those resisting.

As discussed in Chapter 16, people resist change for many reasons. Some feel insecure leaving their comfort zones. Some mistrust any change in policy or practice that wasn't their idea. Others balk for the special attention they get by resisting. It's usually a waste of time trying to force change on these folks. In fact, resistance hardens as more pressure is applied.

It saves time to prioritize. First focus on those who "want to dance." The ones willing to get involved. Then turn these leaders loose on the folks who "came to the dance" but are reluctant to get involved. At least these people are willing to consider a change proposal. Peer instruction can cultivate change champions (Principle 3), as well as increase participation. When a critical mass of individuals gets involved and achieves success as a result of change, many initial resisters will join in—out of choice, not coercion.

PRINCIPLE 6: Giving people opportunities for choice can increase commitment, ownership, and involvement.

A basic reason for preferring the use of positive over negative consequences to motivate behavior (Chapter 10) is that people feel more free, they perceive more choice, when working to achieve rewards (Skinner, 1971). As illustrated in numerous laboratory experiments and field applications, increasing perception of choice leads to more motivation and involvement in the process (Chapter 14).

Personal choice also implies personal control—enhancing empowerment and willingness to actively care for others (Chapter 14). It's important to realize that eliminating the perception of choice—by imposing a top-down mandate that restricts or constrains work behavior, for example—can sap feelings of ownership, commitment and empowerment, and inhibit involvement.

PRINCIPLE 7: A Total Safety Culture requires continuous attention to factors in three domains: environment, behavior and person.

Early on I stressed the "Safety Triad," with behavior and person sides representing the psychology of safety. That's the focus of this book. But don't overlook the need for environmental change. The environment includes physical conditions and the general atmosphere or ambiance regarding safety. For example, safety can be considered a top-down "condition of employment" or a bottom-up "opportunity to get involved." Which of these perspectives you choose influences your safety-related attitudes and behaviors.

Environment, behavior, and person factors are dynamic and interactive; a change in one eventually impacts the other two. For example, behaviors that reduce the probability of injury often involve environmental change, and lead to attitudes consistent with the safe behaviors. In other words, when people choose to act safely, they act themselves into safe thinking and this often results in some environmental change.

PRINCIPLE 8: Don't count on common sense for safety improvement.

Most common sense is not common. It's biased by our subjective interpretation of unique experiences. As a researcher of psychological principles for more than 35 years, I have become quite committed to this basic principle (discussed in Chapter 2). Indeed, I've dedicated most of my career to discovering principles of human behavior through systematic application of the scientific method. You've probably noticed by now that I get quite disturbed when I read or hear "pop psychology" based on unfounded intuition or "common sense." Many statements I've read or heard relating to the psychology of safety sound good but are incorrect. Profound knowledge comes from rigorous research and theory development, and often runs counter to common sense.

PRINCIPLE 9: Safety incentive programs should focus on the process rather than outcomes.

One of the most frequent common-sense mistakes in safety management is the use of outcome-based incentive programs. Giving rewards for avoiding an injury seems reasonable and logical. But it readily leads to covering up minor injuries and a distorted picture of safety performance. The basic activator-behavior-consequence contingency (see Chapter 8 and Principle 19) demonstrates that safety incentives need to focus on process activities, or safety-related behaviors.

I recently consulted with a chemical plant well known for exemplary safety performance. The annual number of OSHA recordables among approximately 550 employees had varied from 3 to 10 over several years. At the start of 1995, management initiated an outcome-based incentive program to reach a "step improvement" in safety. Specifically, 20 percent of a year-end bonus, amounting to $800 per employee, hinged on having six or fewer OSHA recordables at the end of the year.

By mid-August 1995, the plant had experienced seven OSHA recordables. Everyone lost that $800. Needless to say, morale for safety plummeted to an all-time low. To boost spirits, employees were promised a significant surprise reward "that will warm your hearts" if they could go the rest of the year without a single OSHA recordable. When I visited this plant no OSHA recordable had been reported for 117 days. The plant was on it's way to achieving the goal.

These outcome-based incentive programs were clearly well-intentioned. But I hope you're suspicious about the result. My discussion with employees at this facility verified that minor injuries were being covered up. In fact, one line worker remarked, "It's only common sense, isn't it, that when you put so much pressure on a person to not have an injury, they'll be motivated to conceal it if they can."

Yes, some common sense is correct.

PRINCIPLE 10: Safety should not be considered a priority, but a value with no compromise.

As discussed in Chapter 3, this is the ultimate vision: Safety becomes a value linked to every priority in the workplace, or wherever we find ourselves. Priorities change according to circumstances; values are deep-seated personal beliefs beyond compromise. Establishing safety as a value won't happen overnight. It will take time, and there is no "quick fix."

PRINCIPLE 11: Safety is a continuous fight with human nature.

I know people who meet the behavioral criteria for holding safety as a value—they practice safety, teach it, go out of their way to actively care for the safety of others. But their numbers are few. Why? Because human nature (or natural motivating consequences, Chapter 10) typically encourages at-risk behavior. The soon, certain, positive, and natural consequences of risky behavior are hard to overcome. We're talking about comfort, convenience, and expediency. Now consider safe alternatives, which often mean discomfort, inconvenience, and inefficiency. The inconveniences involved in safely locking out equipment are illustrated in Figure 17.1. When you compete with natural supportive consequences in order to teach, motivate, or change behavior, you're fighting human nature.

Figure 17.1. Safe behavior often requires us to resist natural motivation for comfort, convenience, and efficiency.

Principle 12: Behavior is learned from three basic procedures: classical conditioning, operant conditioning, and observational learning.

Through naturally occurring consequences and planned instructional activities, we learn every day, and we develop attitudes and emotional reactions to people, events, and environmental stimuli. The mechanisms for learning voluntary and involuntary behavior and emotions were reviewed in Chapter 8.

The critical aspect of this principle is that our actions and feelings result from what we learn through experience, both planned and unplanned. Basic learning principles can be applied to change what we do and feel. But experience and practice develop habits that are hard to break. Plus, it's possible that natural contingencies and social influences support a "bad habit" or "negative attitude." So, learning new behaviors and attitudes often requires another fight with human nature.

PRINCIPLE 13: People view behavior as correct and appropriate to the degree they see others doing it.

Since personal experience often convinces us that "it's not going to happen to me," we need a powerful reason to perform safely when personal injury is improbable. So consider this: Everyone who sees you acting safely or at risk either learns a new behavior or thinks what you're doing is OK. Now consider the vast number of people who observe your behavior every day. Our influence as a social model gives us special responsibility to go out of our way for safety.

PRINCIPLE 14: People will blindly follow authority, even when the mandate runs counter to good judgment and social responsibility.

This principle was discussed in Chapter 5 as a potential barrier to safe work practices. The fact that people often follow top-down rules without regard to potential risk is alarming. This puts special responsibility on managers and supervisors who give daily direction. These front-line leaders could signal, even subtly, the approval of at-risk behavior in order to reach production demands. And people are apt to follow even implicit demands from their supervisor to whom they readily delegate responsibility for injury that could result from at-risk behavior.

PRINCIPLE 15: Social loafing can be prevented by increasing personal responsibility, individual accountability, group cohesion, and interdependence.

This principle was introduced in Chapter 14 when discussing ways to increase group productivity and synergy. Giving up personal responsibility for safety to another person (Principle 14) could be due to the social mechanisms presumed to influence social loafing. It's possible to decrease blind compliance to rules that foster at-risk behavior by manipulating factors found to decrease social loafing. Thus, workplace interventions and action plans need to be implemented with the aim of increasing people's perceptions of personal responsibility and accountability, as well as their sense of group cohesion and interdependence.

PRINCIPLE 16: On-the-job observation and interpersonal feedback is key to achieving a Total Safety Culture.

Critical behavior checklists (Chapter 8) and communicating the results of checklist observations (Chapter 11) put this principle to work. The penguin in Figure 17.2 does not have sufficient equipment nor a method to fly. "Try harder" won't work without the right process. Interpersonal observation and feedback are basic tools to use when the safety director says, "Try harder."

Figure 17.2. Best efforts fail without the right equipment and method.

PRINCIPLE 17: Behavior-based safety is a continuous DO IT process.

D = Define target behaviors, ***O*** = Observe target behaviors, ***I*** = Intervene to improve behaviors, and ***T*** = Test impact of intervention. Motivation to "try harder" cannot work without appropriate method. As Dr. Deming (1991) put it, "Goals without method—what could be worse?" This four-step process enables continuous improvement through an objective behavior-focused approach. As detailed in Chapter 8, people need to decide on critical target behaviors to observe. After baseline observations are taken, an intervention is developed and implemented. By continuing to observe the target behaviors, the impact of the intervention program can be objectively evaluated. Results might suggest a need to refine the intervention, carry out another one, or define another set of behaviors to work on. The next four principles provide guidance for designing behavior-change interventions.

PRINCIPLE 18: Behavior is directed by activators and motivated by consequences.

As explained in Chapters 8, 9 and 10, external or internal events (referred to as activators) taking place before behavior only motivate to the extent that they signal or specify consequences. Intentions and goals can motivate behavior if they stipulate positive or negative consequences. Understanding this principle is critical to developing effective behavior-change techniques. Chapter 9 showed how this principle guides the development of more effective activators, and Chapter 10 outlined procedures for improving the motivational power of consequences.

PRINCIPLE 19: Intervention impact is influenced by: amount of response information, participation and social support, and external consequences.

In Chapter 11, I discussed ways to maximize the immediate impact of an intervention. Interventions that give specific instructions (response information) and get participants actively involved are likely to influence behavior and attitude change. And, if the intervention facilitates support from others, such as co-workers or family members, it can have lasting effects.

The role of external consequences is a bit tricky. Pioneering research by Professor Jonathan Freedman (1965) demonstrated the need to limit external consequences if we want people to develop internal motivation. Dr. Freedman used a mild or severe threat to prevent seven- to nine-year-old boys from playing with an expensive battery-controlled robot. In the Mild Threat condition, the boys were merely told "It is wrong to play with the robot." Alternatively, the boys in the Severe Threat condition were told, "It is wrong to play with the robot. If you play with the robot, I'll be very angry and will have to do something about it." Four other toys were available for the boys to play with when the experimenter left the room.

From a one-way mirror, researchers observed that only one of 22 boys in each condition touched the robot. About six weeks later, a young woman returned to the boys' school and took them out of class one at a time to perform in a different experiment. She made no reference to the earlier study, but instructed a boy to take a drawing test. While she scored the test, she told the boy he could play with any toys in the room. The same five toys from the previous study, including the robot, were available. Of the boys from the Severe Threat condition, 17 (or 77 percent) played with the robot, compared to only 7 (33 percent) from the previous Mild Threat condition. Presumably, more boys in the Mild Threat condition developed an internal rationale for avoiding the robot, and as a result avoided this toy when the external pressure was absent.

Other researchers have followed up this study, and demonstrated that people are more apt to develop internal motivation when external rewards or threats are relatively small and insufficient to completely justify the target behavior (Riess & Schlenker, 1977). This phenomenon has been referred to as the "less-leads-to-more effect" (Baron, 1995), and is most likely to occur when people feel personally responsible for their choice of action and the resulting consequences (Cooper & Scher, 1990; Goethals *et al*, 1979; Lepper & Green, 1978).

PRINCIPLE 20: Extra and external consequences should not overjustify the target behavior.

The various examples of positive consequences presented in Chapters 10 and 11 (from thank-you cards to the privileges, commendations, and small tangibles listed in Figure 11.2) are not large nor expensive. For reasons discussed above, rewards should not provide complete justification for desired behavior. We don't want people complying with safety rules only to gain a reward or avoid a penalty. If that's the case, what happens when we take away the consequence, good or bad? We take away the reason to comply. This is why people wear PPE at work, but rarely at home.

PRINCIPLE 21: People are motivated to maximize positive consequences (rewards) and minimize negative consequences (costs).

This principle offers another reason why people are not likely to follow safe operating procedures in the absence of external controls, or behavior-consequence contingencies. As reflected in Principle 11, natural external consequences usually support risk-taking at the expense of safe alternatives, which are usually more inconvenient, uncomfortable, or time consuming.

Of course, this principle relates to many behaviors. In Chapter 12, it was used to explain why people often don't rush to help in a crisis. If there are more perceived costs than benefits to intervening, actively caring behavior is unlikely. Therefore, a prime strategy for increasing safety and actively caring behaviors is to overcome the costs (negative consequences) with benefits (positive consequences). Various kinds of consequences are defined by the next principle.

PRINCIPLE 22: Behavior is motivated by six types of consequences: positive vs. negative, natural vs. extra, and internal vs. external.

Understanding these characteristics (as detailed in Chapter 10) can enable significant insights into the motivation behind observed behavior. Appreciating these various consequences can also suggest whether external intervention is called for to change behavior, and what kind of intervention to implement. It's possible the natural external consequences supporting ongoing at-risk behavior cannot be overcome with extra external consequences. In this case, long-term behavior change requires the modification of the natural consequences or the application of techniques discussed in Chapter 14 to alter internal consequences. Throughout this text, I have downplayed the use of negative consequences, and the reasons are reflected in the next principle.

PRINCIPLE 23: Negative consequences have four undesirable side effects: escape, aggression, apathy, and countercontrol.

How did you feel the last time you received a reprimand from a supervisor? Maybe you felt like slinking away, or taking a swipe at him. Chances are you didn't go back to the job charged up. Perhaps you wanted to do something to make him look bad. These and other undesirable side effects of using negative consequences are discussed in Chapter 10.

PRINCIPLE 24: Natural variation in behavior can lead to a belief that negative consequences have more impact than positive consequences.

As detailed in Chapter 11, behavior fluctuates from good to bad for many reasons. Peak performance seldom can be sustained, and poor performance is almost bound to get better at some point. So if you praise someone and their performance falters, don't swear off positive feedback. And don't overestimate the power of your reprimand if it gets some immediate results. Keep things in perspective.

PRINCIPLE 25: Long-term behavior change requires people to change "inside" as well as "outside."

The psychology of safety requires us to consider both external behavior and internal person factors. Chapter 13 focused on the role of person states in influencing people to actively care for another person's safety and health. Chapter 14 showed how outside factors can be manipulated to influence these person states, and thus increase actively caring behavior. A Total Safety Culture requires integrating both behavior-based and person-based psychology. The next several principles focus on understanding "inside" factors.

PRINCIPLE 26: All perception is biased and reflects personal history, prejudices, motives, and expectations.

Appreciating this principle is key to understanding people, and realizing the importance of actively listening to others before intervening. It also supports the need to depend on objective, systematic observation for knowledge rather than common sense (Principle 8).

It's important to realize the reciprocal relationship between perception and behavior. Perceptions influence actions, and in turn actions influence perceptions. If we perceive risk, we'll act to reduce it; by acting to reduce risk, we'll become more aware of other risks.

The current popularity of safety perception surveys in industry reflects an increased awareness of how perceptions impact safety performance. These surveys can help pinpoint issues that need attention, and activities in need of an intervention. They can also be used to assess the person factors influenced by a particular intervention program.

PRINCIPLE 27: Perceived risk is lowered when a hazard is perceived as familiar, understood, controllable, and preventable.

When people perceive a new risk, they adjust their behavior to avoid it. Call it "fear of the unknown." The reverse is also true. As discussed in Chapter 5, research has shown that hazards perceived as familiar, understood, controllable, and preventable are viewed as less risky. This is why many hazards are underestimated by employees.

PRINCIPLE 28: The slogan "all injuries are preventable" is false and reduces perceived risk.

Frankly, I believe that telling people all injuries are preventable insults their intelligence. They know better. It's difficult enough to anticipate and control all environmental and behavioral factors contributing to injuries, but controlling factors inside people is clearly impossible.

Such a slogan can make it embarrassing to report an injury, and could influence a cover-up. "If they think all injuries are preventable, they will think I was really stupid to have this injury, so I'd better not report it." The most critical problem with this popular slogan is that it can reduce the perception of risk. Hazards considered controllable and preventable are perceived as relatively risk free (Sandman, 1991; Slovic, 1991).

PRINCIPLE 29: People compensate for increases in perceived safety by taking more risks.

As reviewed in Chapter 5, researchers have shown that some people will compensate for a decrease in perceived risk by performing more risky behavior. In other words, some people increase their tolerance for risk when feeling protected with a safety device (Wilde, 1994). As shown in Figure 17.3, high technology safety engineering can give a false sense of security. This is not the case for people who hold safety as a value (Principle 10).

Figure 17.3. Sometimes high technology causes reduced perception of risk and at-risk behavior.

PRINCIPLE 30: When people evaluate others they focus on internal factors; when evaluating personal performance, they focus on external factors.

As discussed in Chapter 6, this principle is termed "The Fundamental Attribution Error." It contributes to systematic bias whenever we attempt to evaluate others, from completing performance appraisals to conducting an injury investigation. Because we are quick to attribute internal (person-based) factors to other people's behavior, we tend to presume consistency in others because of permanent traits or personality characteristics. To explain injuries to other persons we use expressions like, "He's just careless," "She had the wrong attitude," and "They were not thinking like a team."

On the other hand, when evaluating our own behavior, we point the finger to external factors. This should make us stop and realize the many observable variables that can be measured and often changed to increase everyone's safety-related behavior and reduce injuries throughout a culture.

PRINCIPLE 31: When succeeding, people over-attribute internal factors; but when failing, people over-attribute external factors.

This research-based principle is referred to as the "self-serving bias" (see Chapter 6), and is sure to warp injury investigations. Placing blame for a mistake on outside variables is just a basic defense to protect one's self-esteem. In most organizations, even a minor injury is perceived as a failure. As a result, the victim is sure to avoid discussing inside, person factors contributing to the mishap. Statements like "I was fatigued," "I didn't know the proper procedure," or "My mind was on other things" are far less probable than "The work demands were too severe," "The trainer didn't show me the correct procedure," or "Excessive noise and heat distracted me." My advice is to accept the self-serving bias and allow people their ego-protecting excuses. Then search for measurable external factors (including behaviors) that can be changed to reduce the probability of another injury.

PRINCIPLE 32: People feel more personal control when working to achieve success than when working to avoid failure.

The sense of having control over life events is one of the most important person states contributing to our successes and failures. When we feel in control we're more motivated and work harder to succeed. We're also more likely to accept failure as something we can change. Thus, the value of increasing people's sense of personal control over safety is obvious. One way to do this is to develop scoring procedures for safety achievements, rather than focusing on the number of reported injuries as a measure of success. This puts the emphasis on measuring process activities that can lead to loss control or injury prevention, as detailed in Chapter 15.

PRINCIPLE 33: Stressors lead to positive stress or negative distress, depending on appraisal of personal control.

When we believe we can do things to reduce our stressors—work demands, interpersonal conflict, boredom—we're more motivated to take control. As discussed in Chapter 6, this is positive stress, an internal person state not nearly as detrimental to safety as distress. We feel distress when we believe there is little we can do about current stressors. This state can lead to frustration, exhaustion, burnout, and dangerous behavior.

It's important to recognize states of distress in others and attempt to help them. After actively listening to another person's concerns, you might be able to offer constructive suggestions. Sometimes it's useful to help people distinguish between the stressors they can control and the ones they can't. We can be concerned about a lot of things, but we can only control some of these. Helping people focus on the stressors they *can* reduce builds their sense of self-efficacy, personal control, and optimism. These are the person states that imply empowerment and increase one's willingness to actively care: "I can make a difference."

PRINCIPLE 34: In a Total Safety Culture everyone goes beyond the call of duty for the safety of themselves and others—they actively care.

Here we have a primary theme of this book. While behavior-based psychology provides methods and techniques to improve the human dynamics of safety, principles from person-based psychology need to be considered to ensure that behavior-based tools are used. The ultimate aim is to integrate behavior-based and person-based psychology so everyone participates in efforts to achieve a Total Safety Culture. In the ideal culture, everyone actively cares for the safety and health of others.

PRINCIPLE 35: Actively caring should be planned and purposeful, and focus on environment, person, or behavior.

Part Four of this book is all about actively caring for safety, from understanding why people resist it (Chapter 12) to implementing strategies that increase it (Chapter 14). We need to plan ways to enable and nurture as much actively caring behavior as possible, rather than sit back and wait for "random acts of kindness."

By considering the three domains of actively caring focus, we can sometimes get more benefit from an act of kindness. In particular, including behavior-focused actively caring can often result in the most benefit. For example, it's often possible to include specific behavioral advice, direction, or motivation with a donation (environment-focused actively caring) and with crisis intervention and active listening (person-focused actively caring).

PRINCIPLE 36: Direct, behavior-focused actively caring is proactive and most challenging, and requires effective communication skills.

Some acts of caring are relatively painless and effortless—contributing to a charity, sending a get-well card, or actively listening to another person's problems. But telling someone how to change their behavior can be confrontational and challenging, especially when it's direct. Think of the parent telling the child: "I want to discuss your behavior." This is the type of active caring we're most likely to avoid, which is unfortunate because it's the most beneficial. But even parents will pass up chances to talk about behavior with their kids. Why?

You often hear parents lament that they had no training in how to raise children. Our resistance is partly due to lack of confidence in our communication skills. Proper training and practice as a safety coach increases our ability to actively care for safety in this most beneficial way.

PRINCIPLE 37: Safety coaching that starts with Caring and involves Observing, Analyzing, and Communicating, leads to Helping.

The basic components of effective safety coaching were presented in Chapter 11, with the letters of COACH signifying labels for the sequential events in the process. The coaching process should start with an atmosphere of interpersonal Caring and an agreement that the coach can Observe an individual's performance, preferably with a behavioral checklist. Then the coach Analyzes the observations from a fact-finding, system-level perspective. Subsequently, the results are Communicated in one-to-one actively caring conversation, with the sole purpose to Help another individual reduce the possibility of personal injury.

PRINCIPLE 38: Actively caring can be increased indirectly with procedures that enhance self-esteem, belongingness, and empowerment.

This principle reflects one of the most innovative and critical theories discussed in this book (Chapter 13). Substantial research is available to support each component of this principle, but prior to my journal editorial (Geller, 1991), no one had combined these components (or person states) into one actively caring model. Procedures that enhance a person's sense of self-esteem ("I am valuable"), belongingness ("I belong to a team"), and empowerment ("I can make a difference") make it more likely that person will actively care for the safety or health of another person. Nourishing each of these person states leads to the actively caring belief that "We can make valuable differences."

PRINCIPLE 39: Empowerment is facilitated with increases in self-efficacy, personal control, and optimism.

This principle was mentioned earlier when reviewing the distinction between stress and distress (Principle 33). When people's sense of self-efficacy ("I can do it"), personal control ("I am in control"), or optimism ("I expect the best") is increased, they are less apt to experience distress and more likely to feel empowered ("I can make a difference"). In addition, empowerment increases one's inclination to perform actively caring behaviors.

Note that empowerment does not necessarily result from receiving more responsibility. In order to truly feel more empowered, people need to perceive they have the skills, resources, and opportunity to take on the added responsibility (self-efficacy), believe they have personal impact over their new duties (personal control), and expect the best from their efforts to be more responsible (optimism).

PRINCIPLE 40: When people feel empowered their safe behavior spreads to other situations and behaviors.

In a Total Safety Culture people go beyond the call of duty for safety. This means they perform safe behaviors in various situations. More specifically, they show both stimulus generalization—performing various particular safe behavior in various settings—and response generalization—performing safe behaviors related to a particular target behavior. Both types of generalization occur naturally when safety becomes a value rather than a priority (Principle 10). Obviously we need to intervene in special ways to promote safety as a value. As reviewed in Chapter 16, our applied research has shown that facilitating empowerment is one special way to increase generalization and cultivate safety as a value.

PRINCIPLE 41: Actively caring can be increased directly by educating people about factors contributing to bystander apathy.

In Chapter 14, I discussed strategies for encouraging actively caring behavior directly. This principle expresses the most basic procedure for doing this. Research has shown that educating people about the barriers to helpful behavior can remove some obstacles and increase the probability of actively caring behavior. Similarly, I have found that discussing the barriers to safe behavior can motivate people to improve safety, provided they also learn specific techniques for doing this.

PRINCIPLE 42: As the number of observers of a crisis increases, the probability of helping decreases.

This principle, supported with substantial behavioral research, is probably the first barrier to actively caring behavior that should be taught. It's strange but true, and means that people cannot assume that someone else will intervene in a crisis. In fact, the most common excuse for not acting is probably something like, "I thought someone else would do it" or "I didn't know it was my responsibility." This principle reflects the need to promote a norm that it's everyone's responsibility to actively care for safety. We can never assume someone else will correct an at-risk behavior or condition.

PRINCIPLE 43: Actively caring behavior is facilitated when appreciated and inhibited when unappreciated.

Making an effort to actively care directly for someone else's safety is a big step for many people and deserves genuine recognition. Then, if advice is called for to make the actively caring behavior more effective, corrective feedback should be given appropriately. Be sure to make your deposits first. All actively caring behavior is well-intentioned, but not frequently practiced with the kind of feedback that shapes improvement. A negative reaction to an act of caring can be quite punishing, and severely discourage a person from trying again. Consequently, much of the future of actively caring behavior is in the hands of those who receive people's attempts to actively care.

PRINCIPLE 44: A positive reaction to actively caring can increase self-esteem, empowerment, and belongingness.

This is a follow-up to Principle 43, and supports the need to sincerely recognize occurrences of actively caring behavior. Although research in this area is lacking, it is intuitive that feeling successful at actively caring behavior should lead to more active caring. Success should enhance self-esteem, empowerment, and belongingness, and so indirectly increase the probability of more caring acts. Thus, we have the potential for a mutually supporting cycle of actively caring influence, provided the reactions to actively caring behavior are positive.

PRINCIPLE 45: The universal norms of consistency and reciprocity motivate everyday behaviors, including actively caring.

These social influence norms have powerful impact on human behavior. Sometimes people apply these norms intentionally to influence others. At other times, these norms are activated without our awareness. Regardless of intention or awareness, behavior-change techniques derived from these norms can be very effective. In Chapter 14, I discussed how these social influences can be used to directly increase actively caring behavior. The next three principles reflect that information.

PRINCIPLE 46: Once people make a commitment, they encounter internal and external pressures to think and act consistently with their position.

This is why I say you can act people into thinking differently, or think people into acting differently. If people act in a certain way on the "outside," they will adjust their "inside"—including perceptions, beliefs, and attitudes—to be consistent with their behaviors. The reverse is also true, but throughout this text I've recommended targeting behavior first because it's easier to change on a large scale. As presented in Chapters 9 and 10, we know much more about changing behavior than perceptions, beliefs, and attitudes because behavior is easier to measure objectively and reliably.

PRINCIPLE 47: The consistency norm is responsible for the impact of "foot-in-the-door" and "throwing a curve."

As detailed in Chapter 14, the "foot-in-the-door" and "curve ball" techniques of social influence succeed because of the consistency norm (Principle 46). When an individual agrees with a relatively small request, for example, to serve on a safety committee, you have your foot in the door. To be consistent, the person is more likely to agree later with a larger request, perhaps to give a safety presentation at a plant-wide meeting. Similarly, when people sign a petition or promise card that commits them to act in a certain way, say to actively care for the safety of others, they experience pressure from the consistency norm to follow through.

The technique of "throwing a curve" occurs when a person is persuaded to make a particular decision because there's not much at stake. Then the stakes are raised. Due to the consistency norm, the individual will likely stick with the original decision. Here's a safety application: An employee is asked to serve on a safety committee. No big deal—the committee meets just once a month. But then the employee is asked to attend more meetings because a special project has come up. To remain true to his first decision, the employee will probably stay on the committee and take on the additional work.

PRINCIPLE 48: The reciprocity norm is responsible for the impact of the door-in-the-face technique.

The reciprocity norm is a powerful determinant of human behavior. Its influence is reflected in the popular expression, "one good turn deserves another" and in the well-known Golden Rule. This is another reason to actively care for safety. One good act will likely lead to another.

The success of the "door-in-the-face" technique depends on the reciprocity norm. If an employee shuts the door on a major request, he's more likely to be open to a lesser request. If you ask for something less imposing, costly, or inconvenient after the initial refusal, your chances of being accepted are greater than if you started with the minor request. Your willingness to withdraw the larger request sets up an obligation to reciprocate and accept the smaller request. In Chapter 14, I discussed applications of this principle to promote actively caring behavior.

PRINCIPLE 49: Numbers from program evaluations should be meaningful to all participants and direct and motivate intervention improvement.

The last two principles relate to the critical issue of program evaluation (Chapter 15). In safety, the total recordable injury rate (TRIR) is the most popular evaluation number used to rank companies for safety rewards. It is calculated by multiplying the number of workplace injuries by 200,000 and dividing the answer by the total person-hours worked in that time period (U.S. Department of Labor, 1994). What an obvious example of an abstract number with little meaning. The most direct measure of ongoing safety performance comes from behavioral observations, and in Chapters 8, 11, and 15, I recommended ways to obtain meaningful feedback numbers from such process evaluation.

Throughout this text I presented various questionnaires that assess particular person states to gauge reactions to interventions. Such evaluation tools are not as objective and directly applicable to process improvement as feedback charts from behavioral observations. But results of surveys to measure perceptions, attitudes, or person states can be meaningful to program participants if explained properly. If given before and during an intervention process, questionnaires can reflect changes in the "inside" factors that impact program acceptance, participation, and future success.

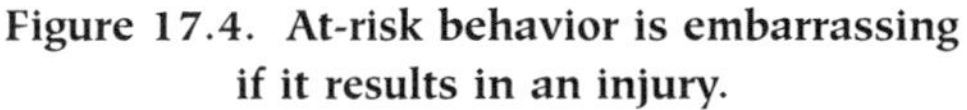
Figure 17.4. At-risk behavior is embarrassing if it results in an injury.

Figure 17.5. Personal testimony is a more powerful motivator than group statistics.

PRINCIPLE 50: Statistical analysis often adds confusion and misunderstanding to evaluation results, thereby reducing social validity.

Complex statistics are appropriate and often necessary for research journals. But if the purpose of a program evaluation is to improve a safety process, we need to provide numbers that give the most meaningful feedback to program participants—the people in the best position to improve the process.

Recall also the lesson from Sandman (1991) and Slovic (1991) that group statistics have minimal impact on risk perception (Chapter 5). If your goal is to increase risk awareness and motivate safe behavior, the most influential evaluation tool you can use is actually anecdotal. The most stimulating feedback usually comes from the personal report of an injured employee. However, as illustrated in Figure 17.4, the victim might want to cover up at-risk behavior leading to an injury. The culture needs to support reporting personal injuries, as well as discussing ways to prevent future incidents.

As illustrated in Figure 17.5, when people give personal testimony, the presentation is more influential than a statistical analysis. We should probably spend less time calculating summary injury statistics and more time eliminating the barriers to the personal reporting and investigating of safety-related incidents—from near hits and first-aid cases to lost-time injuries.

In Conclusion

This chapter reviews the principles of human dynamics discussed throughout this book. Founded on research published in scientific journals, they enable profound understanding of the psychology of safety. Use them as guidelines to develop, implement, evaluate, and refine safety-improvement programs and you'll make a positive difference in the safety of your organization, community, or culture.

Champions are needed to lead this process. Some are easy to find; others will evolve when the principles reviewed here are taught. Give potential champions opportunities to teach these principles and help develop interventions. Active participation increases both belief in the principles and empowerment to apply them to achieve a Total Safety Culture.

There is no quick-fix to culture change. The journey is not to be without bumpy roads, forced detours, and missed turns. These principles are your map to reach an enviable destination, but be prepared to blaze new paths and traverse difficult terrain. And please don't forget to take a break now and then to appreciate journey milestones. Recognize behaviors that contribute to a successful journey.

At the end of the second Deming workshop I attended, a participant raised his hand to ask one final question. When acknowledged, he stood and walked to the nearest microphone and stated, "Dr. Deming, you have taught us many important principles to consider when designing procedures to transform a culture. But frankly, the challenge seems overwhelming. Can we really expect to make a difference in our lifetime?"

W. Edwards Deming, at age 92, replied, "That's all you've got!"

REFERENCES

Baron, R. A. (1995). *Psychology* (Third Edition). Boston, MA: Allyn and Bacon.

Cooper, J., & Scher, S. J. (1990). Actions and attitudes: The role of responsibility and aversive consequences in persuasion. In T. Brock & S. Shavitt (Eds.), *The psychology of persuasion.* San Francisco, CA: Freedman.

Deming, W. E. (1991, May). *Quality, productivity, and competitive position.* Four-day workshop presented in Cincinnati, Ohio by Quality Enhancement Seminars, Inc.

——. (1992, January). *Instituting Dr. Deming's methods for management of productivity and quality.* Two-day workshop presented in Washington, D.C. by Quality Enhancement Seminars, Inc.

Freedman, J. L. (1965). Long-term behavioral effects of cognitive dissonance. *Journal of Experimental Social Psychology,* 1, 145-155.

Geller, E. S. (1991). If only more would actively care. *Journal of Applied Behavior Analysis,* 24, 607-612.

Goethals, G. R., Cooper, J., & Naficy, A. (1979). Role of foreseen, foreseeable, and unforeseeable behavioral consequences in the arousal of cognitive dissonance. *Journal of Personality and Social Psychology,* 37, 1179-1185.

Lepper, M., & Green, D. (1978) (Eds.). *The hidden cost of reward.* Hillsdale, NJ: Erlbaum.

Riess, M., & Schlenker, B. R. (1977). Attitude changes and responsibility avoidance as modes of dilemma resolution in forced-compliance situations. *Journal of Personality and Social Psychology,* 35, 21-30.

Sandman, P. M. (1991). *Risk = Hazard + Outrage: A formula for effective risk communication.* Videotaped presentation for the American Hygiene Association. Environmental Communication Research Program, P.O. Box 231, Cook College, Rutgers University, New Brunswick, NJ.

Skinner, B. R. (1971). *Beyond freedom and dignity.* New York: Alfred A. Knopf.

Slovic, P. (1991). Beyond numbers: A broader perspective on risk perception and risk communication. In D. G. Mayo & R. D. Hollander (Eds.), *Deceptable evidence: Science and values in risk management.* New York: Oxford University Press.

U.S. Department of Labor (1994). *Recordkeeping guidelines for occupational injuries and illnesses* (September). Washington, D.C.: Bureau of Labor Statistics.

Subject Index

Name Index